Introduction to
Industrial Hygiene

Ronald M. Scott

LEWIS PUBLISHERS
Boca Raton   Ann Arbor   London   Tokyo

Cover photo courtesy of Lab Safety Supply, Janesville, WI.

Library of Congress Cataloging-in-Publication Data

Scott, Ronald M.
  Introduction to industrial hygiene/Ronald M. Scott.
      p.   cm.
  Includes bibliographical references and index.
  ISBN 1-56670-140-6
  1. Industrial hygiene.   I. Title.
RC967.S36   1995
613.6'2—dc20   94-45567
                                                                    CIP

© 1995 by CRC Press, Inc.
Lewis Publishers is an imprint of CRC Press

No claim to original U.S. Government works
International Standard Book Number 1-56670-140-6
Library of Congress Card Number 94-45567
Printed in the United States of America 1 2 3 4 5 6 7 8 9 0
Printed on acid-free paper

Ronald M. Scott received a BS in Chemistry from Wayne State University and PhD in Biochemistry from the University of Illinois. Since 1959 he has lectured in biochemistry and, more recently, in toxicology at Eastern Michigan University. Dr. Scott has consulted on problems of metal ion toxicity, and has been a visiting professor at the University of Warwick, England. He has published papers in professional journals on chromatographic methods of trace analysis, techniques for clinical analysis, and work relating to the use of freshwater clams as environmental monitors of industrial effluents. Dr. Scott has written three chromatography reference books, a biochemistry textbook, and two industrial toxicology references, and was technical editor of a reference book on trace analysis.

This book is a non-encyclopedic introductory textbook of industrial hygiene, the field dedicated to the protection of workers. It is the outgrowth of two influences. First, the book is based on years of teaching a one-semester industrial hygiene course. It is a broad survey of the field, and addresses the typical student. Extra discussion is provided where experience has shown that some students will not be strongly prepared, given the variety of backgrounds of students enrolling in such a class. At the end of each chapter, material covered is summarized in a KEY POINTS section. Problem sets often have a practical basis, and lead students into the CFR to familiarize them with the contents and the manner of locating information in this source. References are provided both to material that helps the student with little background in the topic of that chapter and to sources that expand beyond the scope of the chapter. Extensive appendices provide practical information to support the instructional message and to allow the text to be a reference of value to the student later.

Second, the book is an expansion and updating of a more specialized book, *Chemical Hazards in the Workplace,* written originally to serve the person faced with managing or supervising in a workplace where chemicals are in use and who is not prepared by background for such an endeavor. *Chemical Hazards* covers only regulation of chemicals, and completely omits many topics needed by an industrial hygienist, but it has been used with supplements as a text in the author's class, providing valuable insights into what a good text should include. Further, written discussions of the missing sections became the basis of the added chapters and provided an opportunity to experiment with the approach while using the material.

The United States is experiencing a shortage of people trained to serve as industrial hygienists. The maintenance of high standards for health and safety in the workplace involves controlling a variety of working conditions, including control of noise level, radiation level, temperature, and potential for physical injury. Very importantly, industrial toxicology focuses on the threat of injury due to contact with chemicals used in the workplace. This threat is not restricted to workers in the chemical industry, since the use of chemicals is a ubiquitous part of modern industry. Many commercial processes involve painting or packaging a product, degreasing or otherwise cleaning machinery, assembling with adhesives, dyeing or printing with dissolved pigments, and a host of other processes that expose workers to chemicals. Furthermore, it can be argued that the need for regulation is greater outside chemical manufacturing, since management there is less likely to include chemistry specialists.

The book is divided into three sections. The first section focuses on chemical hazards. The basics of toxicology are presented, problems arising from skin contact or inhalation of chemicals are described, the detection and control of airborne contaminants are outlined, and finally the threat of fire or explosion is discussed. As early in Section I as is reasonable—given the need for some specialized terminology—the government regulations and agencies enforcing them are described. The second section is concerned with injury from physical causes, including sound, radiation, heat, and accident. Ergonomics, the study of problems due to the manner of performing the job, is introduced. Finally, the third section describes a range of industries, ones which are major sources both of jobs and of potential injury, and applies the principles presented in Sections I and II illustratively.

ACKNOWLEDGMENTS

The author wishes to recognize and express appreciation to Ralph Pedersen, an engineer specializing in noise problems, and Dr. Krish Rengan, a radio-chemist, for their help with the manuscript.

INTRODUCTION

The Historical View

The first occupational injury victim may well have been a caveman with a chip of stone in the eye from trying to form a spear point or an axe head. The development of civilized society led to a division of labor, with individuals providing specialized services. In these ancient civilizations, the variety of ways to injure oneself increased with the sophistication of jobs. In classical societies, however, citizens had a disdain for manual labor. High-risk work such as mining and construction was done by slaves, provided in plentiful supply by wars, and there was a minimum of concern with job injuries or illnesses. The embryonic medical professions were directed toward cure of the citizen, not the slave. Plinius Secundus described the use of a bladder across the face to filter out dusts from mercury or lead processing. Galen, a physician of the second century A.D., observed the hazards of mining copper, but had no suggestions for protecting or treating the workers.

During the Middle Ages medical knowledge waned in Europe, and earlier gains would have been lost had it not been for the Arab societies. Other than farming, work tended to be largely cottage industries, and little advance was made toward protection of the workers.

In 1473 Ellenbog published a pamphlet about occupational diseases, and in 1556 Agricola, a mining town physician, wrote a book that included descriptions of injuries and sicknesses to which miners were subject, describing both treatments and preventions. The first real comprehensive work concerned with occupational diseases was an effort by Bernardo Ramazzini, published in 1700. A few other advances were made in the 18th Century, particularly the observation by Percival Pott that the scrotal warts of chimney sweeps were caused by the soot they rarely washed away. This was the first linkage of cancer to chemical exposure.

Government finally became concerned with the welfare of workers with the English Factory Acts of 1833. These acts focused on workers' compensation rather than safe practices, but were beneficial in improving safety because they gave employers a financial incentive to prevent accidents.

In the U.S. during the first decade of the 20th Century, Alice Hamilton, a physician dedicated to the welfare of workers, raised public consciousness about the safety of workers and caused laws to be passed providing workers' compensation. At about that time the U.S. Public Health Service and the Bureau of Mines began studies of hazards in the workplace. The 1930s saw significant advances in promotion of worker safety.

Although the U.S. moved rapidly to improve the health and safety of workers through regulation and research, public expectations grew even faster. The social consciousness of the 1960s led to three significant pieces of legislation. The Metal and Nonmetal Mine Safety Act of 1966 created advisory committees with representatives of mine owners, miners, and mine inspectors, required reporting of accidents and occupational diseases, and regularized mine inspection. The Federal Coal Mine Health and Safety Act of 1969 allows the federal government to close unsafe mines. Standards were set for respirable dust, roof supports, and electrical distribution systems, and medical examinations of miners were required. Finally, the Occupational Safety and Health Act of 1970 (OSH Act) provided sweeping national standards for safe working conditions, an agency to set standards with powers of enforcement (OSHA) and a research agency to develop new standards (NIOSH). OSHA has greatly altered the workplace and vastly expanded the need for occupational safety professionals.

Industrial Hygienists

An industrial hygienist is one of a group of health professionals whose concern is the health and safety of workers in their work environment. The focus of the industrial hygienist is the workplace itself, and this person is expert in the identification and correction of a wide range of hazards. These include harm from chemicals, whether due to their short- or long-term toxic characteristics, corrosive or other threats to the skin or eyes, or narcotic or other behavior-altering properties. Physical hazards involve exposure to harmful radiation (ionizing or nonionizing), excessive sound levels, or extremes of temperature. Biological hazards arise from the possibility of infection or other damaging effects from organisms or their products found in the workplace, such as bacteria, fungi, or plant products. Remember, many jobs are performed outdoors, and others bring the indoor worker into contact with living or once living materials. Finally, the worker may be harmed by the nature of the work itself, for example, by repetitive hand or arm movements, or by lifting objects that are too heavy or in a fashion that leads to back injury. There is a wide variation in the level of hazard of each type of job (Table I.1).

Once the potential hazard is identified, it is often necessary to make measurements to determine whether or not the levels of hazard are within tolerable limits. Some sound is unavoidable around machinery, but is the sound above or below levels that might injure hearing? Workers are likely to have some contact with a chemical in use in a factory, but are levels encountered below those that cause harmful effects. Has exposure to X-rays reached a level such

Table I.1. Incidence Rates in Selected Industries. Date from 1982.

Industry	Loss of Work Time; Death Incidents/100 workers
Chemical	0.64
Textile	0.69
Communication	0.79
Agri-chemicals	1.01
Electrical equipment	1.12
Primary metals (nonferrous)	1.18
Steel	1.32
Auto manufacturing	1.36
Furniture building	1.75
Cement	1.88
Average—all industries	2.11
Polymers	2.32
Metal fabrication	2.33
Paper	2.65
Foundries (iron and steel)	2.66
Printing	2.75
Construction	2.92
Mining (metals)	3.02
Mining (coal)	3.40
Food processing	3.65
Lumber	4.66
Meat processing	5.18
Trucking	6.54

Data from National Safety Council.

that an airport security guard should not check luggage for a while? Such analysis may be done in part, or entirely, by an industrial hygienist.

Once hazards are identified, the industrial hygienist understands corrective measures that can be taken. These may focus on changes in the workplace. Airborne contaminants are removed by properly designed ventilation. The job process may be changed to isolate the worker from the threat. Proper shielding may be installed to block harmful radiation. Sound levels are lowered by sound barriers or changes in the design of machinery. Such changes are called "engineering solutions," and are the preferred means of providing a safe work environment.

In cases where an engineering solution is not employed or is still not adequate, the worker utilizes protective devices. Examples include safety glasses, gloves and other protective clothing, or hearing protection. If such devices are required, the industrial hygienist ensures that employers provides them, instruct the worker in their importance and proper use, and enforce their use on the job.

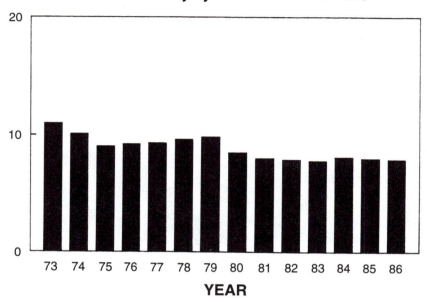

Illness and injury/100 full-time workers

Figure I.1. Job-related illness and injury.

Increasingly, the conditions necessary for worker safety are defined by law. In the United States a major step toward establishing detailed national safety standards was taken with the passage of the OSH Act in 1970. This act has the goal to "assure so far as possible every working man and woman in the nation safe and healthful working conditions and to preserve our human resources." Since its passage, a succession of new regulations have defined aspects of safe conditions for workers in quantitative terms. Employers are required to meet these standards, and for many industrial hygienists the daily task is to determine if a specific workplace is "in compliance."

Other safety professionals work to assure safe working conditions. Most important is the occupational physician. Just as the focus of the industrial hygienist is on the workplace, that of the occupational physician is on the worker. The two cooperate, with the industrial hygienist providing a description of the work environment of the worker that can be used by the occupational physician to determine if the worker is avoiding excessive workplace exposures, if observed problems have an occupational origin, or if a particular worker is or is not well suited to a specific job. Records of worker health are required by OSHA, and these are useful in identifying unsafe conditions that might not have been recognized otherwise.

Absolute safety for the worker is as unattainable as absolute safety for motorists. The objective is to approach that goal as closely as possible, which is the purpose of laws and their enforcement. Problems exist, but data shows that progress has been made (Figure I.1).

CONTENTS

SECTION I
CHEMICAL HAZARDS IN THE WORKPLACE

SECTION II
PHYSICAL HAZARDS IN THE WORKPLACE

SECTION III
SPECIFIC HAZARDS

SECTION I

CHEMICAL HAZARDS IN THE WORKPLACE

In the first section of the book, attention is focused on health and safety problems associated with chemicals. The first three chapters are a brief discussion of the principles of toxicology. We see that chemicals present a variety of health problems, learn how they enter and leave the body, and become acquainted with the terminology used to describe relative toxicity and toxic hazard. The history and practice of regulation by OSHA are reviewed. Then a detailed discussion of the chief problems arising from the use of chemicals in industry is presented. This begins with damage resulting from skin contact, i.e., dermatosis. The next three chapters deal with inhalation of chemicals, from the toxicological standpoint through methods of monitoring the plant atmosphere, and ending with corrective and protective measures. Finally, the section deals with fire and explosion hazards of chemicals.

CHAPTER 1

KINDS OF TOXIC EFFECTS

WHAT IS TOXIC?

In this opening section of the text, attention is directed toward damage to workers from contact with toxic chemicals. First we must answer the question: What is toxic? Chemicals are toxic. Does this mean every chemical is toxic? At the risk of raising the levels of chemophobia already loose in the land, there is a sense in which the answer to that is: Yes. When addressing nonscientific groups about concepts of toxicology, the author often asks the group to name a *completely safe* chemical. Almost always, water is named, at which time a newspaper clipping is presented reporting the death of a British man who, believing quantities of water were good for his health, drank so much that he died.

If you stop drinking water as a result of reading this, you are missing the point. The key words in this statement are "drank so much." Water is normally a *very safe* chemical because the quantities we take into the body are well below levels that cause us harm. Although it may take some of the pleasure away from sitting down to a favorite dish in a good restaurant to think of it in these terms, we must recognize that all of our food is a mixture of chemicals. Any of these individual chemicals taken *in excess* could cause us harm, but we are unlikely to damage ourselves with the quantities consumed at dinner. One responsibility of toxicologists, then, is to determine what level of exposure to a given chemical is safe, and what level can cause harm.

CLASSES OF TOXICITY

In order to assess the toxic hazard of a chemical, it is very useful to know how the chemical causes damage. We must put aside the "Snow White" picture of toxicity, in which a poison is an oily green liquid which, when placed on an apple, causes the apple eater to crumple to the floor with the first bite. A specific chemical can cause one or more of a variety of toxic effects. We place toxic chemicals in such classes as *systemic poisons, mutagens, carcinogens, teratogens,* or *behavioral toxins,* according to the nature of the damage caused. In some cases *the damage is apparent immediately,* while in others *the effect is only seen later,* perhaps even years later. The symptoms may be obvious, but they also may be subtle and easily missed. It is necessary to understand these classes of toxicity and to be familiar with the terminology associated with them.

First, some ground rules. Some chemicals are so highly reactive they cause damage just by contacting the skin, eye, or mouth. This sort of damage would be caused by a corrosive, such as an acid or alkali. These problems may be classed as local effects, and are not the object of this discussion. In this chapter we concentrate on chemicals that do harm after entering the body and being distributed by the circulatory system. The subdivisions of this chapter, then, are the classes of these effects.

SYSTEMIC POISONS

Body cells contain a vast array of specific catalysts, called enzymes, which regulate a complex pattern of chemical changes that convert dietary material to energy, repair and replace cell components, lead to cell growth and reproduction, and perform specialized processes that are the contribution of that particular cell type to the overall coordinated functioning of the multicellular organism. A specific chemical may reach a concentration in the body at which its presence begins to interfere with one or more of these normal cell processes. As a result, the cell may alter its functions, reduce its contribution to the organism, or die. This chemical is now acting in a toxic fashion, and is termed a *systemic poison* or *systemic toxin.*

Poisoning is not normally the result of a general damaging of all body cells in a uniform fashion. Most compounds damage one specific organ, the *target organ,* more than others. Symptoms that appear result from malfunction of that organ, even though cells in other organs are also affected, and if the person dies, it is the result of failure of that organ. The identity of the target organ generates a "class name" for the substance. For example, a liver poison is called an *hepatotoxin,* nervous system damage is produced by a *neurotoxin,* and a compound affecting kidney function is a *nephrotoxin* or *renotoxin.* Once the target organ is known, medical testing to assess the health of that organ monitors the degree of damage.

Particular tissues or organs become targets for a variety of reasons. An airborne toxin may primarily effect the lungs simply because it contacts those tissues first. Some cells employ chemistry directed toward a service provided to the rest of the body, and this unique chemistry is the target of the toxin. The liver metabolizes nonpolar substances into more polar compounds to facilitate their removal from the body, and it may be damaged when, during this process, it converts a relatively innocuous chemical into one that is highly toxic. Kidneys filter blood into a tubular system, recapture desired substances, then reduce the amount of water in this filtrate. The kidneys may be damaged as the toxin concentration rises to a dangerous level during the removal of water.

Damage to an organ generally involves cell death (necrosis), although it could be caused by interference with the normal function of particular cells, or result from excessive intake or loss of cell fluids. Where capillary wall cells are destroyed, hemorrhaging results. In some tissues, notably the liver and the kidney, dead cells are replaced fairly rapidly, while in other tissues, replacement occurs more slowly. In the most extreme case—neurons of the nervous system—cell loss is permanent.

When cell replacement is rapid, there is minimal risk from a single exposure below the level at which the degree of damage seriously interferes with organ function. Such a statement must be qualified to include such considerations as the general health of the subject. Specific health problems can place an increased burden or reliance on the organ in question, and habits or practices of the subject—such as the use of prescription drugs, alcohol, or tobacco—can create special risk situations. Although the risk may be low at a given level for a single exposure, that does not mean that repeated exposure to the same level of the compound carries a similarly low risk. Because a portion of the dose may remain in the body until the time of the next dosage, the intake level may remain constant while the concentration in the target organ is increasing. In a similar fashion, if exposure produces the need to repair tissue, and the repair is not complete at the time of the next dose, damage to the body accumulates. Under conditions where tissue repair is constantly required, another process involving enlargement or irreversible scarring of the organ occurs. Cirrhosis of the liver, common in long-term alcoholics, is an example of the latter. The healthy tissue is gradually replaced by nonfunctional tissue, eventually resulting in an organ inadequate to the needs of the victim.

When the dose of toxicant is sufficiently high that the target organ is immediately damaged beyond its ability to function, the person dies. This level of exposure, termed an *acute lethal dose,* is a fundamental measure of the toxicity of the substance. Because of individual differences, the minimum dose of a particular chemical that causes death is not the same for every individual. Published values that communicate acute lethal dose levels are therefore based on mathematical manipulation of data from several observations.

TOXINS AFFECTING DNA

DNA: The Blueprint of the Organism

In order to understand some other toxic effects of chemicals, it is necessary to acquire some biochemistry background. With very few exceptions, each cell contains blueprints for the complete organism in the form of giant polymeric molecules called *DNA.* These blueprints, the genetic message, tell how to build all the functional molecules of the organism, which are the *proteins* and *RNA molecules.* The message for a single protein is called a *gene.* The message is in coded form, the code being the order of assembly of the four kinds of subunits or monomers into the DNA molecule. Simple life forms such as bacteria may have the entire genetic message on a single DNA molecule. In more complex life forms, the message requires a number of separate DNA molecules, each termed a *chromosome*. In humans and other higher life forms, *two complete sets* of chromosomes are present in each cell, one set from the male parent and one from the female parent. Such cells are termed *diploid*. Because humans are very complex, and because each cell carries two complete messages, the amount of DNA per human cell is relatively large. In fact, if the DNA molecules from a single cell were laid end to end, they would stretch approximately the height of the person.

It is extremely important that the genetic message remain undamaged. A first level of protection is provided by the attachment to every DNA molecule of a second DNA molecule called the *complement*. Complement DNA is a "carbon copy" of the actual message molecule. There are four kinds of subunits in DNA: A,T,G, and C. Everywhere on the *sense* (message carrying) *strand* of DNA that an A unit appears, the complement strand has a T, and everywhere the sense DNA has a T, the complement has an A. G and C have the same relationship (Figure 1.1). The A units uniquely hydrogen bond to the T units, and the G units uniquely hydrogen bond to the C. The two strands of DNA cling to one another by these hydrogen bonds. The paired DNA molecules spontaneously form a spiral assembly called the *DNA double helix*. Just before a cell divides, these two strands separate. New monomer units assemble next to the message strand pairing A with T and G with C in response to the hydrogen bonding forces, and as they assemble they are joined together to form a new molecule of DNA. The DNA now attached to the sense strand is a new complement strand. In the same fashion, the original complement strand serves as the template for the formation of a new sense strand. We now have two double helix DNA molecules, each one identical to the original—one for each of the cells resulting from the division. This process, called *replication,* assures that each of the cells has a complete genetic message (Figure 1.2).

Mutagens

A mutagen is a substance or force that can cause damage to the DNA message. The most important mutagens are either specific chemicals or such radiation as X-rays, radioactive emissions, and ultraviolet light. The machinery of replication includes systems to check the accuracy of the copy and to correct mistakes. This same machinery can locate and correct alterations of the genetic message that occur at times other than replication. Since we are continuously exposed to both chemical mutagens and radiation, we depend on our DNA repair mechanisms to correct the damage caused. Our concern is that some damage may escape notice. If the level of exposure to a mutagen is raised, that likelihood is increased. At high levels of damage, the capacity of the repair system may be exceeded.

If genetic damage has escaped repair before the cell divides, the altered message becomes a permanent feature of the DNA of every descendant of that cell. This results in one of a variety of possible responses. In the best case, the

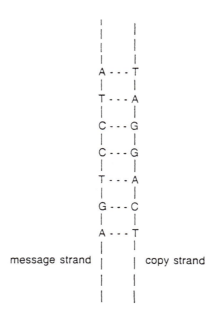

Figure 1.1.   The relationship between two strands of DNA.

new blueprint produces cell components that are functional, and the cell suffers no ill effects. However, we can observe a whole spectrum of damage, ranging from the production of cell components that are less efficient, to complete failure to produce a functional version of a particular component. A cell so crippled may or may not survive. If it survives, all its descendants will be similarly crippled. Other permanent changes in the cell can result, such as progression down the path to becoming a cancer cell.

There are situations where damage caused by mutagens is a greater concern. This is why we wear a lead apron during routine medical X-rays. Alterations to the DNA of either sperm or egg cells, followed by the damaged cell joining with its counterpart to produce a fertilized egg, will be replicated in every cell of the developing embryo, with potentially crippling or lethal effects. It was such occurrences in earlier times that left human populations with the hereditary problems we term genetic diseases, including such difficulties as phenylketonuria (PKU) and sickle cell anemia. Exposure to either radiation or chemical mutagens is more serious in women than in men, because a woman carries the lifetime supply of egg cells in her ovaries. By contrast, men generate a new set of sperm cells in a period of perhaps two weeks. Hazard from exposure is therefore short term in men, but lasts for the entire period of fertility in women.

Exposure to mutagens is also particularly serious during pregnancy. The developing child displays very rapid cell growth and frequent cell division, each cell in early pregnancy being the ancestor of vast numbers of cells by the time the child is born.

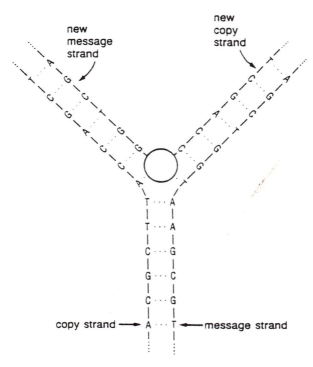

Figure 1.2.   Replication of DNA. Here we see DNA in the process of being replicated. The lower arm is DNA that has not yet been acted on by the enzyme, represented by the circle at the fork. Each of the upper arms is one of a pair of identical DNA strands. When replication is complete, there will be two identical sets of DNA, one for each of the cells that will result from cell division. Each of those sets will have one strand of old DNA and one of newly replicated DNA.

Carcinogens

The relationship between cancer and certain chemicals was not always recognized. The original observations were made in the nineteenth century by a British physician, Sir Percival Potts, who noted that young chimney sweeps were prone to get scrotal warts, some of which later transformed into malignant tumors. Suspecting that something in the soot that these young men wore almost as a badge of office was responsible for the warts, he proposed that bathing after work should be a regular practice of the sweeps. The subsequent success of his proposal established the beginning of our present understanding of the relationship between cancer and chemicals. The general recognition of the hazard these compounds present and the development of usable testing methods followed slowly, so it is only more recently that testing for carcinogenic properties has become a routine procedure in toxicology.

Not all tumors are cancerous. Warts, for example, are common examples of tumors. People can develop warts and wear them for most of their lives with only cosmetic damage. Such tumors are termed *benign*. A benign tumor in the wrong location, for example in the brain, can be a serious medical problem, but often such tumors are nonthreatening. They may, however, at some time change character and become cancerous. Unrestrained growth, accompanied by the tendency to metastasize (to spread to other organs and tissues), characterizes a malignant tumor, hence cancer.

Exposure to a carcinogen increases the likelihood of the disease occurring, but even relatively heavy exposure does not assure tumor development. Consider the occurrence of octogenarian cigarette smokers. Furthermore, exposure to a specific chemical is only one aspect of the initiation of cancer. Radiation is also a causative agent. Additionally, there are differences in susceptibility to the disease based on the age, sex, and genetic makeup of the individual.

The classification of carcinogens is an area of discussion and change at this time. The more traditional classification serves our purposes well enough, however. Primary carcinogens are chemicals that cause cancer upon contact with the tissues. Procarcinogens, or secondary carcinogens, do not cause cancer in the chemical form entering the body. They are metabolically converted in the body into active causative agents for cancer. Finally, cocarcinogens, or promoters, do not cause cancer by themselves. They do stimulate cells that have been converted to latent cancer cells to begin the rapid cell division characteristic of cancer.

There is a strong correlation between mutagenesis and the causing of cancer. Although more research is needed before we fully understand the cause of cancer, there are presently some very important clues. There are genes that code for proteins that regulate cell growth and trigger cell division. Damage to these genes may result in production of proteins that fail to properly regulate cell growth and division, and the resulting tissue displaying unrestrained growth is a tumor. Damage to more than one gene seems to be a prerequisite to malignancy.

Carcinogens display a degree of target organ specificity. One compound induces bladder tumors; another causes growths in the colon. This specificity may be the result of the manner of exposure, as in the generation of lung tumors by smoking. In other cases, a compound always targets a particular organ. The herbicide paraquat targets the lungs, whatever its path of entry to the body.

Of two carcinogens, one is more potent than another if, at the same concentration, it causes a higher percentage of animals to produce tumors. Assessing the carcinogenic potential of a compound is challenging. If a compound is a hepatotoxin, animals exposed to it at high levels consistently show liver damage. By contrast, dosing with a carcinogen is like playing Russian roulette. Not every heavy cigarette smoker gets lung cancer, but the percentage of smokers that do is much higher than the percentage of nonsmokers. Studies must therefore be performed using a large enough test group that the result is statistically valid. Furthermore, tumors develop for reasons other than exposure to a particular compound, so there is a normal or baseline occurrence of cancer without dosing. To identify a carcinogen, we must observe an increase in occurrence of tumors above this baseline level, so an undosed control group must be maintained for comparison. In summary, the percentage of animals lightly dosed with a carcinogen that develop cancer is higher than that of non-dosed animals; more heavily dosed animals develop the disease even more frequently; and the frequency at a given level of exposure differs with the carcinogen used.

Response to a carcinogen is not observed for an extended period of time. For example, in a society where women who smoked were less socially acceptable, movies portrayed glamorous women as being sophisticated when they lit a cigarette. Twenty years later, the resultant increase in women smoking was followed by an increase in women with lung cancer. The effect of some chemicals is to convert a normal cell to a precancerous cell. Perhaps months or years later, this precancerous cell may undergo a second change into a malignant cell. The transformation into a cancer cell may appear to be spontaneous, or it may be the response to a second chemical. Following such a long incubation, it is very difficult to establish unequivocally the cause of the original cell transformation. Studies to determine the carcinogenic potential of a chemical must extend for the life of the animal, so the introduction of carcinogenicity testing into the toxicologic assessment of compounds has greatly increased both the effort and expense of such procedures.

REPRODUCTIVE TOXINS

Teratogens

Teratogens are compounds that cause offspring to be born with abnormalities. We typically see skeletal faults such as improperly formed limbs or vertebral column, soft tissue flaws such as cleft palate, heart malformation, closed segments of intestinal tract, or nervous system imperfections. The hazard arises when the pregnant woman is exposed to a teratogen when the embryo is first forming an organ, and at that time the teratogen interferes with proper development of that organ. Exposure of the male or female parent to teratogens before conception is of no significance. Mutagens, by contrast, can cause birth defects by altering sperm or egg cell DNA perhaps long before the egg is fertilized—quite a different mechanism.

Concern about teratogens is a recent phenomenon. Rubella (measles) infections during pregnancy had long been associated with birth defects, but strictly chemical agents were not generally of concern until 1960, when the drug thalidomide, a potent teratogen in humans, was widely prescribed to pregnant women in Europe and Japan. The defect frequently caused by thalidomide was malformation, even to the point of complete absence, of arms and legs. Approval for the drug had not yet been obtained in the U.S., sparing that population from this catastrophe.

Testing for teratogenic potency, as for carcinogenic potency, presents special difficulties. Animal testing must be performed while the animal is pregnant. Traditional test animals, such as rats and mice, are very different from humans in aspects of the reproductive process. Development in rodents moves forward much more rapidly. Small rodents bear litters rather than single offspring, and differences in exposure with position in the uterus is a variable. The placentas of humans and small rodents differ in their effectiveness as barriers to the entry of foreign substances into fetal blood.

Extrapolating animal results to humans, always a problem, becomes even less dependable when assessing teratogenic potential. Aspirin, for example, is a potent teratogen in small rodents, while thalidomide is not. Had thalidomide been tested before release for teratogenicity in mice, it might well have been evaluated as safe.

Exposure of a pregnant woman to a teratogen does not assure birth defects. In addition, a very high level of malformed human offspring occur naturally. About 20% of all pregnancies result in spontaneous abortion. Very little information is available to determine what percentage of these abortions is the result of malformation. Of the children carried to birth, about 2% are physically malformed, and another 8% must be added to this figure if mental handicap is included in the statistics. A chemical that caused 1% of the exposed population to develop cancer would not only be highly visible, but would be looked on as a serious threat. However, against such a high background level of childbirth abnormalities, an agent causing 1% teratogenic malformations might escape detection.

Regardless of their failings, animal teratogenic testing is all that is available before human exposure to a compound is allowed. After humans are exposed to the chemical, its teratogenic potency is likely to be detected only if it is a very effective teratogen.

Toxins Affecting Fertility

Damage to reproductive potential is a special type of systemic toxicity. Here the target organ is the male or female reproductive system. Processes essential to reproduction, such as sperm production, normal ovulation, and implantation of the fertilized egg, can be affected by exposure to chemicals. Interference with these functions does not produce illness or death either in test animals or humans, and is therefore less likely to be detected. It was only when male shipping room workers at an agricultural chemicals plant in California became aware of their common fertility problems that the potential of dibromochloropropane to inhibit sperm production was discovered. Anticipating this sort of problem by means of animal testing involves the use of different kinds of testing, and adds to the difficulty and expense of screening chemicals thoroughly.

NEUROTOXINS

As our most sensitive and least understood organ, the brain presents yet another type of difficulty to toxicologists. In the sense that we judge systemic poisons by the presence or absence of visible damage to organs like liver or kidney, we could look for cell death or tumors appearing in the nervous system following exposure to a chemical. However, this would take into account only one aspect of the possible damage chemicals could cause to the human nervous system.

The very sensitive balance maintained between the various chemically transmitted signals in the brain that represents normal human behavior can be disrupted by the entry of foreign chemicals without a single cell dying or being transformed. Such an alteration could cause mood and personality changes, loss of intellect, or memory impairment. These same changes might also occur without associated chemical exposure. As a result, changes resulting from a toxic episode might be attributed to another cause, and not stimulate the search for a neurotoxin.

The situation in the central nervous system is complicated by the blood/brain barrier, a term describing the fact that many substances cannot be transferred from the blood into the brain as easily as into most other organs. The effectiveness of this barrier varies among individuals and depends greatly on age. Fetuses and neonates (newborns) in general do not have an effective barrier. This means that potentially harmful substances circulating in the blood of a pregnant woman are blocked from entering her central nervous system by the blood/brain barrier. However, if these substances transfer from the mother's to the child's blood in the placenta, they might freely enter the unborn child's brain. Variability of the blood/brain barrier among individuals adds further randomness to the already difficult task of evaluating potential neurotoxins.

Testing a chemical with animal surrogates for its potential to cause harm in the central nervous system presents special difficulties. If cell death or stunted development is the problem, the animal will display these difficulties. However, how does animal testing assess damage to intelligence, effects on sanity, or alteration of behavior? Some progress has been made in these areas, but the obstacles to full understanding are imposing.

OTHER TOXIC EFFECTS

In addition to substances that cause serious damage, the worker must be guarded against exposure to substances that are simply irritating. Of particular significance are compounds that cause irritation to the skin, eyes, or respiratory tract. Characterizations by animal tests are generally easily performed, and since we are not concerned with lethal consequences, a degree of experimental human exposure can sometimes be an acceptable option.

Clearly, the establishment of a profile of the hazard of chemicals in the workplace requires a variety of types of testing and a complex array of decisions. The result of this decisionmaking is a set of standards for the upper limits of acceptable worker exposure.

KEY POINTS

1. All chemicals are toxic at sufficiently high concentrations. Toxicologists determine safe levels.

2. Chemicals cause a variety of toxic effects, and are classified by such terms as systemic toxins, mutagens, carcinogens, teratogens, and behavioral toxins.

3. Toxic damage may be immediate or may require an incubation period.

4. Systemic toxins interfere with normal cell function, usually most severely in a specific organ or tissue.

5. DNA carries the genetic information of the organism as a sequence of monomers on a very large polymeric molecule. Higher organisms such as humans have more than one molecular type of DNA, and have two copies of each type, one from each parent.

6. DNA is a double-stranded helical structure, and the sequence on one strand correlates with that on the other in a fashion termed complementary. One strand is the genetic information, and the other is useful when duplicating DNA at the time of cell division or when repairing damage.

7. Mutagens damage DNA monomer sequences. When the damage is to a sperm or egg cell that then contributes to forming a fertilized egg, the damage is copied in every cell of the developing fetus.

8. Carcinogens increase the likelihood of cancer occurring. Most carcinogens are also mutagens.

9. Teratogens damage a fetus during development, producing birth defects.

10. Some toxicants interfere with sperm production in men or in the processes involved in ovulation or implanting a fertilized egg in women, reducing levels of fertility.

11. Compounds that affect the nervous system—neurotoxins—produce complex effects that are sometimes difficult to assess.

PROBLEMS

1. A wise early investment for a person in training for a career in industrial hygiene is a medical dictionary, since many references use terminology with which lay persons or even many science majors are not familiar. Following are direct quotes from an excellent but sophisticated toxicology reference. Translate these statements into less technical language, then describe in terms of the major topics of this chapter what sort of toxicant each compound is.

Discussions in L. S. Andrews and R. Snyder, "Toxic effects of solvents and vapors," in C. D. Klaassen, M. O. Amdur and J. Doull, *Toxicology: The Basic Science of Poisons,* 3rd ed. Macmillan, New York, 1986.

A. "The renal histopathology, after 6 to 13 days, varied depending on the amount of ethylene glycol consumed. At high dose levels, deposition of calcium oxalate crystals occurred in the proximal renal tubules, and necrotic areas of tubule epithelium occurred adjacent to the crystals."

B. "Thus, bromobenzene…yields massive hepatic necrosis in animals pretreated with phenobarbital."

C. "In humans the adverse effects of benzene are variants of either aplastic anemia or leukemia. It is likely in each case that metabolites of benzene initiate the disease process and also appear to be implicated in mutational events…."

D. "Reports of hematologic effects or peripheral neuropathies (from glue sniffing) are more likely due to the use of glues containing benzene or hexane."

E. "Although there were no indications of maternal toxicity, 100 percent and 50 percent of the fetuses died in the 200- and 100-ppm exposure (to ethylene glycol monomethyl ether) groups, respectively. Cardiovascular and skeletal malformations were increased above unexposed control values in both 50- and 100-ppm exposure groups."

F. "…in both rats and rabbits exposed to 300 ppm; every male exhibited degeneration of the testicular germinal epithelium. At the end of the exposure period, male rats were mated to unexposed females and were found to be infertile."

G. "Guinea pigs demonstrated…hyperplasia of lung cells, as well as squamous cell carcinoma in the nasal cavity."

2. Let's understand DNA function and mutation a little better. Below is a representation of a segment of DNA—the sense strand. Below is a translation of the genetic message, indicating which amino acid (as a three-letter abbreviation) each three-base sequence designates. (People who have had a biochemistry class will recognize that this is not the genetic code as presented in texts, which translates three-base sequences on *RNA* into specific amino acids, and reads in the opposite direction.)

```
TTTTTTTTTTTTTTTTTTTTTTTTTTTTTTTTT
AATGTTTGTGGGTCTGCCTAGCTTACTCCC
```

	A	G	T	C
A	AAA Phe	AGA Ser	ATA Tyr	ACA Cys
	AAG Phe	AGG Ser	ATG Tyr	ACG Cys
	AAT Leu	AGT Ser	ATT STOP	ACT STOP
	AAC Leu	AGC Ser	ATC STOP	ACC Trp
G	GAA Leu	GGA Pro	GTA His	GCA Arg
	GAG Leu	GGG Pro	GTG His	GCG Arg
	GAT Leu	GGT Pro	GTT Gln	GCT Arg
	GAC Leu	GGC Pro	GTC Gln	GCC Arg
T	TAA Ile	TGA Thr	TTA Asn	TCA Ser
	TAG Ile	TGG Thr	TTG Asn	TCG Ser
	TAT Ile	TGT Thr	TTT Lys	TCT Arg
	TAC Met	TGC Thr	TTC Lys	TCC Arg
C	CAA Val	CGA Ala	CTA Asp	CCA Gly
	CAG Val	CGG Ala	CTG Asp	CCG Gly
	CAT Val	CGT Ala	CTT Glu	CCT Gly
	CAC Val	CGC Ala	CTC Glu	CCC Gly

Proteins are long polymer chains in which all the monomer units (amino acids) are in a specific order. The characteristics of the protein are determined by the amino acid sequence. In an individual, every protein molecule serving a particular function, for example hemoglobin in red blood cells carrying oxygen, may well have exactly the same sequence. At most there will be two sequences, one from the person's father and one from the mother.

A. First look over the table. STOP means that this is the end of the particular protein chain being assembled. The rest of the DNA sequences are associated with a specific amino acid. Thus the appearance of CGG in the DNA sequence says the amino acid Ala [alanine] belongs next in the protein molecule. How many different amino acids may be incorporated into protein?

B. Write the protein sequence this DNA sense strand describes. (Recognize that real proteins typically have hundreds of amino acids, so this is an unrealistically short message.)

C. Suppose the 11th base from the left is altered by ionizing radiation so it is now read as A. What happens to the protein sequence?

D. Suppose the 18th base from the left is altered chemically to read as T instead of C. What happens to the protein sequence? Why do you think this is termed a silent mutation?

E. Suppose the 4th base from the left is altered chemically to read as A instead of G. What happens to the protein sequence?

F. Suppose the 27th base from the left (4th from the right) is altered chemically to read as C instead of T. What happens to the protein sequence?

G. All of the above are called point mutations because a single base is changed to another. A base could also be deleted by mutagenic action. Suppose the 4th base from the left were deleted. What happens to the protein sequence?

H. The example in G is termed a frame shift mutation. Why? Sometimes when a point mutation changes a single amino acid in the protein sequence, the protein can still perform its designated function because the particular amino acid is not critical to protein function. Would this happen with a frame shift mutation?

I. Write the DNA sequence of the complementary strand to this sense strand using A with T and G with C as the complementary pairings, as described in this chapter.

J. If the 5th base from the left were damaged by a mutagen, how would repair enzymes know how to fix it?

3. A workplace discrimination issue has arisen because women have been banned from certain jobs because the chemicals involved pose a special risk to women. Chemicals displaying what toxic characteristics might

generate such an issue? Can you provide any examples of jobs from which men might be banned?

BIBLIOGRAPHY

D. Brusich, *Principles of Genetic Toxicology,* Plenum Press, N.Y., 1980.

T. H. Corbett, *Cancer and Chemicals,* Nelson Hall, Chicago, 1977.

B. D. Dinman, *The Nature of Occupational Cancer,* Thomas, Springfield, Illinois, 1974.

L. Fishbein, *Chemical Mutagens,* Academic Press, N.Y., 1970.

L. Fishbein, *Potential Industrial Carcinogens and Mutagens,* Elsevier, N.Y., 1979.

W. G. Flamm, *Mutagenesis,* Hemisphere Publishing Co., N.Y., 1978.

G. Flamm and M. Mehlman, Eds., "Mutagenesis" in *Adv. in Modern Toxicology,* Vol. 5, Halsted Press, N.Y., 1978.

Health and Welfare-Canada, Ottawa, *The Testing of Chemicals for Carcinogenicity, Mutagenicity, and Teratogenicity,* 1975.

M. Kirsch-Volders, Ed. *Mutagenicity, Carcinogenicity, and Teratogenicity of Industrial Pollutants,* Plenum Press, N.Y., 1984.

P. Lehmann, *Cancer and the Worker,* New York Academy of Science, N.Y., 1979.

J. L. O'Donoghue, *Neurotoxicity of Industrial and Commercial Chemicals,* CRC Press, Boca Raton, FL, 1985.

N. I. Sax, *Cancer Causing Chemicals,* Van Nostrand-Reinhold, N.Y., 1981.

T. H. Shepard, *Catalog of Teratogenic Agents,* 3rd ed., Johns Hopkins Univ. Press, Baltimore, 1980.

CHAPTER 2

MEASURING AND REPORTING RELATIVE TOXICITY

This chapter examines two related questions. First, what are the techniques used to gather information about the health effects of chemicals? We are interested in both how the testing is done, and in how reliable the results are. Second, how do we communicate the findings once we have studied toxic effects—what is the language used to express relative toxicity? In the process of answering these questions, terminology unique to toxicology is introduced.

FACTORS IN TOXICITY MEASUREMENT

TOLERANCE TO TOXIC SUBSTANCES

The body can tolerate exposure to toxic substances, as long as the level is reasonable. This is necessary to human survival, since contact with toxicants is unavoidable. Plant materials in our diet, for example, contain a broad variety of substances that are of necessity consumed along with the protein, starch, lipids, vitamins, and minerals that constitute nourishment. Many of these substances would seriously threaten our health if taken in large quantities. After all, the chemistry of the plant was designed to serve the needs of the plant, not to serve those who plan to kill and eat it. Safe food need not be completely free of toxicants—such food may not exist. It is only necessary that the toxicants be at a low enough level that they are cleared from the body and whatever problems they caused are repaired before we take in any more such toxicants. The function of toxicology is to determine what levels of exposure are safe for those who are exposed—the *target population*—for the length of time of contact, and then to set standards to assure that these levels are not exceeded.

THE TARGET POPULATION

In order to set realistic exposure guidelines, it is necessary to consider the population affected by the findings. Some branches of toxicology are concerned with hazards to fish, birds, or perhaps every human being. For example, the target population when setting standards for the quality of urban air includes persons of every age and state of health. In this case, a level of carbon monoxide that can pose a threat to an elderly anemic individual with emphysema is of concern. Industrial hygiene is concerned with workplace exposures. This translates into concern for a group of male and female adults roughly 18 to 70 years of age who are in reasonably good health. Problems relating to the health of the fetus are pertinent, since the workforce includes pregnant women. However, special problems associated with the newborn, the very old, or those with serious health disabilities are not a prime focus. The allowable levels of carbon monoxide in the plant atmosphere might therefore be higher than would be recommended in urban air.

LENGTH OF EXPOSURE TIMES

It is important to consider the length of time the worker is in contact with the chemical. A dishwashing product that contains a detergent and ammonia might be perfectly safe for household use, given that exposure would only occur when the dishes are washed, but might be a serious health threat to a person whose skin is continuously exposed at work. On the other hand, urban air standards must include the consideration that residents breathe that air 24 hours a day 7 days a week. Exposure at the workstation only occurs for an 8-hour day and a 40-hour week.

ANIMAL TESTING

What do we do to determine how fast a horse can run? We get a device for measuring how fast an animal is running. Then we get a horse and convince it to run. Better yet, since our original horse is anywhere on a continuum from being a potential triple crown winner to being a lame and hopeless nag, we would repeat the experiment with a number of horses and average the values obtained. Would it make sense to do the experiment with mice instead, measuring how fast a series of mice ran? All sorts of arguments are immediately presented—mice have very short legs, mice move very quickly for their size (metabolic rate, perhaps?), mice don't have hooves, and so on. Even if we tried to adjust the data by saying a horse is 100,000 times larger than a mouse, we would certainly hesitate to say it should run 100,000 times faster.

As solid as these arguments are against using mice to gather information about the running of horses, extending the concept that "predictive testing should not be done on dissimilar animals" to all types of testing should be done cautiously. If the information we seek concerns the potential of a new industrial chemical to harm workers, how do we get the answer?

> VOLUNTEERS FOR EXPOSURE TO THE NEW
> CHEMICAL WILL PLEASE FORM A LINE
> TO THE RIGHT AND LEAVE THEIR NAMES AND
> PHONE NUMBERS WITH THE SECRETARY

Although it is true that data concerning toxicity to humans is most valid when exposure is to humans, we retreat from actually running such tests on ourselves, and seek an alternative procedure. Now maybe the mice can do the job for us.

In recent years, the use of animals in research has been strongly, sometimes even violently, opposed by "animal rights" groups. Hate-filled extremists in these groups have raided laboratories, destroying facilities and records, setting important research back years and causing other facilities to be turned into fortresses in an attempt to prevent further such destruction. The action of such crazies gives a movement that includes reasonable individuals with valid concerns a bad name. Positive results of animal rights pressure include formulation and enforcement of standards for the care and housing of research animals that have eliminated some abuses of the past. This movement has also stimulated thinking about alternatives to the use of test animals as *surrogates* or substitutes for humans, and some new methods have been developed. If we are concerned about the effects of a chemical on a specific tissue, we may raise isolated cells of that tissue in a tissue culture medium. For example, a colleague of the author studies the effect of chemicals on sperm production, using cultured sertoli cells (the sperm-producing cells) as a test system. Screening tests to determine if a compound is a mutagen are now done quickly and cheaply using special strains of bacteria. However, for some testing, for example an initial broad screening for any kind of systemic toxicity, the use of test animals is still necessary. A favorite cartoon of the author depicts a physician sitting at the bedside of a patient with a cage containing a small animal in his lap and a hypodermic in his hand. The patient, in great agitation, is saying, "You mean you're just going to test it on a Guinea pig *now?*"

If the use of animal surrogates is necessary, we need to select an appropriate species. It is then important to obtain meaningful values for the toxicity of the substance *to these test animals,* and postpone our concerns about using these data to predict toxicity to humans.

What Animal Shall We Use?

Most toxicologic data is collected using small rodents (rats, mice, Guinea pigs) as surrogates for human beings. There are numerous advantages to using such animals as rats or mice. They are small and relatively easy to maintain in good health in the laboratory. With a life expectancy of typically two years, we can see the effect of a lifetime of exposure in a relatively short time. They propagate rapidly. Because we have used them extensively in the past, we are very familiar with their anatomy and physiology, their normal and abnormal states of health, and their response to a large number of other substances.

First a species of animals is chosen, for example Sprague-Dawley white rats. Notice that here the choice is more restrictive than using just "rats." Pure strains of laboratory rats have been developed and maintained for decades. They are bred to minimize differences between individuals. There are many varieties of rats, and experience teaches that there can be variation in response to a chemical among these varieties. In the hope of obtaining reproducible results, we select a single strain that has been bred genetically pure for many years. To whatever degree it is possible for individuals to respond uniformly, it will happen with these rats. Furthermore, by using this variety repeatedly we learn much about its characteristics, and we become practiced in the transfer of toxicity information into predictions of human response.

It is an important lesson that even in this population the exact dose to produce a particular effect may well be different for every rat. We would not expect the members of the human population, who are far from genetically identical, each to respond in the same fashion to a chemical. However, if we obtain a more reproducible response from our test group, and if they are studied well enough that we understand many of their differences from humans, we can feel more assurance about the dependability of the extrapolation of the animal data to humans.

Extrapolating Animal Data to Humans: The Size Difference

In spite of all these advantages, using rats to predict toxicity to humans involves many risks. In the estimation of the speed of a horse, the rat or mouse is not a miniature horse. Neither in measuring the toxicity of a chemical is it a miniature human. There are differences in metabolism, dietary patterns, basic anatomy and size between mice and humans. As an example, if something is eaten by a human and then causes nausea, the human vomits it. Rats cannot vomit, so that same material would remain in the rat's stomach. If the compound were toxic at the level given, it would be much less threatening to the human, who would remove it from the stomach automatically.

The difference in size between humans and small rodents is a major problem. The observation that 1 mg of a compound makes a white rat ill does not at all mean that the same 1 mg would make a human ill; in fact, we would be very surprised if it did. Spreading that 1 mg around a 250-g rat is altogether different from diluting it into a 70,000-g human.

This difference is handled by employing units that indicate the weight of compound given to the animal divided by the weight of the animal. Most commonly, we list doses in mg (milligrams of compound given the animal) per kg (kilograms of animal body weight). Thus our dose of 1 mg per 250-g rat is a dose rate of 4 mg/kg. We assume that the same dose rate should affect the human, and that 280 mg would have roughly the same effect on a 70-kg (154-pound) adult human. Because of differences between rats and humans, this is almost certainly not exactly true, but it does provide a very useful first approximation of toxicity to humans.

Sample Problem:

A. 25 mg of a narcotic substance under study causes a 200-g rat to sleep for one hour. What is this dose rate in mg/kg?

$$\frac{25 \text{ mg}}{200 \text{ g}} \times \frac{1000 \text{ g}}{1 \text{ kg}} = \frac{125 \text{ mg}}{1 \text{ kg}}$$

B. As a first approximation, what dose would be needed to produce the same effect on an 80-kg human?

$$\frac{125 \text{ mg}}{1 \text{ kg}} \times 80 \text{ kg} = 10,000 \text{ mg} \times \frac{1 \text{ g}}{1000 \text{ mg}} = 10 \text{ g}$$

C. What dose would be predicted to put a 31-kg dog to sleep?

$$\frac{125 \text{ mg}}{1 \text{ kg}} \times 31 \text{ kg} = 3875 \text{ mg} \times \frac{1 \text{ g}}{1000 \text{ mg}} = 3.9 \text{ g}$$

For reporting the toxicity of airborne toxicants, *no adjustment is necessary for the ratio of animal to human size.* We assume that the rat and the human each breathe an amount of air that is in proportion to their size. The dose is therefore listed in terms of the concentration in the air, with no reference to animal size.

DESIGNING TOXICOLOGIC EXPERIMENTS

Manner of Dosing

Next, a decision must be made regarding how to dose the animals. There are several pathways for dosing. Most commonly, the test substance is added to the food or water, placed directly into the stomach by gavage (a small tube inserted through the mouth into the stomach), injected into a vein or the peritoneal cavity, coated onto the skin, or added to the atmosphere the animal breathes. The final toxicity values obtained usually differ according to the path of entry into the body. Which path we choose reflects the following concerns:

A. What will be the manner of exposure of workers in the plant? Whatever it is, we should replicate it as nearly as possible in the testing. A solvent to be used for degreasing parts may make contact with the skin of the worker, or its vapors may be inhaled, but it is highly unlikely that the worker would ever swallow any. Testing should therefore concentrate on exposure to the skin and lungs.

B. The technique should provide an accurate measure of the amount of compound taken into the test animal. If the food is dosed, then the amount of food consumed by the animal should be recorded. An advantage of the gavage tube or injection is the high level of accuracy with which the dose level is known.

The actual design of the experiment depends on how the dose is to be administered and what responses are to be studied (irritation of the skin, appearance of cancer, amount constituting a lethal dose, and so on).

Length of Testing Time

Testing can vary in time period. When the animal is dosed one time orally, by injection, or by coating on the skin, or when the dosing is for a limited period of time (as for a few hours one time by inhalation), the study is called *acute*. In such a test, we determine the level of the compound sufficient to achieve the desired measure of toxicity. On the other hand, if the testing exposes the animal daily for months or perhaps the animal's lifetime, the study is termed *chronic*. Chronic testing reveals whether there is an accumulation of either the compound itself or the damage it causes. Knowledge of acute toxicity is always needed, since we may have excessive contact with any chemical in our surroundings at any time. Chronic dosing is particularly important *when workers contact a chemical every day* in evaluating whether damage accumulates.

Variability Among Test Animals

The toxicity of a compound is not a constant term. It has been ypointed out already that differences are sometimes displayed between animals of different species, even when they are very closely related.[1] Males sometimes display a greater or lesser degree of sensitivity to a particular toxicant than otherwise identical females.[2] Some of the enzymes that change the chemical structure of a foreign substance in the body are produced at levels that depend on body levels of such a hormone as estrogen, a female sex hormone. This results in variations in toxic response. Whether the variation is an increase or decrease in relative toxicity is hard to predict. These enzymes convert the foreign substance into a more readily eliminated compound, which would seem to predict that more enzyme activity equates to lower toxicity. However, the new chemical may also be more toxic than its unmetabolized predecessor.

Age is another important factor. Newly born (neonate) mammals are often more sensitive than adults, and animals of advanced age may also show increased susceptibility as body defense mechanisms gradually lose effectiveness. For example, dosing at a constant rate in mg/kg with a barbiturate kills week-old rabbits, puts two-week-old rabbits to sleep, and has no effect on four-week-old rabbits.

Even if species, age, and sex are held constant, the effect of a compound on a group of test animals is still not precisely predictable. Wouldn't it be surprising if we were to dose each of 10 animals with the same level each of morphine, have them all fall asleep, then all wake up exactly 32 minutes later? Think of the differences between people: Joan can't eat raw vegetables, aspirin doesn't seem to work for Ralph, sour cream upsets Betty, I can't take certain antihistamines and drive safely, and those same antihistamines don't bother Bob. Individual test animals show variations in response, but the variations tend to cluster in a statistically predictable way around mean or average values. We would be surprised if 10 identically dosed animals awoke from their mor-

---

[1]Comparing three strains of female rats exposed to the same dosage of 7, 12-dimethylbenz(A)anthracene: 0% of the Marshal strain, 16% of the Long-Evans strain, and 100% of the Sprague-Dawley strain developed mammary tumors.

[2]For example, a dose of 7,12-dimethylbenz(A)anthracene that generates tumors in 79% of male rats causes tumors in only 18% of female rats.

phine-induced sleep in exactly 32 minutes, but not if we ran the study twice and the group *averaged* 32 minutes of sleep each time. Variations can usually be explained in terms of differences in the activity of key metabolic enzymes used to chemically alter the toxicant, of the ease of crossing barriers into particular body compartments such as the brain, and of the effectiveness of the systems used to eliminate the compound from the body. All these variations in the response of individual animals to a chemical don't make it impossible to do meaningful toxicity testing, *but they need to be remembered when we design the experiments and interpret the results.*

The experiment must be designed to take these variations into account. To minimize complexities, the animals used are as nearly identical as possible with respect to age and genetic makeup. Standard strains of test animals are available. Unless we want to focus on male/female differences, we average these out by using equal numbers of each sex. Hopefully, we may now determine reasonable toxicity values *for that narrowly defined sample.* Extrapolating the result to humans is a separate problem, one based on clear parameters. Now to do the actual experiments. Suppose we want to determine what level of exposure puts the animal to sleep. Using a few animals and a broad range of dose levels constitutes a *range finding* experiment that narrows the range of concentrations to study in a more comprehensive test. Once the approximate dose is estimated, a second experiment utilizes probably five or more concentrations in that range, with at least 4–5 animals per dose level. If the experiment is properly designed, few become unconscious at lower levels, but an increasing percentage do as the dose level is increased.

REPORTING THE RESULTS—TOXICITY UNITS

Once we see how much of a compound does what degree of damage to what particular animal, our next problem is to communicate that information to all interested parties. For this purpose a whole dictionary of units has been devised. Most of these units are known by letter or letter and number abbreviations.

ACUTE LETHAL DOSE

Traditionally, the most basic information, generally the first to be measured, is the value that describes how much compound taken in a single dose is likely to cause death. For compounds taken orally, injected, or absorbed through the skin, dosage is reported as a weight of compound—either pure or dissolved in a noninterfering solvent—per unit animal weight. Recently, increasing concern about the use of animals in research has led to reassessment of the importance of this value. However, it is a traditional toxicity value, almost certainly listed for every compound that has been evaluated, and often it is the only value

Table 2.1. Selected LD50 Values.

Compound	LD50 (mg/kg, rats, oral)
Glycerol	25,200
Ethanol	10,300
Ethylene glycol	8,500
Acrylic acid	2,600
Hydroquinone	320
Acrylamide	1 70
Acrylonitrile	93
Nicotine	1
Dioxin	0
Botulinus toxin	0.00001

listed. It provides a valuable measure of relative toxicity, for example, allowing us to compare the relative hazard of a compound new to a process with that of a compound it might replace.

That all animals, no matter how carefully selected, do not respond identically to a given dosage now becomes a problem in communication of the findings. Do we select the lowest dose that caused a susceptible animal to die, the highest necessary to kill the most resistant animal, or arrive at an average dose? To resolve this, the first step is to prepare a dose-response curve from the data (Figure 2.1). The response, expressed here as percent mortality, is plotted on the y-axis, and the dose, usually log dose in mg/kg, is plotted on the x-axis. At the center of this curve is found the dose that is estimated to be fatal to half the recipient animals. This dose, the acute LD_{50}, is predicted to be lethal to 50% of the animals, and is the most common value used to describe the relative toxicity of a compound. Once the curve is established, it is possible to describe other measures of toxicity, such as LD_5, the dose lethal to 5% of the animal sample, or LD_{95}, the dose lethal to 95%. We often see LD_{LO}, the lowest value to kill an animal, listed in summaries of toxicological information. Table 2.1

uses some familiar compounds to illustrate the range displayed by these values, and Table 2.2 describes in everyday terms the relative hazard associated with particular LD_{50} values.

In cases where the dose taken by the animal does not have to be adjusted to the size of the animal (for example, when the intake is through the lungs), the standards are based on the concentration of the substance in the environment of the animal. The toxicity is expressed as the LC_{50}, or concentration lethal to 50% of the test group. The units in this case are usually either parts per million (ppm) or milligrams per cubic meter of air (mg/m^3). For a solid or liquid, the "parts" in ppm normally are in terms of weight,[3] but when a gas is discussed the general usage is volumes of gas per million volumes of air. Expressing a given gas concentration by weight and by volume does not produce the same number.

Probit Plots:

The probit plot, an important graphical method for presenting dose/response information, is based on standard deviation calculations. Standard deviation is a statistical technique for describing the degree of deviation a set of data displays from the average value, and is described in Appendix A. In a large collection of data displaying a normal Gaussian distribution, 68.3% of the responses fall between one standard deviation (σ) less than and one standard deviation more than the average, 95.5% fall within $\pm 2\sigma$, and 99.7% fall within $\pm 3\sigma$. This translates into the following:

A.   The dose that affects 50% of the animals is the LD_{50}.
B.   The dose that affects 2.3% of the animals is 2σ below the LD_{50}.
C.   The dose that affects 15.9% of the animals is 1σ below the LD_{50}.
D.   The dose that affects 84.1% of the animals is 1σ above the LD_{50}.
E.   The dose that affects 97.7% of the animals is 2σ above the LD_{50}.

Data may be considered in terms of how many standard deviations it is from the average, rather than in terms of its numerical value. This was first handled by units called *NEDs* (normal equivalent deviations), where zero is the average value and all values less than the average have negative numbers. In order to simplify handling the numbers by eliminating negative numbers, 5 was added to every NED value, so now the average is 5 and one σ less is 4. These numbers are called *probits*.

---

[3]mg per kg used in describing dosage levels is actually mg per 1,000,000 mg, a variation of ppm.

Table 2.2. Toxicity Classes.

Toxicity Rating	Descriptive Term	LD_{50} wt/kg Single Oral Dose in Rats	4–hr Inhalation LC_{50} in Rats (ppm)	Extrapolated Dose (g) for 70-kg Human
1	Extremely toxic	≤1 mg	< 10	<0.07
2	Highly toxic	1–50 mg	10–100	0.07–3.5
3	Moderately toxic	50–500 mg	100–1,000	3.5–35
4	Slightly toxic	0.5–5 g	1,000–10,000	35–350
5	Practically nontoxic	5–15 g	10,000–100,000	350–1000
6	Relatively harmless	≥ 15 g	> 100,000	> 1000

The advantage of this system is that when log dose (x-axis) is plotted against probits as the response unit, a straight line is produced (Figures 2.2a and 2.2b). Visually estimating the best fit to such a line is relatively easy, and computer programs designed to optimize the fit of a straight line to data can be used. Using probits in dealing with toxicologic data sounds complex, but in practice it is greatly simplified by the use of probit paper, special graph paper drawn so that when data is plotted according to the calibration of the axis, all the calculations are handled (just as the semilog paper takes the log for you of data plotted on the log axis).

THRESHOLD TOXICITY VALUES

The value we might wish to determine next is one that reflects the lowest level producing toxic effects (the threshold). This is needed in order to set standards for allowable maximum exposure. A word of caution is necessary before this discussion proceeds further. Whether we are considering rats, mice, or humans, there are a few individuals in any population whose response is extreme. These individuals exhibit behavior that statistically is highly unlikely. This could be an unusual display of resistance, and although this is interesting, it is of no value for setting standards. Unusually sensitive individuals also occur for a variety of reasons. Perhaps the animal was born with an abnormally high or low level of activity in a metabolic pathway that chemically alters the

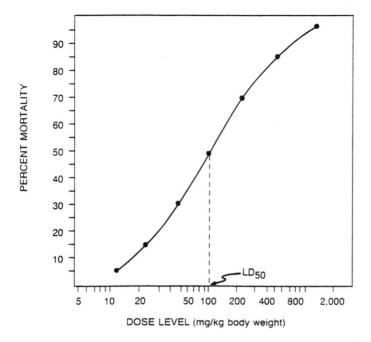

Figure 2.1.   Flow diagram of a chemical into, around, and out of the body.

toxicant. If the pathway detoxifies the compound, low activity produces greater sensitivity. However, if the dangerous substance is one to which the original compound is converted by the pathway, toxic effects increase directly with the activity of this pathway.

Allergic Response

Most commonly, abnormally high sensitivity results when the individual develops an *allergic response* to the compound. This may take the form of the production of itchy welts on the skin, of an asthma response where the breathing passages become restricted, or of some other allergic symptom. This is not the normal toxic response observed for the chemical. Once a person is sensitized to the chemical, extremely low concentrations of the chemical produce the symptoms. A sensitized worker normally must be completely isolated from the chemical thereafter. Some chemicals are prone to produce allergy, and are termed *sensitizers*.

Hypersensitive responses are difficult to predict and are a serious problem when attempting to set standards for exposure. In general, standards are based on the normal distribution of responses, rather than on a few hypersensitive reactions. For some chemicals, such as the isocyanates used to make polyure-

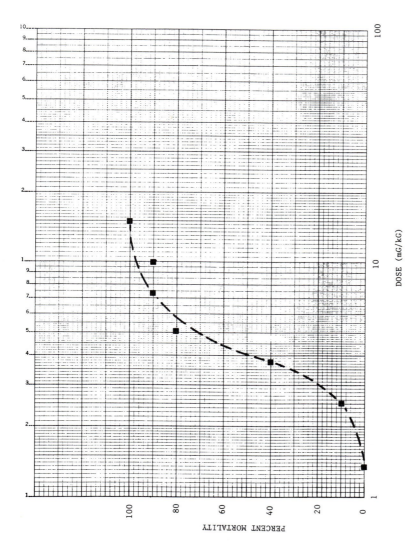

Figure 2.2a.   Here and in Figure 2.2b, the same data is plotted two different ways. Here semilog paper is used, producing the sigmoidal (S-shaped) curve often associated with dose/response studies. Locating a value for LD_{50} is not hard, especially since it is found at the steepest segment of the plot. However, drawing the best sigmoidal curve through the data may present problems.

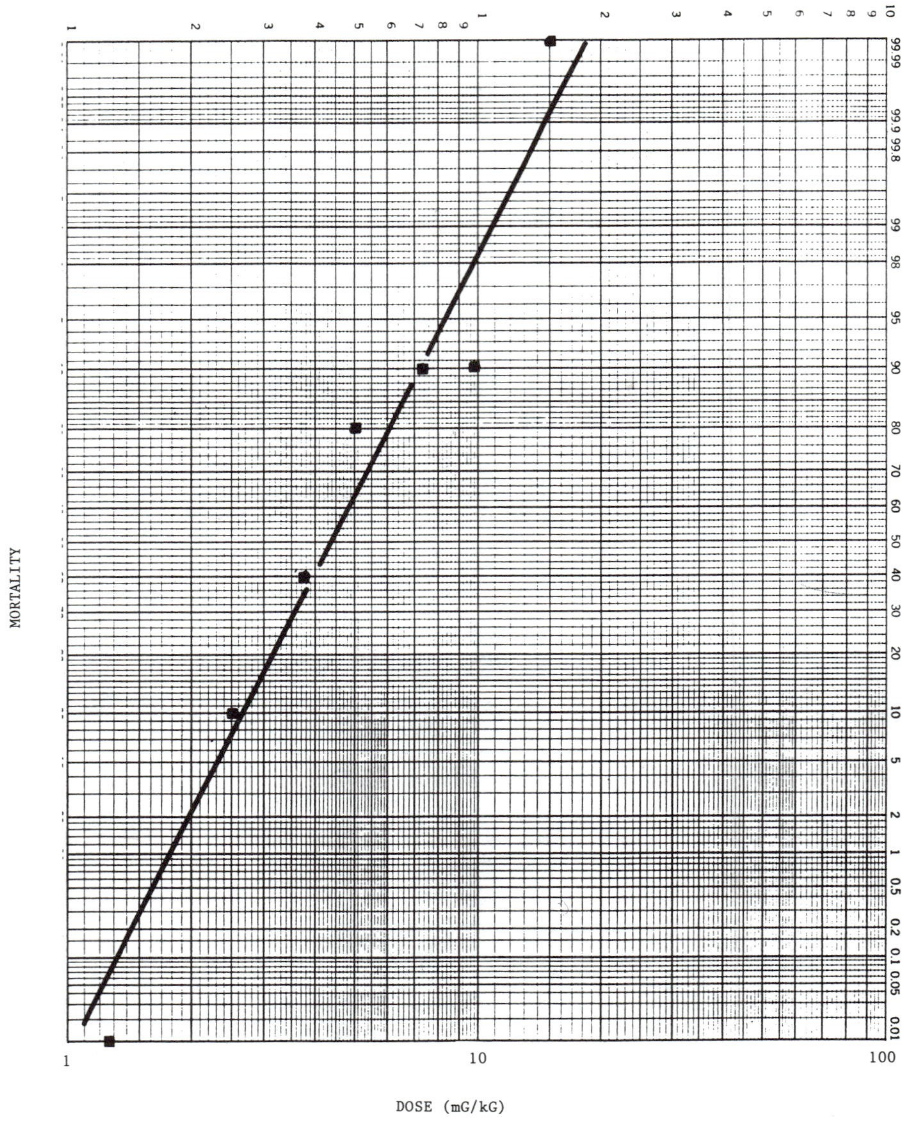

MORTALITY

DOSE (mG/kG)

Figure 2.2b. Here the same data as that in Figure 2.2a is plotted on log vs. probit paper. Once again, the LD$_{50}$ is easy to locate, but drawing the best line through the data is much easier. Furthermore, it is very easy to extrapolate the line to the x axis to provide a threshold value for toxicity, something difficult to do in plot A.

thanes, occurrence of sensitization is so commonplace that the allergic effect becomes the basis for standards.

Animal State of Health

The state of health of the animal is also a factor in response to a chemical being tested. A compound that produces edema (fluid) in the lungs is more deadly to an animal already suffering from a cold or pneumonia. Test animals must be in a good state of health at the start of a test, and symptoms appearing during a test should be the result of exposure to the chemical, not due to an entirely different health problem.

Terms Denoting Threshold Values

Several terms are used to describe a hypothetical dosage that is either the highest level to produce no response or the lowest level to produce a response. These values should be very similar, but often are not because of the variability of animal response and the design of the experiment. Commonly we refer to the lowest dosage to produce a response as a *threshold value* or LD_{LO}. The highest dosage producing no symptoms in test animals generally is termed the *NOEL* or no observed effect level. This may also be abbreviated NEL (no effect level) or NOAEL (no observed adverse effect level). Such a value is of great interest when establishing standards for exposure, but would not itself be used as a standard. We should not deliberately expose ourselves to levels right at the brink of symptomatic response. Possibly, responses actually occurring at that level are not easily observed.

EXTRAPOLATING ANIMAL DATA TO HUMANS

There is considerable uncertainty involved in extrapolating animal test results to the prediction of toxicity to humans. The variability in response within the human population and hypersensitive response must be considered. A "rule of thumb" applied when setting standards is to lower the recommended maximum exposure tenfold to protect against unforeseen higher toxicity in man as compared to the test animals, then to lower it another tenfold to protect the hypersensitive members of the group of humans exposed. The resulting recommendation is dealt with as a permitted level of exposure. The level of the chemical in the workplace should be monitored, and the level found should be no higher than the permitted exposure level. Finally, the health records of the workers in that plant should be monitored in order to learn if the margin of safety was sufficient, or if the chemical has an effect on humans of a type not anticipated by the animal testing.

UNITS FOR RECOMMENDED MAXIMUM EXPOSURE

The recommendations arrived at through animal testing are expressed in a variety of units. Because it is the most likely route of exposure in a factory situation, particular attention focuses on exposure by inhalation. The term MAC (maximum allowable concentration) was used for many years, but is not heard much today. More recently, the TLV (threshold limit value) was established by the American Conference of Governmental Industrial Hygienists (ACGIH).[4] Recommended maximum exposures are expressed on the basis of *time weighted average* (TLV-TWA). Calculation of a time weighted average TLV assumes an 8-hour day and 40-hour week. The exposure levels of the compound are measured at intervals. The TWA exposure is calculated by multiplying the concentration of compound in each analysis by the length of time of exposure to that level. These are summed and divided by the total time to produce an average exposure level.

Sample Problem:

Exposure (ppm)	Time (hr)	Product (ppm x hr)
2	10	20
3	20	60
4	10	40
totals	40	120

$$TWA = (120 \text{ ppm} \times \text{hr})/40 \text{ hr} = 3 \text{ ppm}$$

[4]The ACGIH is a private group headquartered in Cincinnati that for many years has assembled recommendations for maximum exposure to chemicals and for other hazardous exposures in the workplace.

In such a standard, no importance is attached to the occurrence of a short period of high exposure, as long as it is counterbalanced by an adequate period of low exposure. However, if a special hazard exists above a certain concentration in the air, the recommendation will include a *ceiling* value (TLV-C), a concentration the level must not rise above regardless of the duration of that exposure. Limitations on high-level exposure may also be expressed as the acceptable maximum peak, which describes as the exposure limit both a concentration level and a period of time at that level. To use benzene vapor as the example, the TLV-TWA is 10 ppm. On top of this regulation, there is an acceptable ceiling of 25 ppm, so that during this 8-hr period, levels higher than 25 ppm are not allowed for extended periods, even if they are countered by extended periods of low exposure. Finally, there is an acceptable maximum peak of 50 ppm for not more than 10 minutes. This says that the ceiling of 25 ppm may be exceeded, up to 50 ppm, if the exposure at the higher level is limited to 10 minutes. It must be remembered that if the level reaches 25 ppm for some period, or 50 ppm for under 10 minutes, these exposures must be balanced by extended periods of low exposure to allow the time weighted average to stay below 10 ppm. Another term used to express a standard for intermittent high exposure is the short-term exposure limit (TLV-STEL).

Why is such a complex system used? If workers were exposed to relatively constant levels of chemicals in the plant atmosphere, the time weighted average TLV would be a sufficient protection. However, it is not uncommon in a factory to have levels of a volatile material rise sharply for short periods of time. This would be the case when a process is carried out in a closed vat, but at the completion of the process, the vat is opened and the product removed. During that period, the levels of a chemical in the air rise sharply, then drop again when the vat is reclosed.

OSHA STANDARDS

When the Occupational Safety and Health Administration (OSHA) was established in 1970, it adopted TLV values as the best existing standards. However, in a controversial move, it adopted them as ceiling values, rather than as maximum average exposure values. Some of these have since been modified by OSHA, which now uses the term PEL (permissible exposure levels) for its standards.

Recommendations for maximum worker exposure to a chemical are based on characteristics that go beyond the lethal effects. They include consideration of adverse factors such as eye irritation, odor, skin irritation, or ability to produce narcosis. To the person in a plant charged with the responsibility for maintaining a safe environment, knowing the information upon which a recommendation is based is important.

Other recommendations are provided by such sources as OSHA or the chemical manufacturer. OSHA literature lists a value called the IDLH, which stands for the level that is Immediately Dangerous to Life and Health. It was developed as a part of the Standards Completion Program begun in 1974 by OSHA and NIOSH to establish complete occupational health standards. It represents the maximum level from which a person could escape in 30 minutes without any escape-impairing symptoms or irreversible health effects. The recommendations also include requirements for the use of eye protection, protective clothing, or respirators in handling a chemical.

EPIDEMIOLOGY

Another valuable technique used by toxicologists is *epidemiology,* the study of trends and events in similar populations—one exposed to a chemical and one not exposed. Several compounds have been implicated as carcinogens by observing elevated levels of a particular type of tumor in a specific population, then determining the chemical exposure shared by members of that population. This is how the carcinogenic nature of vinyl chloride was discovered, and how the teratogenic potential of thalidomide was first recognized. In Table 2.3 data from a hospital in Germany illustrate the dramatic effect of thalidomide. Once the rise in birth defects was documented, the task remained to determine what exposure the victims of the toxic effect shared.

KEY POINTS

1. Toxic substances enter the body constantly. Only when concentrations in the body rise to disruptive levels, or when damage occurs that is not repairable—or accumulates because it is not repaired before the next exposure—do we experience problems.

2. When setting standards for exposure to a toxicant, the age, sex, and state of health of those likely to be exposed, and the length of time they will be exposed must be considered.

3. Because of the risks involved, preliminary toxicity testing is performed on animals, usually small rodents. It is then necessary to extrapolate the data to predict risks to humans. There is less emphasis on testing with lethal levels of toxicant, and alternatives to the use of animals are being developed.

Table 2.3. Occurrence of Phocomelia.[a]

Time Period	Number of Cases
1949-1959	0
1959	1
1960	30
1961	154

Source: University Pediatric Clinic, Hamburg, West Germany.
[a]Birth defects in which arms and/or legs are malformed or missing. This is a dramatic example of epidemiological data serving to uncover the existence of an unsuspected teratogen. The next problem was to determine what substance had been used by these women which had not been used in the control period of 1949–1959. It was found to be the drug thalidomide.

4. When expressing results of toxicity studies, differences in size between humans and animal surrogates are accommodated by expressing data in mg (dose) per kg (body weight).

5. Dosing animals should be done so that (1) we know the intake accurately, and (2) the manner of dosing resembles the manner of human exposure as closely as possible.

6. The length of exposure to a toxicant in testing varies with the information sought. Studies of acute exposure reveal immediate toxic effects, and chronic exposure studies show whether the compound or damage accumulates in the body.

7. Response to a toxicant varies according to sex and age. To a degree, the response of each animal or human is unique.

8. Testing is done using several small groups of subjects, each at a different dose level. Results are plotted to produce a dose/response curve. From this curve the LD_{50}, or dose lethal to 50% of the subjects, expressed in mg/kg, is determined. Testing airborne toxicants does not involve consideration of animal weight, is expressed in parts per million (ppm), and is termed the LC_{50}, or concentration lethal to 50% of the subjects.

9. The lowest dose to produce a response is the threshold value, and the highest to produce no effect is the NOEL (no observed effect level).

10. Levels initially recommended as the upper limit for safe exposure are roughly NOEL/100. Such recommendations are termed TLV values by the ACGIH and PEL by OSHA. Such values are average exposures for 8-hr days and 40-hr weeks, and are termed time weighted averages (TWA). Ceiling (C) values may be listed either regardless of the time of exposure, or with a maximum length of time permitted at that level. A short-term exposure limit (STEL) is such a ceiling value used by OSHA.

11. Epidemiology in toxicology is a method of identifying toxic compounds by comparing health statistics for an exposed group with those of a similar but unexposed group.

PROBLEMS

1. The recommended adult dose for an antibiotic is one 100-mg capsule three times a day. Would you give the same dose to a small child? If an average adult is 70 kg and a child who is to receive the antibiotic weighs 23 kg, how much antibiotic should the child be given per day, and in what pattern?

2. A chemical is used as a component in the solvent for automobile finishes sold to bump and paint shops. Many of these shops allow the paint to dry in the workplace air in a building that is closed in winter. Testing with rats established that 75 ppm is a safe level of exposure to this solvent. If the average rat in the test weighed 150 g, and a typical adult worker weighs 75 kg, what would your recommendation be for a maximum level of exposure to be allowed in a bump and paint shop?

3. In a facility that rebuilds used fuel system parts for automobiles, including carburetors, fuel injectors, and fuel pumps, a solvent is to be used to dissolve accumulated gums and fuel additives from the old units. Each unit is disassembled and the parts placed in a fine-mesh copper screen basket. The container and contents are then shaken as they are sprayed with hot solvent in an enclosed machine. Workers then open each basket, remove and inspect all the parts, replace all worn and broken components, and reassemble the unit. When the basket is opened, it is not uncommon to find some solvent still in depressions or chambers of the unit. A series of solvents have been proposed, all of which have satisfactory properties for the cleaning operation, and you are asked, as the industrial hygiene consultant to the company, to chose the one that is safest for the workers. Here are measures of toxicity you have assembled for the solvents:

Solvent (boiling point, °C)	LD_{50} (acute, oral, rat, mg/kg)	LD_{50} (acute, dermal, rabbit, mg/kg)	LD_{50} (90-day, oral, rat, mg/kg)	LC_{50} (24-hr, rat, ppm)
A (85°C)	920	3600	880	68
B (79°C)	870	3500	330	220
C (145°C)	125	2900	118	225
D (91°C)	780	850	699	256

As each solvent is considered for use, think of the pathway of worker exposure. Which would you recommend?

4. Table 2.4 displays oral and dermal LD_{50} values displayed by white rats dosed with a series of organophosphate insecticides. Using these values, answer the following questions:

A. Which compound taken orally is known to be the most toxic to male rats? To female rats?

B. Which compound taken orally is known to be the least toxic to male rats? To female rats?

C. Name three compounds where oral toxicity to male and female rats is very similar.

D. Is it true that the relative oral toxicities, male vs. female, are closer in the case of Phorate, where the difference in LD_{50} is only 1.2, than for Trichlorfon, where the difference is 70?

E. Name all the compounds more than twice as toxic orally to males than to females. To females than to males. As a good generalization, does this family of compounds tend to be more toxic to males or females?

F. Are these compounds typically more toxic by the oral or the dermal route?

Table 2.4. Acute Oral and Dermal LD$_{50}$ Values of Organic Phosphorus Insecticides for Male and Female White Rats.

Compound	Oral LD$_{50}$ (mg/kg)		Dermal LD$_{50}$ (mg/kg)	
	Males	Females	Males	Females
Carbophenothion	30	10.0	54	27
Chlorthion	880	980	<4500	4100
Co-Ral	41	15.5	860	–
DDVP	80	56	107	75
Delnav	43	23	235	63
Demeton	6.2	2.5	14	8.2
Diazinon	108	76	900	455
Dicapthon	400	130	790	1250
Dimethoate	215	–	400	–
Di-syston	6.8	2.3	15	6
EPN	36	7.7	230	25
Ethion	65	27	245	62
Fenthion	215	245	330	330
Guthion	13	11	220	220
Malathion	1375	1000	>4444	>4444
Methyl parathion	14	24	67	67
Methyl trithion	98	120	215	190
NPD	–	–	2100	1 800
Parathion	13	3.6	21	6.8
Phorate	2.3	1.1	6.2	2.5
Phosdrin	6.1	3.7	4.7	4.2
Phosphomidon	23.5	23.5	143	107
Ronnel	1250	2630	–	–
Schradan	9.1	42	15	44
TEPP	1.05	–	2.4	–
Trichlorfon	630	560	>2000	>2000

G. Let us look at the dermal toxicity of Diazinon to males and females, of Schradan to females, and of Phosdrin to males. For each insecticide, what do the values tell us? What does high dermal toxicity imply?

    H. Is it a good generalization to say that compounds that are more toxic orally are also more toxic dermally?

5. Recall that 1 liter (L) is 1000 cm^3, and that 1 mol of a gas at STP (standard temperature and pressure, or 0°C and 1 atm) occupies 22.4 L.

    A. How many liters are there in a cubic meter of air?

    B. How many liters does 1 mole of gas occupy at 25°C and 1 atm pressure?

    C. Now, we sometimes see concentrations in air expressed as ppm (1 volume per 10^6 volumes) and sometimes as mg/m^3. How many mg/m^3 are there of a gas whose concentration is 1 ppm and whose molecular weight is 100 g/mol at 25°C?

    D. We have a sample of air at 25°C that has two vapors in it whose concentrations are 2 ppm and 4 ppm, and whose molecular weights are 320 and 150 g/mol respectively. Which has the highest concentration in mg/m^3?

6. You are on a committee trying to minimize the use of animals in toxicity testing. The suggestion is made that if you start by knowing the level of concentration in the air that a compound in question reaches in the workplace, you could simply expose an animal to that level to see if it is safe. You object to that approach because exposure recommendations should have a safety margin. (A) What is the safety margin described in the text, and (B) how would this "one animal, one level" test be modified to incorporate that margin? A toxicologist on the committee says, "Suppose we used that approach, let us consider what the results would tell us. There are two possibilities: the animal lives or the animal dies. If the animal lives, we propose to say that this means the compound is safe at that level. (C) Is there any problem with that supposition?" Another participant then says, "I see where you are going with that, and of course the other side of it is: (D) what information do we lack if the animal dies?"

7. A chemical used in cutting oils has been found to resist removal by preliminary treatment in industrial wastewater and is going to the local sewage treatment facility. On checking, it is found that its oral acute toxicity is unknown, and tests are immediately undertaken. One proposed experimental design is to dose test animals by way of their drinking water, and another is to inject the animals intraperitoneally (in the abdominal cavity, where blood vessels readily absorb chemicals). (A) What are the advantages and disadvantages of each approach? Injection is selected as the method of study, and 90 Sprague-Dawley rats are purchased. They are divided into

9 groups of 10 rats each, each group receiving a different dose level. Assume for simplicity that each of the rats is very close to 200 g in weight. The following data are collected:

Group Number	Concentration of Compound (g/mL)	Volume Injected (mL)	Number of Rats Surviving
1	0.0010	0.25	10
2	0.0010	0.50	9
3	0.0010	0.75	6
4	0.010	0.10	2
5	0.010	0.15	1
6	0.010	0.20	1
7	0.010	0.30	0
8	0.010	0.40	0
9	pure water	0.20	10

(B) From this data, what is the LD_{LO}? (C) What is a value for NOEL? Now you are going to plot the results. (D) Decide first what units your x-axis should have in your dose/response plot, and convert this data to those units. Then plot the data on three types of graph paper: conventional linear paper, semilog paper and probit paper. Determine a value for LD_{50} from each. One of these plots is particularly good for estimating a threshold value. (E) Which plot is that and what is the value?

BIBLIOGRAPHY

E. J. Calabrese, *Principles of Animal Extrapolation*, Lewis Publishers, Chelsea, MI, 1991.

D. Dollberg, *Analytical Techniques in Occupational Health Chemistry*, American Chemical Society, Washington, 1980.

Filov, Golubev, Liublina, and Tolokontsev, *Quantitative Toxicology*, Wiley, N.Y., 1979.

W. J. Hunter, Ed., *Evaluation of Toxicological Data for the Protection of Public Health*, Pergamon Press, N.Y., 1977.

M. Kapis and S. Gad, Non-Animal Techniques in Biomedical and Behavioral Research and Testing, Lewis Publishers, Chelsea, MI, 1993.

C. Klaassen, M. Amdur and J. Doull, *Toxicology*, 3rd. ed., Macmillan, N.Y., 1986. The first five chapters provide excellent basic coverage regarding the measurement of toxicity.

F. W. Mackinson, R. S. Stricoff, and L. J. Partridge, *NIOSH/OSHA Pocket Guide to Chemical Hazards*, U.S. Government Printing Office, Washington, 1978. A compact summary of PEL and IDLH values, physical properties, precautions in handling, and health hazards for 375 compounds used in industry.

NIOSH, *Registry of Toxic Effects of Chemical Substances*, U.S. Government Printing Office, Washington.

CHAPTER 3

TOXICOKINETICS

A chemical must contact us in some fashion in order to cause injury. Some substances cause damage simply by contacting the skin, or the most sensitive part of our outer surface, the eyes. In this chapter, however, we are concerned with substances that enter the body and are distributed by the bloodstream. Once a toxicant is in the body, we are concerned about how long it stays, what organs it damages, and how it is removed. The overall pathway is shown in Figure 3.1.

ENTRY OF TOXICANTS INTO THE BODY

When we say that a chemical has entered the body, we usually mean it has gained access to the circulatory system (the bloodstream). *Three important routes to the circulatory system are absorption through the stomach or intestines (the G.I. tract), the skin, and the lungs.*

THROUGH THE G.I. TRACT

If a person is described as having been poisoned, one immediately thinks of a person who has swallowed a chemical. Once swallowed, the chemical can cross the mucosal barrier of the stomach or intestine (the layer of cells lining the lumen of these organs) and enter the blood. This is the most important route of entry for toxicologists concerned with the safety of food, pharmaceuticals, or the water supply, so their studies focus on the hazards of specific ingested toxicants.

However, this is the least important pathway into the blood threatening the worker. Few workers would voluntarily place industrial chemicals in their mouths. Chemicals can inadvertently enter the G.I. tract if the worker does not carefully wash up before a meal break or at the end of the day, and later transfers the chemical to food or a beverage. The prevention of such accidental intake is a matter of stressing personal hygiene to the worker whenever possible.

THROUGH THE SKIN

Absorption through the skin is a more important problem in the workplace. Workers are likely to have contact with solvents, wash solutions, coolant solutions, lubricants, and other liquids in the course of performing their jobs. Beyond the question of irritation to the skin (Chapter 5), the hazard to the worker from such contact depends on the answers to two questions. First, how effective is the skin as a barrier to the entry of the compound into the body? Second, if the compound does enter the body, how much can the body tolerate without ill effects? Remember that if the worker is close enough to a chemical for skin contact, he or she may also be inhaling vapors of that chemical, adding to the level of intake.

The skin functions to prevent the loss of body fluids, and the resultant structure effectively blocks random entry of many chemicals. This is accomplished by an outer layer of tightly interlocked surface cells, the *keratin layer* of the *epidermis*. As these cells develop, they move farther away from the blood supply, replace normal cell contents with protein fiber, and cease to be living cells (Figure 3.2).

However, skin has blood vessels below this keratin layer, and some compounds are able to penetrate to these blood vessels. Several solvents belong in this group. Passage through the skin is easier where the skin is thinner, so it is more threatening to spill a chemical on the forearm than on the palm of the hand. Warm water, especially water containing detergents, and some organic liquids loosen and soften the keratin layer, reducing its effectiveness as a barrier.

Pores and hair follicles, which pass through the keratin layer, represent a potential alternative pathway for toxicant entry. However, they are sufficiently infrequent in occurrence that they are considered to be of minor importance. Damage to skin—as by cuts, abrasions, or burns—increases the likelihood of a compound gaining entry. For this reason, when dermal irritation and the potential for entry of a chemical through the skin are tested with animals, half the skin in the contact area is often lightly abraded.

THROUGH THE LUNGS

Finally, compounds may enter the body by way of the lungs, the route of greatest concern in the workplace. Users of illegal drugs that "smoke" the drug

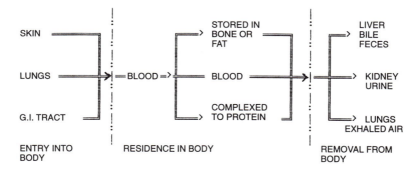

Figure 3.1. Flow diagram of a chemical into, around, and out of the body.

(*crack* cocaine or *ice* amphetamines, for example) take advantage of the lungs being a rapid route of entry to the blood that bypasses metabolic mechanisms of protection in the liver. Most standards for permitted maximum exposure to a chemical (TLV, PEL) are standards for concentration of the chemical in the workplace atmosphere. Overall, the largest single effort in plant inspection for health and safety is directed at potential worker exposure to chemicals in the air.

Lungs consist of a highly branched set of air passages—the *bronchial tree*—terminating in little sacs called *alveoli* (Figures 3.3 and 3.4). The barrier between blood and inhaled air is very thin in alveoli in order to facilitate the exchange of oxygen and carbon dioxide, and it is chiefly here that inhaled chemicals gain entry to the blood. More details about this route of entry are given in Chapter 6.

DISTRIBUTION OF TOXICANTS THROUGHOUT THE BODY

Compounds are transported by the blood to all parts of the body. Their immediate fate depends on two factors:

1. *Binding to protein.* Blood contains a variety of proteins, some of which bind to specific substances to transport them. Hormones, oxygen, and metal ions are examples of substances normally transported in this fashion. Some toxicants also bind to protein, which prevents them from passing through the capillary walls and entering the surrounding cells. Although this prevents them from causing damage, it also means they cannot enter the kidney tubular system to be removed from the body with theurine, so their stay in the body is extended. That is the good news and the bad news.

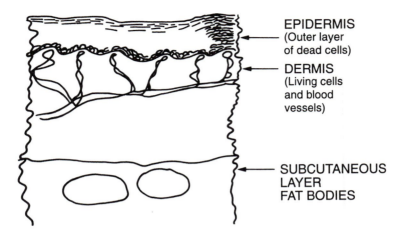

EPIDERMIS
(Outer layer
of dead cells)

DERMIS
(Living cells
and blood
vessels)

SUBCUTANEOUS
LAYER
FAT BODIES

Figure 3.2.   Cross section of the skin.

2. *Polarity of the compound.* Cell membranes are very nonpolar; they are essentially grease barriers. Chemicals that are very polar, such as most nutrients and minerals, would not be able to cross these barriers into the body cells without specific transport systems built into membranes for that purpose. Polar toxicants are similarly blocked by membranes, although a few are moved by the nutrient transport systems because they resemble substances the cell needs.   By contrast, nonpolar toxicants dissolve through the membranes and move relatively freely into body tissues. They may then enter body fat much as they would dissolve into a nonpolar solvent and remain stored there to be released later.   For example, storage in body fat of nonpolar environmental toxicants such as DDT, PCB, and PBB has become an important concern.   Levels in the body can become relatively high when exposure to those compounds continues over a period of time, even when the level of exposure at any given time is relatively low.   For nonpolar toxicants that is the bad news and the bad news.

METABOLISM

Throughout the body, but particularly in the liver, enzyme systems convert nonpolar foreign substances into different chemical species. Readers familiar with the enzymes of normal metabolism would find these systems surprising. There are only a few different enzymes, and rather than being highly specialized, these few collectively are able to chemically alter almost any organic

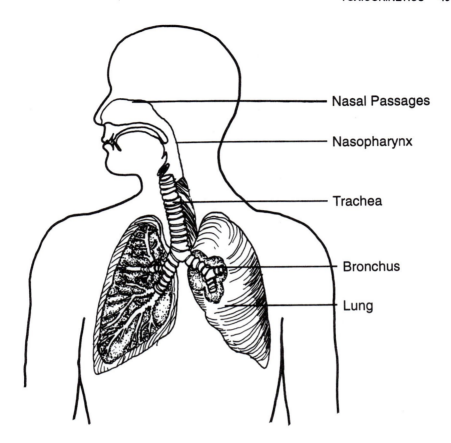

Nasal Passages

Nasopharynx

Trachea

Bronchus

Lung

Figure 3.3. The respiratory system. All the air passages up to but not including the terminal air sacs are covered with a sticky mucous surface
that traps inhaled particles. Hairlike cilia move the mucus toward
the nasal passages, carrying the trapped particles along. *Source:*
Godish, T. *Air Quality* (Chelsea, MI: Lewis Publishers, Inc., 1985).

structure. However, they catalyze changes at a much slower rate than do
conventional enzymes.

Most of the changes promoted by these enzymes fall into two broad classes.
One class is *oxidation*, and in oxidation reactions an important generalization
is that the substances become more polar (Figure 3.5). The other class, *conjugation*, involves coupling foreign substances to such structures as sulfate ions
or charged derivatives of carbohydrate (Figure 3.6). Here, too, the effect is to
increase the polarity of the compound. This increased polarity reduces access
of the compound to cells, and facilitates removal from the body.

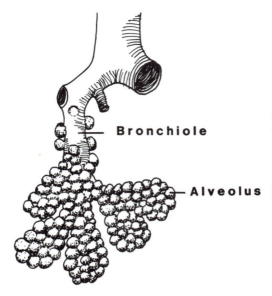

Figure 3.4.   This figure shows the alveoli—thin-walled air sacs found at the ends of the finest subdivisions of the air passages. In these sacs, the barrier between inspired air and the blood is very thin; the exchange of gases takes place through these thin walls. *Source:* Godish, T. *Air Quality* (Chelsea, MI: Lewis Publishers, 1985).

It does not follow necessarily that this metabolism reduces toxicity at the same time. It may, but there are also cases where just the opposite occurs. For example, combustion produces polycyclic aromatic hydrocarbons (PAHs), which are relatively harmless compounds. However, they potentially could be retained in the body for a long time because they are nonpolar. Oxidation by metabolic enzymes converts them into products that are more polar, but these polar products are potent carcinogens. Once again we have a good news/bad news situation.

REMOVAL FROM THE BODY

KIDNEYS

The chief pathway for removal of toxicants from the body is their transfer in the kidneys from the blood into the urine. Each time the heart contracts, about 25% of the blood pumped out to the body passes through the kidneys. The basic functional unit of the kidney is the *nephron,* a structure that starts with the *glomerulus,* where fluid is removed from the blood, extends through a long

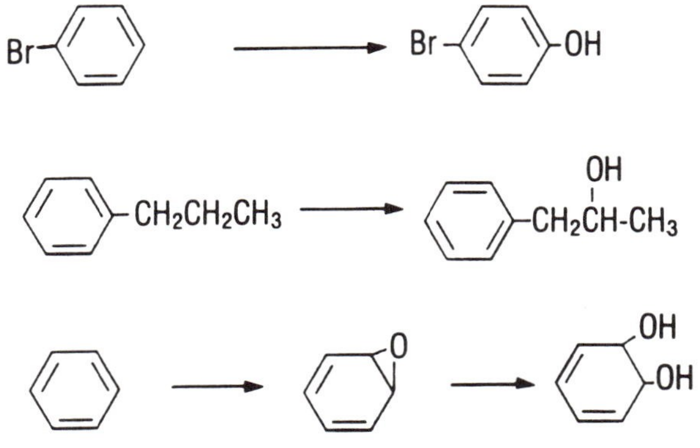

Figure 3.5.   Oxidation of toxicants in the body. The products of oxidation, which goes on largely in the liver, are more polar, making it easier for the kidney to take them out of the blood. Oxidation also provides attachment points for conjugation, shown in Figure 3.6.

tubular system in which the content and concentration of the fluid is adjusted, and ends with the fluid (now urine) being passed into a collecting system that leads to the bladder.

In the glomerulus, blood is literally filtered through a porous membrane so that a water solution of small molecules enters the tubular system and the relatively large blood cells and proteins remain in the blood.[1]   The filtrate passes down a tube, which has blood vessels closely associated with its walls. Selective transport proteins that are built into this tube bind desirable compounds such as nutrients and important minerals and move them back into the blood vessels. When the fluid reaches the end of this tube, glucose, amino acids, sodium ion, and about 99% of the water have been scavenged (Figure 3.7). As water returns to the blood, the concentration of solutes remaining in the fluid increases. Polar solutes are trapped in the absence of transport structures specific for them. However, nonpolar molecules are now much more concentrated in the urine than in the blood, and driven by this concentration gradient, dissolve through membranes of the tube walls and return to blood without the help of a transport structure.

The term *clearance* is used to describe the efficiency of the kidney at removing substances from blood. Clearance is *the volume of blood completely cleared*

---

[1]One of the symptoms of kidney damage is the appearance of blood or protein in urine, inferring that the glomerulus has been damaged and is allowing larger structures to pass into the nephron.

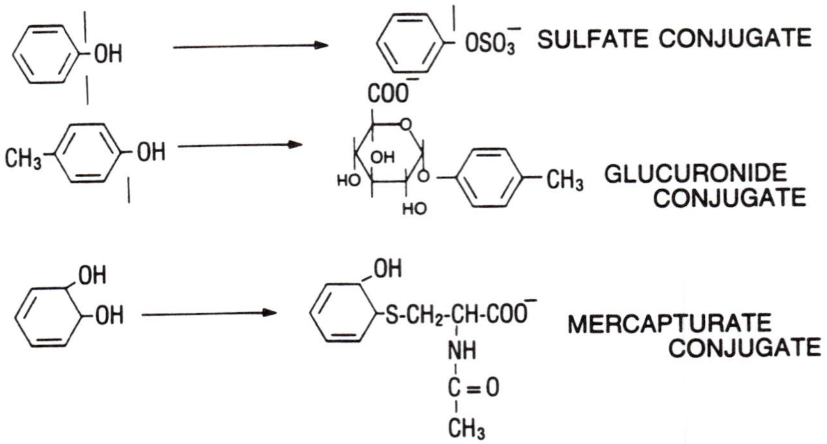

Figure 3.6.   Conjugation of toxicants in the body. In this process, the foreign substance is attached to another molecule. Notice that in each of these cases, the attached molecule is charged, making the combination much more water-soluble. This assists in the elimination of the substance from the body via the kidney.

of a substance per unit time, and has units of mL/min.[2]   For a given compound (t), at a time when t is not entering the body, the amount leaving the blood equals the amount added to the urine:

$$U_t V = Cl_t B_t$$

where

U_t =   concentration of t in urine
B_t =   concentration of t in blood
V =   volume of urine excreted per minute
Cl_t =   clearance of t

$$Cl_t = U_t V / B_t$$

---

[2]The definition of clearance is misleading, since it seems to infer that the kidney takes a volume of blood and completely cleanses it of toxicant or waste product. In fact, compounds add to urine as the result of equilibria involving all of the blood filtrate, some equilibria being influenced by transport structures.

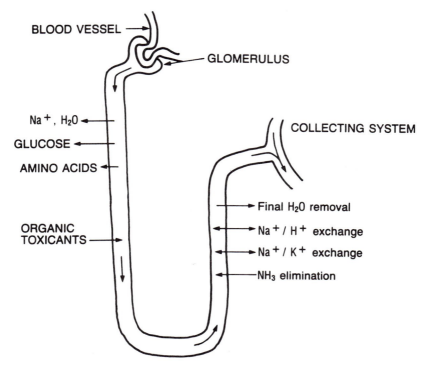

Figure 3.7. Kidney function. The fluid portion of the blood is filtered into the tubular system at the glomerulus. As this fluid (which contains the wastes of metabolism and the toxicants that have entered the blood) moves through the tubes, those parts that the body wishes to save are moved back into the blood, whose vessels line the tubular system. In particular, glucose and amino acids–important nutrients–and sodium ion are saved. As the sodium ion is moved back into the blood, water follows to preserve the osmotic pressure balance. Organic toxicants are pumped into the urine from the blood, if they passed the filtration step earlier. Finally, at the end of the tubular system, exchanges are made that adjust blood acidity, and the last of the water retention is performed.

Sample Problem:

If 25 mg/mL of toxicant t are excreted in urine at a rate of 2.0 mL/min, and the blood concentration of t is 0.5 mg/mL, what is the clearance of t?

$$Cl_t = U_t V / B_t$$

$$Cl_t = (25 \text{ mg/mL})(2.0 \text{ mL/min}) / 0.5 \text{ mg/mL}$$

$$Cl_t = 100 \text{ mL/min}$$

A high clearance value is displayed by a substance that is cleared rapidly from the blood. A comparison of the expected clearance of a compound (such as creatinine) regularly appearing in urine with its clearance in a particular subject provides a measure of the operating efficiency, and thus the health, of the kidney in that subject.

Basically, waste products and polar foreign substances are concentrated in the urine and are sent to the bladder. (Other adjustments to urine are also performed by the kidney that are of less importance to this discussion.) When compounds in the urine are carcinogens, such concentrating raises the dosage of surrounding tissues and is the cause of many kidney and bladder tumors.

LIVER

The liver removes a few foreign substances from the blood, transfers them into a fluid called bile, and sends them for storage to the gall bladder. Bile is primarily a solution of detergent-like molecules. When food enters the small intestine, the gall bladder contracts and sends the bile into the intestine, where the detergents of bile, called *bile salts,* stabilize the dietary fats as droplets (micelles). The toxicants placed in bile by the liver pass along the G.I. tract with food and eventually are eliminated with the feces, unless during transit they are reabsorbed through the intestinal wall. At best, this pathway eliminates only a small fraction of the amount and number of foreign substances removed by the kidney.

HALF-LIFE OF TOXICANTS

One of the studies performed on potential toxicants is the determination of their half-life in the body. By half-life we mean the length of time required, when a known amount of a compound is taken into the body, for half of it to be removed again. Half-life depends on all the factors presented in this chapter. Compound polarity is an important factor. Polar compounds, restricted in their ability to cross membranes, tend to distribute less well into cells from blood, and remain trapped in urine, facilitating their removal from the body. They display shorter half-lives because of these factors. Nonpolar compounds are more difficult to eliminate since they distribute throughout the body cells, store in the fat, and transfer spontaneously back to the blood in the kidney by dissolving across membranes. Their half-life is generally much longer.

If a compound is bound to a blood protein, its half-life is greatly extended. It does not enter cells readily, but it is not filtered out of the blood in the kidney either. Such binding is not a permanent attachment. During periods when the foreign molecule equilibrates off (is temporarily freed from) the protein, it acts like any similar unbound molecule and may then exert toxic effects or be eliminated.

Not mentioned earlier is the ability of a few toxic metal ions, because of their similarity to calcium ion, to be incorporated into bone structures. Cadmium and lead ions are significant examples. Bone is not a static tissue, but exchanges its ions with fluids surrounding it. There is a fluid bathing the bone surface that contains dissolved ions, and this fluid is separated by a membrane from surrounding body fluids. Transport structures to move ions either way are found in this membrane, and the regulation of blood calcium concentrations is accomplished by transport at these sites under the control of a series of hormones. Ions resembling calcium are moved at the same time. Toxic ions incorporated into bone do not permanently remain there, but their entry into it greatly extends their half-life.

KEY POINTS

1. Toxicants enter the blood through the G.I. tract, skin, and lungs, to be distributed throughout the body. The G.I. tract is a minor route of entry in the workplace, since workers would seldom swallow industrial chemicals. Passage through the skin, involving crossing the keratin layer of protective cells, is more likely. The lungs are the pathway of greatest concern, since they provide a very large surface across which airborne toxicants may transport.

2. In blood, toxicants may bind to proteins, which prevents them both from entering tissues to cause damage and from being removed from the blood by the kidneys.

3. Nonpolar toxicants cross membrane barriers easily, and so are hard to remove from the body.

4. Many cells, but particularly liver cells, metabolize toxicants to facilitate their removal from the body. Oxidation makes toxicants more polar and conjugation attaches them to polar structures.

5. The chief exit route from the body is transference to urine in the kidneys. A few compounds transfer to bile in liver and, when bile is added to the intestinal contents, may leave with the feces.

6.   The time required to remove half the dose of a toxicant from the body is the toxicant's half-life.

PROBLEMS

1.   Workers get cutting oils on their hands during a gear cutting operation in a transmission plant. A chemical used in cutting oils has an acute oral LD_{50} of 195 mg/kg. What is the likelihood of a worker being injured by eating an apple on break without washing his or her hands first? If the worker did this as a regular practice, what characteristics of the chemical could raise the level of hazard from this careless practice?

2.   Workers in the repair and maintenance facility of a highly automated plant building two-cycle chain saw engines and drive systems have borrowed some solvent from the painting operation to clean grease from parts before they work with them. A part is dipped briefly in the solvent, then carried to a bench. The solvent has a dermal LD_{50} of 90 mg/kg. When challenged that this is hazardous, they point out that no one has shown any ill effects from the practice, and the solvent does an excellent job of degreasing. You counter by saying, "You have been lucky so far. Here are circumstances that just haven't happened yet that would raise the toxic potential of skin exposure to this chemical: ..."

3.   On testing, the following compounds have shown potential as mold growth inhibitors to be added to water-based paints:

A.
$$CH_3-CH-CH_2-CH-CH_2-CH_2-NH_2$$
with CH_3 on the third carbon and $O-CH_3$ on the fifth carbon

B.
$$CH_3-CH-CH_2-CH-CH_2-CH_2-NH_2$$
with CH_3 on the second carbon and OH on the fourth carbon

C.
$$CH_3-CH-CH_2-CH-CH_2-CH_2-N(CH_2-CH_3)_2$$
with CH_3 on the second carbon and $O-CH_3$ on the fourth carbon

D.
$$CH_3-CH-CH_2-CH-CH_2-CH_2-N^+(CH_2-CH_3)_3$$
with CH_3 on the second carbon and OH on the fourth carbon

E.

$$CH_3-CH-CH_2-CH-CH_2-CH_2-N(CH_2-CH_3)_2$$

with substituents: CH_3 on the second carbon, $O-CH_3$ on the fourth carbon, and $CH_2-CH_2-CH_3$ below the second carbon.

A. Predict the relative half-lives these would display in the body based on relative polarity.

B. As a first assumption would you expect values for clearance by the kidneys to be in the order you would have predicted for half-life?

C. Using test animals to study compounds A and C, the following data was obtained:

Compound	Volume of Urine (mL/2 hr)	Conc. in Blood (µg/mL)	Conc. in Urine (µg/mL)
A	1.5	9.1	38
C	1.4	7.2	41

What are the clearance values?

D. Do these clearance values correspond to predictions based just on polarity?

E. What factor other than polarity affects half-life?

BIBLIOGRAPHY

E. J. Calabrese, *Pollutants and High Risk Groups: The Biological Basis of Increased Human Susceptibility to Environmental and Occupational Pollutants*, Wiley-Interscience, N.Y., 1978.

C. D. Klaassen, M. O. Amdur, and J. Doull, *Toxicology: The Basic Science of Poisons*, 3rd ed., Macmillan Publishing Co., N.Y., 1986. A very thorough and complete presentation of the material in chapters 1-3 is given here in Unit 1.

CHAPTER 4

GOVERNMENT REGULATION

THE OCCUPATIONAL SAFETY AND HEALTH ACT

Regulation of occupational safety in the workplace is primarily the responsibility of the Occupational Safety and Health Administration (OSHA), an organization within the Department of Labor of the federal government. This agency came into being to fulfill the requirements of the Williams-Steiger Occupational Safety and Health Act of December, 1970, (OSH Act; Public Law 91-596). In the House of Representatives, the Health and Safety Subcommittee of the Education and Labor Committee, and in the Senate, the Labor Subcommittee of the Labor and Human Resources Committee have jurisdiction.

BEFORE THE OSH ACT

Before passage of the OSH Act, regulation of health and safety matters in the workplace was a state responsibility, with a few exceptions, such as contractors to the federal government. Standards were not uniform across the country, nor was the level of enforcement. The American Conference of Governmental Industrial Hygienists (ACGIH) provided some common ground in setting standards, since their threshold limit values (TLV) were widely, though not uniformly, employed. In discussing the history of this legislation, Mary Worobec *(Toxic Substances Controls Primer)* describes the situation as it existed before passage of the OSH Act, quoting these statistics:

Each year 14,000 workers died and 2.2 million were disabled by accidents in the workplace, according to former Labor Secretary George P. Schultz.

Work-related deaths and injuries were causing annual losses of $1.5 billion in wages and $8 billion in gross national product.

About 65 percent of U.S. workers were being exposed to harmful physical agents, yet only about 25 percent of them were adequately protected.

A total of 390,000 new cases of occupational disease occurred each year.

Chemical agents, including lead, mercury, asbestos, and cotton dust, posed known health threats to workers, but were not controlled.

A new, potentially toxic chemical was being introduced in industry every 20 minutes, according to a Public Health Service estimate.

PROVISIONS OF THE OSH ACT

The OSH Act has as a goal the establishment of standards to prevent injury or illness among workers. It applies to workplaces with more than 10 employees, but exempts employers already regulated by another federal agency (farms, mines), the activities of governmental agencies, and the self-employed. Although federal, state, and local government employees are not protected by OSHA, the act requires that these governmental units maintain a program consistent with that of OSHA.

In brief, the powers and responsibilities of OSHA include the following:

A.   To establish safety and health standards.

B.   To conduct inspections and issue citations.

C.   To require records of safety and health to be kept by employers, and in conjunction with Health and Human Services, keep occupational health and safety statistics.

D.   To train employers, employees, and personnel employed to enforce the act.

The act also establishes a research agency, the National Institute of Occupational Safety and Health (NIOSH), organized under the Department of Health and Human Services, to perform health studies and to recommend new standards. NIOSH studies may be laboratory centered, or they may focus on the workplace.

An Occupational Safety and Health Review Commission was established to serve as a board of review to investigate and rule on the appropriateness of actions of OSHA, such as penalties and citations, when complaints are lodged. The commission has three members who are appointed by the president for six-year terms and confirmed by the Senate. The commission is the highest authority in these matters within OSHA, but its rulings can be reviewed by federal courts.

The OSH Act is intended to protect workers from either physical or chemical agents in the workplace.[1] Physical agents include noise, extreme temperature, and types of radiation. Chemical agents are regulated by setting limits on exposure, and by an OSHA cancer policy. Hazards from chemicals must be indicated by labeling, and workers must be trained regarding safe use of chemicals. Recordkeeping of worker exposure is required, and inspections of the workplace are authorized. Safety standards are intended to protect the worker from such accidents as fires, cuts, falls, electrocution, and construction collapse.

THE SOURCE OF STANDARDS

The standards for safety and health were initially set up utilizing existing federal standards or consensus standards that had been recommended at that time for about 400 chemicals. For example, the 1968 ACGIH TLV values were adopted as official exposure standards,[2] as were American National Standards Institute acceptable concentration values.[3] Such standards for exposure were now referred to as *permissible exposure levels* (PEL), and became not recommendations, but law. PEL values are not carved in stone, but may be changed as new information becomes available, for example through advice from NIOSH. Further, substances can be added to the original list at any time.

The decision to consider a new substance or working condition for regulation may be made by OSHA, or it may originate outside OSHA. For example, labor or employer groups may request the establishment of a standard. NIOSH, since it is authorized to conduct research on problems related to health and safety, is the most likely external source of such a request. Recommendations by NIOSH, since they represent the results of NIOSH research and may foretell new regulatory directions, are often included in references listing standards, but they are only enforced when they are adopted by OSHA as legal standards.

When OSHA plans to create a new standard, it may first publish an early warning notice in the *Federal Register* to allow time for public reaction. There then may be a public hearing. This early warning approach allows OSHA to gather information and assess public reaction, without itself being under pressure of a deadline. Whether or not an early notice is published, if OSHA wishes to set a new standard, it is *required* at some point to *publish* the potential standard as a *proposed rule*. Then, in no less than 30 days, or in no more than 60 days after the completion of a public hearing, the final standard is

---

[1]The agency is challenged to protect 90,000,000 workers in 6,000,000 places of employment with a budget of $294,000,000 (1994).
[2]See 29 CFR 1910.1000, Table Z-1.
[3]See 29 CFR 1910.1000, Table Z-2.

published. Disagreement with the standard from this time on is a matter that must be handled by the courts.

Occasionally, where evidence has been presented of a serious threat not adequately dealt with by the existing standards, and delay caused by the usual mechanisms to introduce new standards is risky, the Labor Department has instituted an *emergency standard*. Such standards go into effect at the time of publication in the Federal Register. Emergency standards were set for vinyl chloride, asbestos, and benzene. These were replaced later by standards passed in the conventional fashion. Such replacement must occur within 6 months.

ENFORCEMENT OF STANDARDS

Enforcement of standards begins with an inspection of the workplace. Inspectors normally visit an establishment without advance notice, and the inspectors have the right to inspect not only the workplace itself, but also the employer's records of injury and illness. The decision to inspect a particular facility may be based on a worker's complaint, occurrence of a serious incident involving the health of the workers, a fatality, or a catastrophe wherein five or more workers are hospitalized. Otherwise, inspections are scheduled in a pattern that favors more frequent visitation of higher-risk facilities.

Personnel involved in inspections are called industrial hygienist compliance officers. These individuals are trained to recognize and evaluate health hazards. They interview workers, collect dust and vapor samples, measure noise levels, and watch for such evidence as eye irritation, odor, or visible emissions.

An OSHA industrial hygiene inspection is more complex and formal. There is a conference with the plant management, a walk-through inspection, scrutiny of records, the collecting of samples and measurements according to a standard format modified to be appropriate to the particular plant process, and a closing conference with the employer. In addition to the results of the inspection at this conference, the employer's occupational health program is discussed.

In certain standards, OSHA defines half the PEL as the *action level*. A number of provisions of the safety standards are not required if levels in the plant are below the action level, including some aspects of monitoring, training of employees, and medical surveillance.

If the inspection uncovers violations, citations may be issued and penalties may be imposed. Violations can reflect several levels of seriousness. "Imminent danger" reflects a condition expected immediately to cause death or serious injury. "Serious violation" describes conditions which probably would cause death or serious injury under certain conditions. Anything less than the above is classed as "Other violations."

The employer may contest citations, penalties, time allotted to correct a violation, or notices of failure to correct a violation. However, this must be done to the area director in writing within 15 federal working days of notification. Such a contest *must be reported to the employees.* The appeal may be handled by the area director, or in extreme cases it is heard by the Occupational Safety and Health Review Commission, an adjudicatory commission independent of the Department of Labor. At such a hearing, the Department of Labor must prove its case against the employer. The decision of the judge in such a hearing may be sent for review by the United States Court of Appeals.

The Department of Labor, if it finds a very dangerous situation, can post imminent danger warnings. However, beyond informing the employees of the hazard, the Department has no further authority and must seek the assistance of the courts if the employer refuses to close down the operation.

THE CODE OF FEDERAL REGULATIONS

Every weekday, information about all the new rules, meetings, and other congressional proceedings is printed in the *Federal Register.* This is a huge collection of volumes that is indexed so that original submissions can be located and read. Often a new rule is accompanied by extensive information and documentation, which can be useful. However, it is unwieldy to use, at best.

Once a year the regulations are transferred to the Code of Federal Regulations (CFR), adding them to the accumulated existing regulations. Statements of the regulations now are presented in compact form. The organization is not changed from year to year, so new rules are simply fitted into the established framework at the appropriate location. The organization is not as straightforward as these comments make it seem, and some practice is needed to learn to locate information in the CFR.

When in the library facing the CFR, you notice first that there is a series of 50 titles, each dealing with a specific concern of the federal government (Table 4.1). Everything having to do with OSHA is found in Title 29, Department of Labor. Specifically, OSHA regulations are in the two volumes that include parts 1900 to the end of 1910 (Figure 4.1). The numbers 1900 to 1999 are reserved for OSHA regulations. The standard reference to locate a specific rule might start out as 29 CFR 1910, which means Title 29 of the CFR, which gets you to the Department of Labor section, then 1910 leads you to the volume containing the OSHA rules.

If you take the volume including parts 1900 to 1910 off the shelf and turn the pages starting at the beginning, you come to a page that indicates what each of the major divisions (1900, 1901, 1902, ...) includes. Most of the document is found in section 1910. Turn to the first page of 1910 (Occupational Safety and Health Standards) and you find a listing of the contents of each major subdivision of 1910. This list would lead you to the appropriate section to find the area in which you wish to see the standards.

Table 4.1. Code of Federal Regulations Titles.

Title 1	General Provisions
Title 2	Reserved
Title 3	The President
Title 4	Accounts
Title 5	Administrative Personnel
Title 6	Reserved
Title 7	Agriculture
Title 8	Aliens and Nationals
Title 9	Animals and Animal Products
Title 10	Energy
Title 11	Federal Elections
Title 12	Banks and Banking
Title 13	Business Credit and Assistance
Title 14	Aeronautics and Space
Title 15	Commerce and Foreign Trade
Title 16	Commercial Practices
Title 17	Commodity and Securities Exchanges
Title 18	Conservation of Power and Water Resources
Title 19	Customs Duties
Title 20	Employees' Benefits
Title 21	Food and Drugs
Title 22	Foreign Relations
Title 23	Highways
Title 24	Housing and Urban Development
Title 25	Indians
Title 26	Internal Revenue
Title 27	Alcohol, Tobacco Products and Firearms
Title 28	Judicial Administration
Title 29	Labor
Title 30	Mineral Resources
Title 31	Money and Finance: Treasury
Title 32	National Defense
Title 33	Navigation and Navigable Waters
Title 34	Education
Title 35	Panama Canal
Title 36	Parks, Forests, and Public Property
Title 37	Patents, Trade Markets, and Copyrights
Title 38	Pensions, Bonuses, and Veterans' Relief
Title 39	Postal Service
Title 40	Protection of Environment
Title 41	Public Contracts and Property Management
Title 42	Public Health
Title 43	Public Lands: Interior
Title 44	Emergency Management and Assistance
Title 45	Public Welfare
Title 46	Shipping
Title 47	Telecommunication
Title 48	Federal Acquisitions Regulations System
Title 49	Transportation
Title 50	Wildlife and Fisheries

Suppose what you have is instead a reference to a standard as 29 CFR 1910.133(a)(2)(i-vii). In section 1910 you locate 1910.133,[4] which turns out to be standards for eye and face protection. Heading (a) is called "General," and its inclusion does not mean that there has to be a (b), only that the possibility of a (b) is held open. Subsection (2) lists the minimum requirements eye protection must meet and seven points are included, i-vii (Roman numerals).

THE FEDERAL RIGHT-TO-KNOW STANDARD

A new standard was established in 1983 by OSHA to require that all employees likely to come into contact with hazardous materials shall be informed of the nature of the hazard and the steps necessary to protect the employee from harm. Called the Hazard Communication Standard (29 CFR 1910.1200), and referred to as the Right-to-Know law, it applies to chemical manufacturers, distributors, importers, and manufacturing industry employers of standard industrial classification (SIC) codes 20–39 (Table 4.2).

All materials which represent a hazard in at least one of the categories shown on Table 4.3 are covered by the standard, a list of perhaps 70,000–80,000 commercial materials. An important provision of the standard is the requirement that Material Safety Data Sheets (MSDSs) be available at all times to the employees. (An example is Appendix B.) "Available at all times" means posted or on file in a place that is not locked or otherwise inaccessible any part of the working day. These may either be obtained from the supplier of the chemical, or prepared by the manufacturer. The information included on this form includes items in these categories:

A. Identity of the material.

B. Classification of hazard, including types of hazard by name, PEL, TLV, or other measures of the maximum permitted levels of exposure, if available, and other such information.

C. Recommendations for safe usage, including the need for ventilation, protective clothing, and other precautions.

D. Information to guide emergency procedures.

E. Physical Hazard Data.

---

[4]Notice that we are not using the decimal system as a mathematician would. 1910.133 does not belong between 1910.13 and 1910.14, and 1910.99 is followed by 1910.100.

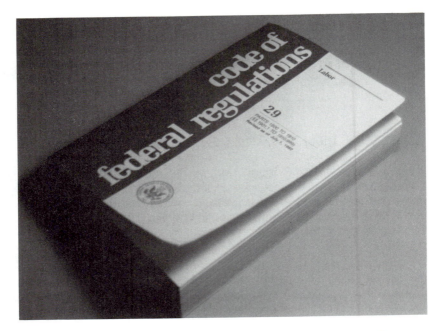

Figure 4.1.   Shown here is the volume of the Code of Federal Regulations
containing parts 1900 to 1910.999, revised as of July 1, 1992.
The CFR is revised frequently, and a more recent version is in
print, containing recent changes and additions. However, changes
in each new edition are inserted into the organization of the previous edition, so that references to unchanged information do not
change.

F.   Chemical Hazard Data.

G.   Physical and Chemical Properties.

H.   Information about the Manufacturer or Supplier.

Where the composition of the mixture is a trade secret, it is not required that
the manufacturer list the composition on the MSDS. However, all of the hazard
information must be provided. OSHA does not supply a standard form, but
leaves it to the individual company to design its own form.[5]

---

[5]Discussions frequently are directed at changing the MSDS format to be more "user
friendly." Consider that obtaining information from such a sheet requires a certain
sophistication from college students, and that the intended beneficiary of this information is a factory worker who may not have completed high school or have
English as a first language.

Table 4.2.   Manufacturing Industries Impacted by the Hazard Communication Standard.[a]

SIC Code	Industry Group
20	Food and kindred products
21	Tobacco manufacturers
22	Textile mill products
23	Apparel and other textile products
24	Lumber and wood products
25	Furniture and fixtures
26	Paper and allied products
27	Printing and publishing
28	Chemicals and allied products
29	Petroleum and coal products
30	Rubber and plastic products
31	Leather and leather products
32	Stone, clay, and glass products
33	Primary metal industries
34	Fabricated metal products
35	Machinery, except electrical
36	Electrical equipment and supplies
37	Transportation equipment
38	Instruments and related products
39	Miscellaneous manufacturing products

[a]Source: Lowry, G. G. and R. C. Lowry. *Handbook of Hazard Communication and OSHA Requirements* (Chelsea, MI: Lewis Publishers, Inc., 1985).

Hazardous substances must be labeled. The label must include the identity of the substance, information regarding every hazard associated with the substance, and the name and address of a person from whom additional information may be obtained. OSHA has an already established list of substances (Table 4.4) for which specific wording is required on the label. In order to reduce the likelihood of lawsuits based on inadequate information on the label, the American National Standards Institute has suggested a nine-item label:

A.   Identity.

B.   CAUTION, WARNING, or DANGER.

C.   A statement of the hazards.

D.   Safety measures, such as protective clothing.

E.   Instructions as to procedures to use if exposure occurs.

F.   Antidotes for poisoning.

G.   Emergency treatment information for physicians.

H.   Instructions as to procedures in the case of fire, spill, or leakage.

I.   Instructions for handling and storing.

If the name and address of a person to contact were added to this, the list would exceed the OSHA requirements.

Each manufacturer is required to set up a training program for the employees who work in the area of the plant containing the hazardous materials. They must be informed about the rules established by the standard and how to use the labeling, MSDS sheets, and other hazards information.

Finally, the manufacturers must have a written hazard communication program. This program must include:

A.   A list of hazardous materials in the plant, by plant area in a large operation.

B.   MSDS sheets.

C.   A description of the warning labels and how other postings are designed and used.

D.   A description of the training program to be used with the employees.

The list of hazardous substances in the plant must be arranged alphabetically and must take account of "slang" names used by plant workers. An excellent source of information about this regulation is provided in a book by G.G. and R.C. Lowry. (See Bibliography.)

TSCA AND THE ENVIRONMENTAL PROTECTION AGENCY

The Toxic Substances Control Act, known as TSCA, was enacted in 1976 in order to prevent unreasonable chemical risks. TSCA gives the EPA authority to act before harmful chemical substances threaten human health or the environment. The act directs the EPA to make judgments on the safety of new chemicals or chemical mixtures produced by manufacturers. The first major action of the EPA under TSCA was to assemble the most comprehensive inventory in existence of chemical substances in U.S. commerce. The law requires the submission of test data by the manufacturer of a chemical relating both to the health effects and to the effects on the environment, should the agency choose to question its safety. The compound or mixture is deemed new (for testing purposes) if it is not already on the agency's list; but, testing may also be required if a new use is to be made of an existing compound or mixture that involves a significant change in the exposure of humans or of the environment. Notification of intent to manufacture must be given to the agency 90 days or more prior to actual production. Action by the agency can range from a requirement for specific labeling to a ban on manufacture. Where such restrictions are suggested, there is a hearing to allow the various interested groups to have input.

To avoid overlapping of enforcement authority, tobacco, pesticides, ammunition, nuclear material, food and food additives, drugs, and cosmetics, all of which are regulated by other agencies, are exempted. For each compound, setting rules for the workplace is left in the hands of OSHA.

In the CFR, regulations of the EPA are found in Title 40. Referencing is done in the same fashion as in Title 29.

Effective February 6, 1986, the EPA and OSHA established new mechanisms for the sharing of information. Twice yearly the EPA sends OSHA a list of chemical substances under consideration for referral under TSCA. Each agency appoints a staff contact person, and meetings are set to exchange information. Such an arrangement helps greatly in reducing the confusion and overlapping of efforts that are a constant problem in our multiagency approach to the regulation of toxic substances.

KEY POINTS

1. OSHA, founded in 1970 as part of the Department of Labor, brought uniformity to health and safety standards.

2. OSHA sets standards, conducts inspections, requires health and safety recordkeeping and trains personnel.

3. NIOSH is the health and safety research arm.

4. Standards for chemical exposure were initially based on existing ACGIH and ANSI standards.

5. On inspection, OSHA can declare a plant to be from *below action level* to *in a state of imminent danger.*

6. Rules are first reported in the Federal Register, then are condensed into the Code of Federal Regulations.

7. The Right-to-Know standard requires that employees be informed of and trained to avoid workplace hazards. MSDSs and hazard labeling are required.

8.   TSCA is an act administered by EPA that passes judgment on the hazards of chemicals.

9.   OSHA and the EPA share information.

PROBLEMS

1.   Contrast the situation in the U.S. before and after passage of the OSH Act. Did it cause the same degree of change to take place in every state? It is clear that a state that has lax worker protection laws will have more frequent accidents and work-related illness. What is the advantage to a state of having lax worker protection laws?

2.   The politically liberal individual tends to be an advocate of the worker, while the politically conservative person takes the side of the employer. Since the OSH Act is designed to protect workers, it is basically liberal legislation. In a democracy, for legislation or anything else to succeed, the needs and interests of both sides must be considered, or an imbalance results that ultimately forces change. Name some ways industry has a say in the provision of worker protections.

   A.   It is possible that a proposed, more rigorous standard, compliance with which could require manufacturers to spend huge sums of money, would not actually increase workplace safety in a significant fashion or would be no better than a somewhat less rigorous standard that could be more easily met. At what point do manufacturers have an opportunity to present this case, and to whom do they present it?

   B.   Plant inspections are done by people who can make mistakes or can be responding to an agenda that is not strictly enforcement of regulations. Can a manufacturer who is charged with noncompliance protest the charge, and if so, how and to whom? [See 29 CFR 1903.17.]

   C.   What are the limits of OSHA's powers: (1) Can OSHA inspect any worksite they choose, and can OSHA inspect without warning? [See 29 CFR 1903.3 and 1903.6.] (2) If an employer refuses OSHA admission to a worksite, what can OSHA do? [See 29 CFR 1903.4.] (3) Can OSHA fine a company for noncompliance, and if so, who decides the size of the fine? [See 29 CFR 1903.15 (a) and (b).] (4) Can an employer send a representative to accompany an inspector during inspection? [See 29 CFR 1903.8.] (5) What can OSHA do for continued noncompliance after an inspection and citation: nothing, levy additional fines, or close a plant? [See 29 CFR 1903.18.] If the time allowed to correct a situation not in compliance seems unreasonable to the employer, does the employer have any recourse? [See 29 CFR 1903.14a.]

3. Learn about the role of the Review Commission by reading 29 CFR 1903.17.

4. How may employees keep informed of the progress and results of an inspection? [See 29 CFR 1903.8, 1903.16, 1903.10.]

5. Radical environmentalists propose banning all chemicals that have chlorine in their structure on the basis that some chlorine compounds are hazardous and in the cases of a few the hazard was not recognized until some damage had been done. In terms of economic cost, contrast that approach to the setting of standards for individual compounds based on laboratory toxicology studies, as is done by OSHA.

6. In section 29 CFR 1905, find out what is meant by variance, exemptions, variations, limitations, and tolerances.

7. Turn to the MSDS for benzene (Appendix B). In this chapter, find the eight classes of information to be included in the MSDS, then locate each of these in the sample sheet.

BIBLIOGRAPHY

E. J. Baier, "Legislation and Legislative Trends," in G. D. Clayton and F. E. Clayton, *Patty's Industrial Hygiene and Toxicology*, 4th ed., John Wiley and Sons, Inc., New York, 1991.

95th Congress, 1st Session Committee Print 95.7, Toxic Substances Control Act, Occupational Safety and Health Act, U.S. Government Printing Office, Washington.

M. Corn, "The Impact of Federal Regulation on Engineers," *Chem. Eng. Prog.,* July, 1978.

G. S. Dominguez, *Guidebook: Toxic Substances Control Act*, CRC Press, Boca Raton, FL, 1977.

Guidebook to Occupational Safety and Health, 1978 ed., Commerce Clearing House, Chicago, 1978.

G. G. Lowry and R. C. Lowry, *Handbook of Hazard Communication and OSHA Requirements,* Lewis Publishers, Inc., Chelsea, MI, 1985.

Subcommittee on Labor, Committee on Labor and Public Welfare, U.S. Senate, *Legislative History of the Occupational Health and Safety Act of 1970* (S.2193, P.L. 91-596), 92nd Congress, 1st Session, June, 1971.

M. D. Worobec, *Toxic Substances Control Primer*, 2nd edition, The Bureau of National Affairs, Washington, 1986.

CHAPTER 5

OCCUPATIONAL DERMATOSIS

To this point the focus of the text has been on damage caused by chemicals entering the body. Absorption across the skin is an important route of such entry, and this aspect was presented in Chapter 3. Dermatosis—irritation of the skin or damage to the surface of the skin—as a result of contact with chemicals is a different and important problem in occupational safety and health, and is addressed in this chapter.

SKIN ANATOMY

We now expand the earlier presentation. The skin is the largest organ of the body, including approximately 18,000 cm^2 (20 ft^2) of surface. The thickness of this covering varies from about 0.5 mm on the eyelids to about 4 mm on the back and such high abrasion areas as the palms of the hands and the soles of the feet. Skin is composed of two layers that are quite different in structure and function.   A small percentage of the surface area includes pores and hair follicles, openings that lead into the interior of the skin.

EPIDERMIS

The *epidermis,* the outermost of the two skin layers, has layers of living cells at its inner extreme. As these cells divide and shift toward the outer surface of the epidermis, they gradually change character. They eventually become scale-shaped nonliving cells filled with the structural protein keratin, producing the *stratum corneum* or *keratin layer*. The keratin layer is resistant to abrasion and is a barrier to the loss of water or entry of water solutions of chemicals. As these outer cells are rubbed off or otherwise lost, they are replaced by new cells moving up from the layers of living cells below. We are generally not aware of this loss of epidermis unless we accelerate it, as when we damage the skin by sunburn, or if we are afflicted by the dread social disease dandruff.

The pigment melanin is produced in the inner portion of the epidermis to reduce damage to the skin by light, especially UV light. People who inhabit tropical regions of the world adapted to greater solar radiation by producing more melanin, and hence have darker complexions. Short-term increase in exposure to light stimulates increased production of melanin, resulting in suntan.

DERMIS

A primary function of the inner skin layer, the *dermis,* is to provide resistance to tearing or puncture. This is accomplished by a meshwork of collagen fibers—ropelike protein structures that are strong without being rigid and unbending. Should damage occur, specialized dermal cells produce new collagen to heal the wound. Scars are the result of heavy collagen production.

The dermis contains blood and lymph vessels, which supply the needs of the dermal and living epidermal cells. The blood vessels also play a role in body temperature regulation, a second major function of the skin. Blood circulating in the skin loses heat through the epidermis to the surrounding air. When body temperature rises, the diameter of the dermal vessels increases, increasing the amount of blood in the skin and consequently the rate of heat loss. When you are overheated, the extra blood in the skin gives it a reddened or "flushed" appearance. In cold weather, the diameter of the vessels shrinks in order to minimize the heat loss.

PORES AND HAIR FOLLICLES

Further temperature control is provided by the approximately 2,000,000 sweat glands of the dermis. These glands are coiled tubes deep in the dermis that extend through the epidermis to the surface of the skin, the opening being termed a pore. Sweat is simply water with dissolved salts collected in the coiled base of the gland, and it is produced at all times. Sweat coats the surface of the skin, and heat is lost as it evaporates. The rate of production of sweat increases as the need to cool the body increases, either because of hot weather or the production of excess body heat due to exercise. The rate of sweat production then exceeds the rate of evaporation, and we are aware that we are perspiring. When the skin remains wet, the moisture loosens and speeds removal of the outer cells of the epidermal keratin layer. Continuous perspiration therefore reduces the protection against abrasion and the entry of chemicals that this layer provides. Skin that is constantly coated with sweat may also develop a "prickly heat" rash.

The surface of the skin grows inward to form a *follicle* or tube, at the base of which the keratin-forming cells construct a shaft of hair. Hair follicles have their roots deep in the dermis. *Sebaceous glands,* generally associated with the hair follicle, produce sebum (skin oil), which travels up the follicle and coats the surface of the skin. Blockage of this flow out of the follicle results in an accumulation of sebum, causing a skin eruption (pimple, or zit). Widespread blockage of pores is termed *acne*.

Pores and hair follicles are passages into the dermis, and might therefore be suspected to be major routes of entry of chemicals past the keratin layer. However, they represent only roughly 0.1% of the surface of the skin and are not major factors in the passage of chemicals.

Skin also contains sensory nerves, which provide touch, pain, and temperature change sensations. Such systems protect us by stimulating removal of our bodies from hazardous situations.

CONTACT DERMATITIS

Most (80–90%) dermatosis is caused by *contact dermatitis,* direct contact at a localized site with a chemical agent. Types of agents and the damage they cause to the skin fall into several categories, with overlap occurring between these classes. Effects are generally restricted to the site of contact with the chemicals, and in the workplace most often the site is the hands, wrists, or arms.

KERATIN LAYER AS A BARRIER

The keratin layer is inert, and, aside from corrosives, must be penetrated by chemicals before their effect is observed. Abrasion to the skin damages or even crosses the keratin layer, reducing the barrier to chemicals provided by this layer. Chapping caused by weather or low humidity in an indoor work environment makes skin more susceptible to damage by chemicals. Physical exertion or a high workplace temperature causes sweating that softens the keratin layer and makes it more permeable, as does emersion of the skin in hot water, such as dishwashing water. Sometimes the skin is exposed simultaneously to several harmful agents, each contributing to weakening the keratin layer barrier.

IRRITATION

Many classes of chemicals have in common the effect of *irritation* of the skin. The irritation may result from the action of the chemical itself, or it may involve an allergic process. These two processes involve different elements, and are considered separately. Mild irritation involves the arousal of inflammatory processes that result in fluid retention (swelling) and increased blood flow (reddening). Stimulation of the sensory nerves leads to itching, burning, or stinging sensations. As irritation worsens, there is cell death, along with further swelling and reddening, resulting in a rash. Damage to blood vessels eventually produces bleeding. Blisters are formed by fluids generated by inflammation in more severe cases.

IRRITANT CONTACT DERMATITIS

Irritation of the skin as the result of direct chemical action on skin tissues is termed *irritant contact dermatitis*. Such irritation may be the result of direct chemical damage to the cell, of physical damage such as might be caused by sharp particles, of the long-term effects of removing skin oils, or by combinations of these causes. Our first assumption and most common model is a liquid chemical into which hands are immersed or which is splashed accidentally on the skin. Reports are not uncommon of skin irritants that are carried to the worker through the air, and this source of contact should not be ignored. Airborne irritants can slip past protective clothing, which seldom presents a continuous airtight barrier.

Irritation by Direct Chemical Damage to the Skin

Intact normal skin is protected from direct toxic effects of chemicals by skin oils and the tight layer of dead keratin cells. Chemicals that do reach living cells cause irritation through a variety of individual toxic mechanisms. Often these chemicals contain relatively reactive functional groups. The extreme case of tissue destruction and erosion by powerful chemical agents called corrosives is considered separately.

Solids That Cause Physical Damage to the Skin

Solids with sharp particles can abrade the skin, directly causing irritation by physically damaging cells and opening routes for other chemicals to penetrate the skin to cause further problems. The thin, sharp fibers of fiberglass are an excellent example of this, and can be contacted both by direct handling and by an airborne route.

Agents That Remove Skin Oils and Damage the Keratin Layer—Solvents

Water is perhaps the most common vehicle used to dissolve substances, but when we use the term "solvent" in industry, we ordinarily are referring to *organic* liquids (structures based on carbon) used to dissolve other organic substances. The chemical characteristics of such liquids vary, but all would primarily dissolve other organic substances and would not dissolve inorganic substances such as metallic salts. A few (such as acetone and smaller alcohols) mix with water, but most do not.

Irritation to the skin by solvents usually requires extensive and regular contact. When this occurs, skin oils are removed from the surface of the skin and keratin cells are then gradually loosened and removed. The result is a loss of water from skin tissue, producing a drying and scaling of the skin. Cell membranes are disrupted, killing the cells. The effectiveness of skin as a barrier to entry into the body of the solvent itself or of other chemicals is reduced. With heavier exposure, skin cracks or splits, producing open sores.

Agents That Remove Skin Oils and Damage the Keratin Layer—Soaps and Detergents

Soaps and detergents (surface active agents, or surfactants) are chemicals with a structural dual character. They have in one molecule both a nonpolar (water-insoluble) and a polar (water-soluble) segment. When water is the medium used to clean objects contaminated by water-insoluble contaminants, the nonpolar part of dissolved surface active agents adheres to these contaminants, coating their surface. Nonpolar structures now have a polar surface and are suspended in water.

Soaps and detergents are formulated differently for different functions, for example for cleaning inert objects versus cleaning the grease-contaminated skin of workers. Commercial cleansing agents contain other chemicals, for example to prevent hard water minerals from forming scum, to make the solution alkaline, or to preserve the agent from bacterial attack. If germicidal action is important, for example in food handling or hospital environments, the agent may contain phenol or phenolic derivatives. Hand cleaners may contain abrasives to mechanically loosen the dirt. Waterless hand cleaners include both a detergent cream and petroleum distillate, the latter being able to dissolve the grease due to its nonpolar character.

Damage to skin by soap or detergent solutions generally requires extended and frequent contact. As with damage by solvents, the skin oils are removed and the keratin cells of the epidermis are loosened. Surface active agents are most effective when the water is hot, which accelerates these effects. Abrasives in mechanics' hand soaps do little damage to the thick skin of the palms, but may have more serious effects on the backs of the hands or the forearms. Damage to the skin by surface active agents makes skin more susceptible to irritation from other agents.

ALLERGIC CONTACT DERMATITIS

An allergic response to chemicals (allergic contact dermatitis) is an important cause of occupational dermatitis, and differs fundamentally from irritations already discussed. Problems with irritation on contact with chemicals previously described depend on the reactivity of the chemicals with components of skin, affect most workers having such contact, and depend in a relatively simple fashion on the amount of contact with the chemical. Allergic response may only affect a small proportion of the workers, but those affected may show severe responses, while those not affected show no response at all. Affected workers may not respond to a chemical on first contact, but at some later time become *sensitized.* Thereafter, those workers respond to very small quantities of the chemical, sometimes in an extreme fashion. The resulting reddening, swelling, and rash are due to an inflammation of the tissue, and look very similar to irritation described for irritant contact dermatitis.

Allergic response begins with the chemical, termed an *antigen,* being absorbed into the skin and *reacting with a skin protein.* This complex must then stimulate an immune response, the production of large numbers of *antibody* proteins which specifically bind the chemical-protein complex. The person is now sensitized.

Antigen-antibody complexes activate body mechanisms whose purpose is to specifically bind and destroy invading bacteria or viruses and so prevent infections, but whose side effect is inflammation. The sensitized individual continues to produce antibody to the chemical. If the antigen had been a bacterium or virus, this aspect of the process would give us "immunity" to reinfection. However, in the case of a chemical antigen, future exposures result in these antibodies binding to the chemical (protein binding is now not necessary) and stimulating the reddening, swelling, itch and other characteristics of inflammation—an allergic response. On reexposure, a dramatic allergic response may occur to even very small amounts of the chemical.

Diagnosis by occupational physicians of the cause of allergic contact dermatitis can be difficult. There is a time lag between exposure to the chemical and appearance of inflammation, complicating identification of the specific exposure that causes inflammation. The work environment is often a complex source of chemical exposures. Furthermore, the cause of the allergic response could be outside of the workplace. There are chemicals that are recognized as being potent allergens or sensitizers, and their use in the plant raises suspicions that they are the source of the problem. Shifting the worker to a workstation that eliminates contact with suspect chemicals may relieve the symptoms, but this is not assured, since only traces of the chemical may be necessary to evoke the response. Patch testing—taping samples of suspected allergens to the skin and observing the appearance of irritation at intervals usually of 48 hours—is a useful diagnostic procedure. It must be done with controls to assure that the worker's skin is not responding just to tape or to being covered.

The most common allergic occupational skin problems are associated with the contact of outdoor workers with plants that produce potent allergens such as poison oak and poison ivy. Similar allergens are found in mango trees, Japanese lacquer trees, and cashew nut trees. Although contact with these latter plants is unlikely in the U.S., products of those plants are found on the U.S. market.   An oil is extracted from cashew nutshells and used in a range of plastics, lubricants, coatings, and other products, while lacquers and varnishes may be based on products of the lacquer trees.   A person already sensitive to poison oak or ivy may respond to those products.   Other plant components that cause problems for nursery workers include chemicals in a number of common plants, and forestry workers may be sensitized to lichens on bark or chemicals in sawdust.

Allergens are found throughout industry.   In polymers, both plastics and rubbers, the monomers, hardeners, accelerators, antioxidants, and diluents of resins and rubbers include some allergens.   A few metals are sensitizers.   Nickel is the most common metallic allergen, and may be contacted in cuttings or solutions such as cutting oils or electroplating baths.   The next most frequent metallic allergen is hexavalent chromium, which is encountered in such industrial applications as tanning, electroplating, etching and photoengraving, and in pigments.   Small amounts of hexavalent chromium may be found in cement. Biocides, agents added to prevent deterioration due to growth of bacteria or fungi, are used in cutting oils, paints, adhesives and hand soaps, and are sometimes allergens.   Dyes, particularly azo dyes, may be allergens.   Azo dyes are used in textiles, color photography and hair coloring compounds.   Agricultural workers can be sensitized to pesticides, particularly to organomercurials, carbamates, captans, triazines, and thiurams.

PHOTODERMATITIS

Some chemicals do not directly cause irritation on contact with skin, but symptoms of irritation appear later when the skin is exposed to sunlight.   Such chemicals absorb UV light, resulting in their being converted into a different and more irritating chemical.

The most common phototoxic agent is coal tar (creosote, or pitch).   Workers who handle railroad ties often complain of "tar smarts."   Symptoms include burning and stinging, and the skin reddens.

A number of agricultural products produce the phototoxic agent furocoumarin, including celery, citrus fruits, parsley, dill, and carrots.   In addition to irritation, skin may darken after exposure.   Workers harvesting these crops are particularly susceptible, but anyone handling such produce extensively can be affected.

Less common causative agents include some dyes, drugs, and other miscellaneous chemicals.   Persons experiencing problems with phototoxic agents who cannot avoid exposure to the agent can obtain relief from the symptoms by placing sunscreen on exposed skin, thus avoiding the photoactivation.

HAIR PROBLEMS

Hair problems in the workplace include loss or color change of hair. These represent a very small proportion of occupational dermal problems. Alopecia (loss of hair) is caused by agents that kill the living and growing cells at the hair root. Three chemicals have most frequently been the cause of such problems: thallium (which was actually used to remove unwanted hair), chloroprene dimers, and boric acid. Loss of hair also occurs on exposure to ionizing radiation, commonly seen in patients undergoing radiation treatment for cancer.

Color change is most often caused by metals, copper and cobalt having been reported as producing green and blue colors, respectively. Picric acid and a few other chemicals also cause such problems.

ACNE TYPE DERMATITIS

Acne (skin eruptions) results when follicles are blocked, so as to prevent the discharge of skin oils. The trapped sebum produces swelling and irritation. A number of working conditions stimulate this condition, particularly in individuals already prone to this problem.

Acne arises for reasons other than chemical exposure. Hot weather or high temperature in the workplace induces perspiration, which results in swelling and softening of the keratin layer, possibly blocking the follicle. Skin friction, as from belts, hard seating, or face masks, also can cause blockage, and heat and friction together have an additive effect.

Hydrocarbon Agents

Acne commonly results from the handling of high-molecular-weight hydrocarbon compounds, particularly on fingers, hands, forearms, and skin in contact with oil-soaked clothing. Lubricating greases and cutting oils are often at fault. The skin develops "blackheads," followed by swelling of the follicle with a burden of skin oil. This, too, is more serious in individuals more prone to acne, and improved personal hygiene reduces the problem.

Chloracne

Chloracne is similar in appearance to—but different in cause from—the acne described above. The follicle is sealed with damaged keratin, rather than oil. Chlorinated and brominated aromatic hydrocarbons cause this problem, with some chemical structures being much more potent than others. Chlorophenols and polychlorinated or polybrominated naphthalenes, biphenyls, and dibenzofurans are recognized causative agents. The problem may clear after contact with the chemical ceases, but improvement is usually slow.

ALTERATION OF SKIN PIGMENTATION

A few chemical agents cause pigmentation changes in skin, either accompanied by irritation or in the absence of other symptoms. Melanin, the skin pigment, is a phenolic structure. Some phenolic chemicals interfere with melanin synthesis in the epidermis, resulting in temporary pigment loss. Heavier exposure may kill melanin-producing cells, inducing a more permanent loss of skin color. Other damage that kills skin cells—for example, exposure to corrosives described in the following section—may result in healing without the regeneration of the cells that produce melanin.

Following skin inflammation, there may be excessive pigment in the skin. Melanin may be deposited in the tissue by damaged melanin cells, and the occurrence is more marked in people with very dark skin, and hence more melanin production. This may occur below the level of cells being carried outward to be shed, and so may persist for a long time.

CORROSIVES

A few chemicals that cause massive tissue damage within short time intervals are termed *corrosives*. Corrosives destroy by direct chemical action, eroding the tissue and potentially producing deep wounds. As with serious burns, tissue damage can be extensive and the healing process produces heavy scarring. Corrosives have such potential for damage that contact with the skin must be avoided completely, and they should not be used without skin protection suitable to withstand the specific agent. They are particularly threatening when used at high temperatures.

Acid

Most commonplace among industrial corrosives are concentrated strong acids, especially concentrated sulfuric acid. This is the highest-volume chemical produced in the U.S., and it is used broadly throughout industry. Sulfur trioxide, a gaseous anhydride of sulfuric acid, is an intermediate in sulfuric acid production. Sulfur trioxide can react with moisture on the surface of the skin to produce concentrated sulfuric acid. By controlling the amount of water reacted with sulfur trioxide, sulfuric acid may be produced for use in concentrated form that contains little or no excess water. Oleum, also called fuming sulfuric acid, has had less water added than the sulfur trioxide requires for complete reaction, and is therefore a mixture of sulfuric acid and sulfur trioxide. The lower the sulfuric acid water content, the more hazardous the preparation, with oleum being the most hazardous. Concentrated sulfuric acid causes more damage by rapidly dehydrating tissue than by action as an acid. Skin contact might occur either with the liquid or with an airborne mist.

Concentrated nitric acid is used to make dyes, plastics, and fertilizer. It adds the ability to oxidize tissue to its effect as a strong acid. As an example of its effectiveness as a corrosive agent, hot nitric acid is used in the laboratory to completely digest tissue samples, destroying all carbon, hydrogen, and nitrogen compounds, for purposes of metals analysis. Skin contact results in ulcers and a reaction with keratin to produce a yellow-colored product.

Phenol (carbolic acid), is a very weak acid that nonetheless seriously damages the skin. Phenol was the original antiseptic, and its derivatives are used in medicinals and in a variety of agents designed to kill or prevent growth of microorganisms. Phenol exposure leaves a thick layer of skin whitened and dead. It then absorbs rapidly into the body, where further harm may occur.

Alkali

In addition to acids, concentrated solutions of strong bases (caustics) are corrosive in nature. The most common strong base in industry is sodium hydroxide (lye, caustic soda, white caustic). It is used in large quantities in production of textiles, paper, soap, petroleum, and chemicals. Once on the skin, caustic solutions are much more difficult to wash off than are acids, which heightens their ability to do harm.

Cement is an alkaline agent. It is prepared at a very high temperature, which drives off even traces of water. When water is added, the components react with and incorporate that water, forming the rocklike product. Mixed with sand and rock, it is called concrete. Typical cement contains aluminum oxide, silicates, and calcium sulfate—none of which are corrosives—and calcium oxide. The latter reacts with water to produce calcium hydroxide, which is a strong base and is capable of skin damage. Cement is not a concentrated alkali like the lye described above, and damage requires long contact. Such contact eventually results in burns and ulceration, however, caused by a combination of the calcium hydroxide, abrasion, and removal of water from skin by the components of cement.

First Aid

With the exception of phenol, which is not very water-soluble, the proper immediate response to skin exposure to a corrosive is thorough washing with water. Because alkali is harder to remove from skin, washing of an alkali exposure should continue for several minutes. A person exposed to a splash or spill of strong acid or alkali should remove their clothing—which may hold some of the corrosive agent—and shower immediately. Corrosives acting on skin can quickly destroy sensory nerves, reducing warning that damage is occurring, and this may go undetected until serious harm is done if it takes place under clothing. Special agents such as polyethylene glycol are recommended for the removal of phenol, but in their absence, soap and water is more effective than water alone.

CANCER

When compared to other industrial skin problems, the development of skin tumors is a relatively uncommon but obviously very serious occurrence. Classes of skin tumors are associated with different classes of cells: squamous cell carcinoma, basal cell carcinoma, and melanoma.

For all cancer types, but particularly for squamous cell carcinoma, a sizable body of epidemiological evidence implicates UV light as a carcinogenic agent. Exposed skin of outdoor workers has a relatively high risk of tumor growth. Such tumors tend to appear in skin that already displays damage generated by the sun. Outdoor workers, particularly those with fair complexions, should wear protective clothing or sunscreen preparations.

Exposure to ionizing radiation also increases the likelihood of skin cancer, particularly squamous cell and basal cell carcinomas. Individuals that operate X-ray machines are at risk and must be careful to avoid exposure.

Polycyclic aromatic hydrocarbons are well established carcinogens, particularly 3,4-benzpyrene and dimethylbenzanthracene. These are found in coal tar, asphalt, and creosote. Workers exposed to these materials may develop "tar warts" after years-long incubation periods, and a small proportion of these become malignant. Heavy exposure to UV light increases the frequency of malignancy.

Heavy and chronic exposure to arsenic results in warts (arsenical keratoses), which occasionally progress to become squamous cell carcinomas. The evidence linking arsenic and cancer is primarily epidemiological.

INFECTIONS

The skin generally provides good protection against infections, but damage to the skin surface, as by abrasions or burns, chapping from weather or low humidity in the workplace, or poor personal hygiene, greatly increase the likelihood of these problems. Infections may be bacterial, fungal, or viral.

It may be difficult to determine whether a particular infection is work-related or not, but some occupations carry an increased risk of a specific infection.

Persons handling animals or animal products are at special risk. This includes then the original producer of the product (farmers and fishermen), transporters of the product (warehouse workers, dockworkers), processors (meat, fish, and poultry packers; tanners and furriers) and distributors (butchers).

Erysipeloid (fish handlers disease) is a bacterial infection causing a spreading erythema and is particularly common among those handling fish but also is found among poultry and meat handlers. Anthrax used to be a serious threat, particularly from contact with goats or sheep. Inoculation programs have sharply reduced this problem in the U.S., but some danger still exists in handling hides imported from countries with less rigorous programs. Sheep or goats may also transmit orf, a viral disease, to ranchers and veterinarians.

PREVENTION OF SKIN IRRITATION OR DAMAGE

Workers may be protected from skin irritation or damage by avoiding skin contact with the offending chemical or chemicals. In extreme cases, for example in the case of a violent corrosive such as concentrated sulfuric acid, the worker is kept away from the chemical. In cases where extensive contact is necessary to cause a skin problem, for example the problems created by detergents or by solvents, contact with the agents by the worker needs to be kept below a hazardous level. It is also possible for the worker to be physically protected from the chemical. Two common protections are barrier creams and protective clothing.

Barrier Creams

Barrier creams are coated on the skin at the start of the workday, and should be washed off and replaced at breaks, as for lunch (Figure 5.1). They block many chemicals from reaching the skin, but the level of protection is limited. Many questions arise: is the coating thick enough overall, was a spot missed, is the cream removed by rubbing against other objects during work? Barrier creams should be used only in low-hazard situations, should be considered to be an adjunct to responsible handling of chemicals and judicious use of protective clothing, and are clearly not a license to work carelessly with hazardous substances.

Protective Clothing

Protective clothing is used when contact with chemicals is either unavoidable—as when manipulating objects freshly removed from a bath—or when there is a possibility of spills or splashes that could be harmful to the skin (Figure 5.2). It may be essential to deal with emergency situations such as cleaning up a spill. Supplying protective clothing should not serve as justifica-

Figure 5.1.   Barrier creams provide a degree of protection, and are especially useful when the chemical is of relatively low hazard. They should not be used as a replacement for gloves or more extensive protective clothing for high-hazard situations. (Photo courtesy of Lab Safety Supply, Janesville, WI.)

tion for a job in which the worker is required to place his or her hands in continuous contact with a dangerous chemical. An engineering solution—the redesign of the task or process to eliminate such contact—is called for instead.

There are two major choices. First there is the question of how much skin must be covered. Often gloves are sufficient, but garments are available that range up to jumpsuits with hoods. The second choice is the material from which the clothing is to be constructed. There is no material that is uniformly resistant to all chemicals. In selecting the material to purchase, refer to tables provided by the manufacturer that list a series of common chemical hazards and indicate the performance of each material in blocking that chemical.

Gloves

Gloves are available in a variety of materials, thicknesses, liners, and prices (Figure 5.3). There are various choices and compromises to be made when selecting gloves; these depend on the particular work situation. The first consideration is the hazard to which the employee will be exposed. The worker

Figure 5.2.   Where exposure to harmful chemicals may occur at any part of the body, protective clothing may be necessary. Here a worker clearing a hazardous waste site is completely protected, including face protection and the use of a respirator. (Photo courtesy of Lab Safety Supply, Janesville, WI.)

may be protected from sharp objects or abrasion using cotton gloves with leather palms and finger pads. Hot objects require heat-resistant and insulating construction. However, neither of these types of gloves provide protection against chemicals.

To protect against chemicals, the choice of material from which the glove is assembled is critical (Table 5.1). For example, hazardous water solutions of acids, alkalies, or strong oxidants are repelled by latex or latex-coated gloves, but latex is rapidly degraded by nonpolar solvents. Conversely, polyvinyl alcohol gloves are of value against some solvents, but become stiff and brittle on exposure to water, so readily that they must have a lining that absorbs perspiration. Manufacturers or distributors provide tables of recommendations about the correct glove construction for specific chemicals. Such tables may

Figure 5.3.   Skin contact with chemicals may be prevented by gloves. It is important to select gloves of the proper thickness and material for a given chemical exposure. (See Appendix E.) (Photo courtesy of Lab Safety Supply, Janesville, WI.)

include *breakthrough times,* i.e, average times in tests of continuous exposure to the chemical required for the glove to fail. Such values are useful for comparison of various gloves, but should not be misused. A worker wearing a glove with an eight-hour breakthrough time for a specific chemical will not actually be safely protected for eight hours of *continuous* exposure to that chemical, at which time the glove is discarded. It is also true that one seldom finds (perhaps *should not* find) a job in which the worker is continuously in contact with a chemical, and a glove that is largely dry all day may not fail in the breakthrough time, or even in a much longer time of use.

Gloves are sold in different thicknesses of protective material, and it is reasonable to suppose that a thicker glove lasts longer than a thin one. This reasoning must not be carried too far, however. Gloves do not remain intact

Table 5.1   Materials Used in Protective Clothing and Recommended Applications.

Material	Protects Well Against	Ineffective Against	Comments
Butyl rubber	Polar solvents	Non-polar solvents	
Cotton	Many solids	Not for solvent protection	Mild abrasion protection
Ethylene Vinyl Alcohol	Most chemicals		Easily damaged, use with other stronger material
Leather		Not for solvent protection	Primarily abrasion protection
Natural Rubber (Latex)	Water solutions, polar solvents	Non-polar solvents	Good cut and abrasion resistance
Neoprene Rubber	Non-polar solvents, acids, caustics		Structurally relatively weak
Nitrile Rubber	Non-polar solvents	Some polar solvents	Good cut and abrasion resistance
Polyvinyl Alcohol	Non-polar and many polar solvents	Water, alcohols	Good cut and abrasion resistance
Polyvinyl Chloride	Water solutions, acids, caustics, some polar solvents	See recommendations of manufacturer	Good abrasion resistance
Vitron	Non-polar solvents. See recommendations	Polar solvents. See recommendations	Structurally weak

Non-polar solvents include hydrocarbons and chlorinated hydrocarbons. Polar solvents include alcohols, ketones, water. Some solvents are intermediate, such as high molecular weight alcohols, esters, aldehydes.

and leak-free right up to some predictable time of failure, then completely dissolve. Glove failure occurs as the appearance of a small hole anywhere on the glove, often at a place where the glove is particularly stressed. Failure could occur at the thumb or finger surfaces where the worker contacts an object, at a flex point on the glove, such as the crease at the base of the thumb, or at a defective place on the glove coating. However, statistically one would predict longer useful life from a heavier gauge glove.

With failure time unpredictable, and even a small percentage of new gloves being defective, it is wise to spot check gloves each time before use. A quick, easy test for small holes is performed by folding the glove opening shut and rolling the material from the wrist toward the fingers to inflate it with trapped air. A change in glove material, such that it becomes more tacky, slippery, discolored, or hardened, may predict forthcoming failure.

As a separate issue, a worker wearing gloves loses a degree of hand dexterity and ability to pick up objects. Some glove materials present more of a problem in this regard than others. Fine manipulation of objects is made increasingly difficult as the glove is made thicker, which can lead to more frequent spills and dropping. It may sometimes be sensible to select thinner, less expensive gloves and discard them frequently.

In order to protect workers dealing with a variety of chemicals so that no single glove is adequate, the worker may "double glove"—wear one glove inside another. In such cases, at least one of the gloves must be very flexible or the combination might immobilize the worker. Double gloving may also be done to improve the worker's ability to handle objects. If the protective glove provides a poor surface (perhaps slippery) for grasping the work object, a thin second glove pulled over it may provide a better surface for grasping.

When gloves are intended to be discarded, it is important that the workers actually do discard them. Leaving gloves saturated with chemical on the hands could actually be worse than not wearing gloves at all. If gloves are reused for multiple workdays, a system to cleanse them of the chemicals in use must be instituted. On final discard it may be necessary to consider contaminated gloves as hazardous waste.

More Comprehensive Protection

More complete protection can be afforded in a number of ways, ranging up to complete body covering. The widespread nature of splashing or other sources of chemical exposure and the degree of hazard of the material may dictate the use of such protection. It should be strongly emphasized that an engineering solution should be sought if a routine job calls for full body covering. Such a protective garment serves as a vapor barrier, blocking the evaporative loss of sweat, which is not only uncomfortable but also interferes with normal body temperature control mechanisms. Whole body garments are intended primarily for emergency situations such as a spill or leak in a plant. However, aprons, lab jackets, coats, and slip-on sleeves may well be sensible and practical routine precautions.

The protection of gloves can be extended by slip-on sleeves that fit under the gloves and extend up the arm. Such commercially available shields have elastic at each end that should fit snugly, and come in a variety of materials. These have seams up the side, and should be checked to see if they are sealed or stitched. Stitched seams may be satisfactory, but they do provide small access holes for chemicals to enter.

Transparent face shields protect the skin of the face from splashing, and might be used alone, along with an apron or coat, or in conjunction with a full body covering, unless a more complete head covering device with a visor is required. Under no circumstances should a face shield be considered as sufficient eye protection (Figure 5.4). Safety goggles should be worn under a face shield to ensure that chemicals do not enter from the sides, top, or bottom of the shield.

OCCURRENCE OF INDUSTRIAL DERMATOSIS

Skin problems are the most prevalent job-related diseases, representing at least 40% and perhaps 2/3 of all reported difficulties, and 70% of industrial claims paid by insurance companies. The rate of occurrence has dropped over the years as a result of both programs to prevent skin contact with chemicals and the changing nature of the workplace (e.g., separation of the worker from the work process, as by automation). Incomplete reporting of dermatosis problems may be common because of a loophole in the requirements for reporting injury, which allows a minor injury to be omitted from the OSHA log by a company if it is classified as an "injury" rather than an "illness" and no time is lost, especially if only first aid is required. The cost in the U.S. of occupational dermatosis, considering medical costs, compensation for lost wages and loss of worker productivity, have been estimated by C. G. T. Mathias (1985) to be in the range of $220 million to $1 billion.

About a quarter of all workers utilize or work near some kind of skin irritant, and around 1% exhibit problems from this exposure. Workers in agriculture and manufacturing run the biggest risks. According to the Standards Advisory Committee on Cutaneous Hazards, very high rates of skin problems are found in leather tanning plants, poultry dressing plants, and meat packing plants. Industries involving contact with irritating chemicals, such as plating and rubber fabrication, also rank high on the list (Table 5.2).

EYE HAZARDS

Eyes are irreplaceable and easily damaged. We have two classes of concerns regarding eye protection: protection against physical damage from objects propelled at them by the work process (for example, metal or abrasive fragments from a grinder), and against chemicals that may splash or be sprayed into the eyes. All eye protection must accommodate vision correction needs of the worker, either by fitting over prescription glasses or by including corrective lenses in their construction. Goggles or face masks that guard against flying objects must be shatterproof, provide adequate coverage, and be comfortable and easily cleaned.

The eye is easily damaged by contact with chemicals. The surface of the eye at its most delicate point consists of the transparent cornea, across the outer surface of which is a layer of cells (Figure 5.5). The cornea is devoid of blood supply in order to improve its transparency, and depends on diffusion of nutri-

Figure 5.4. If full face protection is necessary, a face mask may be worn. Note that this is in addition to safety glasses, not instead of them. (Photo courtesy of Lab Safety Supply, Janesville, WI.)

ents from underlying eye fluids. It is supplied with nerve endings, so damaging contact is painful. A person with a chemical in the eye is usually highly motivated to remove it. The surface layer of cells can be destroyed by chemicals. The water content of the cornea changes in the absence of these cells, causing it to become clouded. However, the cells frequently grow back, so, if the cornea itself is intact, the damage may be temporary.

Safety glasses designed to protect against chemicals should completely enclose the eyes—sides, top, and bottom—to prevent entry of chemicals into the eye from any angle (Figure 5.6). Venting provided to minimize steaming of the glasses should have a baffle design that prevents them from leaking chemicals that strike the glasses at some particular angle. Face masks are often used to protect the entire face, but these do not substitute for safety glasses since they allow chemical entry from the sides.

Table 5.2. Occupational Skin Diseases—High Risk Industries.[a]

Industry	Incidence Rate (per 1000 workers)	Number of Workers (approximate)	Workdays Lost per Year
Leather Tanning	21.2	22,900	1392
Poultry Dressing	16.4	89,800	4405
Boat Building and Repair	11.1	48,000	854
Ophthalmic Goods	8.5	38,000	1390
Chemical Preparations	8.3	36,700	855
Plating and Polishing	8.3	61,400	1270
Meat Packing	7.2	164,300	1561
Frozen Fruits and Vegetables	7.2	43,200	1153

[a]Adapted from: Standards Advisory Committee on Cutaneous Hazards, OSHA.

Although any reactive chemical presents a threat, most eye accidents involve acids, alkalies, solvents, and detergents. If any of these contact the eye, immediately flush the eye with quantities of water. Acids and alkalies kill the surface layer of cells, and can cause corneal structural damage leading to scarring and blindness. Alkalies present the greatest problem, since they are more difficult to remove, and during the extended time required to remove them, they can cross the cornea and do damage inside the eye. Nonpolar acid or base anhydrides are also hard to remove, and they react with water to produce heat. Sulfur dioxide and ammonia, the most common examples of such compounds, are used in a variety of ways in factories. They are gases and are often used under pressure, for example in a refrigeration system. A leak in a line can result in the gas entering the eye under pressure. Lime (calcium oxide) also presents a special problem, since particles from a dusty environment may penetrate the surface of the eye. Once inside, lime reacts slowly with water to produce calcium hydroxide, a very strong alkali.

Organic solvents are generally chemically nonreactive and largely nonpolar chemicals. They damage the surface cells of the eye on contact, and, although this process is painful, it is much less threatening than acid or alkali exposure. The pain on contact drives the victim to remove the solvent (or any other chemical) from the eye as quickly as possible, minimizing the damage. Solvents are unlikely to do structural damage to the cornea.

Detergents function at the interface between nonpolar and polar (usually aqueous) environments, and, in order to do so, generally have a large hydrocarbon component able to interact with the nonpolar component and a charged or very polar structure attractive to water. Soap is the best known detergent:

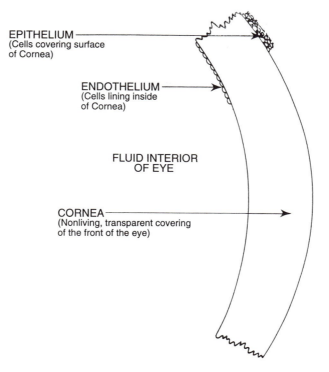

EPITHELIUM
(Cells covering surface
of Cornea)

ENDOTHELIUM
(Cells lining inside
of Cornea)

FLUID INTERIOR
OF EYE

CORNEA
(Nonliving, transparent covering
of the front of the eye)

Figure 5.5.   The cornea of the eye.

— A SOAP MOLECULE —

$$CH_3CH_2CH_2CH_2CH_2CH_2CH_2CH_2CH_2CH_2CH_2CH_2CH_2CH_2CH_2CH_2C\overset{\overset{O}{\|}}{-}O^- \ Na^+$$

nonpolar (hydrocarbon) segment   polar segment

They belong to one of three chemical classes: anionic (generally quaternary amines), cationic (these include soap, a carboxylic acid, and household detergents, which are often sulfonic acids), and nonionic (which have a very polar structure, such as a polyalcohol). All detergents are irritating to the eye, but the three classes vary greatly in the degree of irritation. Anionic detergents, which are used for their bactericidal properties in food processing, hospitals, and other institutional settings, are by far the most irritating, while nonionic detergents are the mildest to the eye.

When safety recommendations call for the wearing of eye protection, the proper safety glasses should be used, and the wearing of this protection should be enforced.

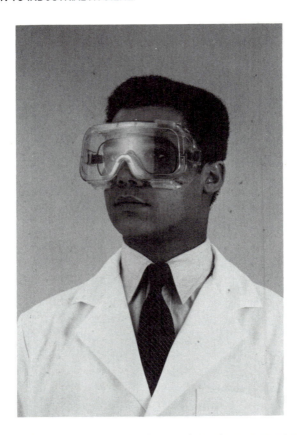

Figure 5.6.   To provide complete protection, safety glasses must protect the eyes from the top, bottom, and sides, as well as from directly in front. Such glasses have baffled vents to reduce steaming in hot, humid conditions. (Photo courtesy of Lab Safety Supply, Janesville, WI.)

KEY POINTS

1. Skin has two layers: the outer epidermis and the underlying dermis.

2. The epidermis has a surface keratin layer of dead, protein-filled cells. These are constantly replaced by cells modifying into keratin layer cells as they grow outward.

3. The dermis has collagen fibers, nerves, and blood vessels. Blood vessels expand when the body is overheated, cooling the blood.

4. Skin is pierced by pores and hair follicles. Evaporating sweat from pores lowers the skin temperature. Sebaceous glands coat the skin with nonpolar sebum.

5. Contact dermatitis is damage caused to the skin by chemicals. Abrasion, chapping, solvents, or extensive contact with water, especially water containing detergents, reduces the effectiveness of the keratin layer as a barrier.

6. Inflammation involves reddening and swelling of skin due to contact with an irritant. This produces an itching or burning sensation.

7. A few agents cause loss or color change of hair.

8. Allergic contact dermatitis is an inflammation of skin due to allergic response to a chemical.

9. In photodermatitis, innocuous agents on skin are converted to irritants on exposure to UV rays in sunlight.

10. Blocking follicles either with grease (acne) or by keratin destruction by chlorinated hydrocarbons (chloracne) produces pustules when sebum is trapped in follicles.

11. Some agents reduce or intensify skin pigmentation.

12. Corrosives chemically destroy skin tissue, generally producing scarring. Corrosives are generally strong acids or alkalies.

13. Direct contact with carcinogens can initiate skin cancer.

14. Workers dealing with animals or animal products are particularly susceptible to skin infections.

15. Protection for skin can vary from simple nonpolar barrier creams through gloves to full protective body coverings. The material chosen for protective clothing should be appropriate to the chemicals encountered.

16. Industrial dermatosis is the most frequent job-related disease.

17. Eyes present a special problem for protection, since the surface is easily damaged, and damage may lead to permanent blindness. Most eye problems concern corrosives, solvents, and detergents.

PROBLEMS

1. Consider the structure of the skin and how it varies with location. If the entire hand is submerged in a chemical, assuming the skin is completely intact on that hand, where will transport of the chemical into the blood be likely to occur first?

2. List conditions that reduce the effectiveness of the keratin layer as a protective barrier.

3. A group of 35 workers in a textile plant is exposed to an azo derivative dye component, and two become sensitized to it.

    A. Of various options (providing protective gloves, etc.), what is the best course to follow with these two sensitized workers?

    B. Why do only these two respond, and not the entire exposed group?

    C. There is no other chemical, of some 15 different chemicals used in that section of the plant, that sensitized a worker. What characteristic of that chemical might explain this?

4. You are employed as the safety consultant to a small industrial plant manufacturing sheet metal warning signs to be mounted in high hazard areas of factories. You are asked to recommend protective gloves to be used by workers in the plant, and are asked to stay within a set budget. You identify these worker tasks:

    There is a small press where worker 1 stamps sheet metal into signs with raised letters. Newly stamped signs often have sharp edges. These are moved on a conveyor to a station where worker 2 places them individually in a machine that grinds the edges. Worker 3 then places the signs in a jig and holes are drilled by a machine. The machine drops them into a solvent bath to clean them before painting. As they leave the bath, with a solvent residue on the metal, they are placed by worker 4 on a holder to be spray-painted. They are turned over after one side is painted by worker 5, who handles them by the holder, which is coated with some paint overspray, and placed back into the paint machine to paint the other side. Worker 6 places the holder and sign on the conveyor into a paint-drying oven. The hot sign emerges from the other side, worker 7 removes it from the holder and places it on a stand where the paint for the raised letters is rolled on. The finished sign rolls under heat lamps on a conveyor to the packing room, and workers 8 and 9 package signs for shipment to distributors.

A. Which workers (by number) need hand protection?

B. Which of these statements are appropriate to this situation?

    (1) A distributor gives a price break on purchases of over 200 pairs of gloves at a time. By getting all required gloves of the same kind, this is feasible and a better grade of gloves can be afforded.

    (2) The tasks require a variety of glove types, all of which must be replaced every day, so only the cheapest grade gloves fit the budget.

    (3) Some gloves can be kept in use for a long time; others need to be replaced frequently.

    (4) If heavy-duty gloves are purchased for worker 4, they need to be inspected before use each day.

    (5) If heavy-duty gloves are purchased for worker 2, they need to be inspected before use each day.

C. How will the gloves be chosen for worker 4?

D. The plant owner asks you about eye protection regulations. Go to the contents section of 29 CFR 1910 and locate where this is found. Does OSHA provide specific or general guidelines? Where are design standards for eyewear found?

5. Heavy-gauge gloves are purchased by an employer to be used several days before discarding by workers in a plant where solvents and moderately strong acids are in use. You are asked to outline a procedure for glove use and daily glove inspection. What should be done?

6. Medical terminology is commonplace in references frequently used by industrial hygienists, and you need to be able to decipher it. As stated earlier, a medical dictionary is a good investment for a person planning a career in the field. The following are from an excellent advanced toxicology book[1]:

---

[1]E. E. Emmett, "Toxic Responses of the Skin," in C. D. Klaassen, M. O. Amdur and J. Doull, *Toxicology: The Basic Science of Poisons*, 3rd ed., Macmillan Publishing Company, New York, 1986.

"The dermis has substantial vascular plexuses, unlike the epidermis, which is avascular....The dermis has a plexus of lymphatics, which drain to the regional lymph nodes and the thoracic duct. The dermis has abundant sensory and sensorimotor nerves."

"The skin and particularly the epidermis is an actively metabolizing organ that is capable of significant biotransformation of xenobiotics."

"Contact dermatitis is manifest by signs of erythema and edema in experimental test animals. In humans more varied responses are seen, and erythema and edema frequently progress to vesiculation, scaling, and thickening of the epidermis. Histologically the hallmark is spongiosis or extracellular edema of the epidermis."

"A number of physiologic and pathologic changes occur in the skin as a result of exposure to the ultraviolet (UV) component of sunlight. These include erythema (sunburn); thickening of the epidermis; darkening of the existing pigment (immediate pigment darkening); new pigment formation (delayed tanning); actinic elastosis (premature skin aging); proliferative and other changes in epidermal cells; suppression of T lymphocytes; actinic keratosis, a precancerous condition; and the development of squamous cell cancers, basal cell cancers, and probably melanomas."

BIBLIOGRAPHY

R. M. Adams, *Occupational Skin Disease,* Grune and Stratton, New York, 1983.

D. J. Birmingham, *The Prevention of Occupational Skin Diseases,* Soap and Detergent Association, New York, 1982.

G. Dupis and C. Benezra, *Allergic Contact Dermatitis to Simple Chemicals,* Basel Dekker, New York, 1982.

A. A. Fisher, *Contact Dermatitis,* 3rd ed., Lea and Febiger, Philadelphia, 1986.

H. I. Maibach , *Occupational Dermatoses,* 2nd ed., Year Book Medical Publishers, Chicago, 1986.

C. G. T. Mathias, "The cost of occupational skin disease," *Arch. Dermatol.* (1985), **121**, 332.

C. G. T. Mathias, "Occupational Dermatoses," in Carl Zenz, *Occupational Medicine, Principles and Practical Applications,* 2nd Edition, Year Book Medical Publishers, Inc., Chicago, 1988.

CHAPTER 6

INHALATION TOXICOLOGY

In Chapter 3 we briefly considered the lung as a route of entry into the body. In the industrial setting this is the most important such route. Consequently, control of the workplace atmosphere is the objective of the majority of industrial hygiene studies and government regulations. This chapter presents more detail on lung structure and function, and describes interaction of the lung with specific atmospheric contaminants.

Inhaled toxicants may consist of dispersed molecules (gases or vapors), aerosols, or particles. By *vapors,* we mean evaporated liquids. *Aerosols* are droplets of liquid small enough to stay suspended in the air.

LUNG STRUCTURE AND THE ENTRY OF TOXICANTS— A MORE DETAILED DESCRIPTION OF THE RESPIRATORY SYSTEM

THE NASAL CAVITY

The first segment of the respiratory system is the nasal cavity, a cavity lined with a moist mucus layer. Passing air through this compartment warms and moistens the air before it enters deeper portions of the system. The sticky mucus surface traps particles, particularly larger particles. Strong irritants initiate sneezing, preventing penetration farther into the system and clearing contaminants out.

Entry into the lungs normally is by way of the nose and nasal cavity. However, when assessing the hazard of a chemical to the complete workforce, such trapping must be discounted as a means of protection, since some workers are constant mouth breathers. Other workers experiencing respiratory congestion, as from a cold, become temporary mouth breathers. Finally, a person strongly exercising resorts to mouth breathing as the most rapid pathway to get more air into the lungs. The alternative path through the mouth is far less effective at blocking progress of toxicants farther into the lungs.

THE PHARYNX AND BRONCHIAL TUBES

Beyond the nasal cavity is the pharynx, which subdivides repeatedly to form the bronchial system. These bronchial tubes have cartilaginous structures in the walls to prevent collapse of the tubes without causing them to be rigid. Smooth muscle is found in these walls, and its contraction narrows and relaxation widens the diameter of the passages. This is regulated by the brain through the sympathetic and parasympathetic nervous systems and the hormone epinephrine in response to physical demands on the body.

The mucus lining found in the nasal cavity extends completely down this bronchial "tree." As in the nasal cavity, the mucus serves to trap particles. The inertia of particles gives them a tendency to travel in a straight line, so they "spin out" onto the mucus when the airways bend at a branch point. Larger particles have greater inertia, so are more likely to be trapped.

Particles settle out of still air, but are held in suspension in moving air. Faster-moving air can suspend larger particles. At each bronchial branch the total cross-sectional area of the system increases because the sum of the areas of the branches is greater than was the area before branching. Visualize a simple model where a duct carrying moving air splits into two branches, each the diameter of the original duct. Each branch is now carrying half the original air, and therefore has half the original flow rate. As the total cross-sectional area of the bronchial system increases, the flow rate correspondingly decreases and larger particles settle out onto the mucus.

Beneath the mucus layer is a forest of hair-like cilia whose motion shifts the mucus layer upward toward the nasal cavity. Trapped particles are carried up this mucus "escalator," and eventually the mucus and its particulate burden reach the nasal cavity and are swallowed. Hopefully the particles will now be removed from the body with the feces, but some components may be absorbed from the G.I. tract.

Gases and vapors pass completely through the bronchial system unless they are water-soluble and dissolve into the moist lining. Those that dissolve and are irritants stimulate coughing, clearing the air out of the lungs.

ALVEOLI

At the end of the smallest subdivisions of the bronchi are the alveoli, little sacs in which oxygen and carbon dioxide are exchanged with the blood. Before age 10, human lungs generate the adult population of about 300,000,000 alveoli, each about 200 µm in diameter, which provide about 70 m^2 for gas exchange. There is no mucus barrier on the alveolar walls. Phagocytic cells populate the inside surface of the alveolus and are thought to be responsible for depositing a detergent-like chemical.

The outside surface of the alveolus is almost completely covered with capillaries, and the thickness of tissue separating air in the alveolus from blood in the capillary is in the micrometer range. Gases and vapors entering the alveolus travel only a short distance across this barrier to enter the blood. Chemicals on particles that reach the alveolus may also cross this barrier and enter the blood.

Phagocytic Cells of the Alveolus

Phagocytic cells serve to engulf foreign substances and bacteria, and normally cleanse the alveolus of inhaled particles. They then migrate toward the bronchi to be moved out of the lungs on the mucus. Exposure to dust stimulates expansion of this cell population. If components of the particles are toxic to phagocytic cells, or if the population grows very large so more are dying at any given time, the dead cells release digestive enzymes that damage lung tissue.

Unfortunately, components of cigarette smoke narcotize phagocytic cells, with the result that both particulate and cells accumulate in the alveolus. This is an important reason cigarette smokers are at greater risk from exposure to toxicants than are nonsmokers.

THE NATURE OF ATMOSPHERIC CONTAMINANTS AFFECTS THE LIKELIHOOD OF ENTRY INTO THE BLOOD

GASES OR VAPORS

Gases or vapors are not trapped on sticky mucus as are particles, so they can penetrate with the inhaled air all the way to the alveoli. Such molecules may then cross the thin barrier into the blood readily. The chemical nature of a gas or vapor influences how effectively the compound crosses this, or any other, barrier. Nonpolar gases or vapors are more capable than polar compounds of dissolving through the grease barrier of cell membranes between the interior of the alveolus and the inside of the capillary.

We have described the alveoli as prime sites for transfer of contaminants from inhaled air to blood, but gases and vapors may transfer to blood or do damage to some degree anywhere in the lungs. A large fraction of more polar, hence more water-soluble, compounds dissolve into the watery mucus lining before they reach the alveolus, causing them to remain longer in the lung. Once dissolved, the gas or vapor must pass through the mucus lining in order to reach the wall of the lung. This lining is relatively thick in the upper airways, but decreases in thickness toward the alveoli. Once reactive chemicals penetrate to the wall of the lung, they may do structural damage to the lung.

If the toxicant is a strong irritant, the individual ceases to inhale, either voluntarily because of distress, or involuntarily because of bronchial contractions or spasms. Irritants stimulate the production of mucus, increasing the thickness of the layer and its effectiveness as a barrier. Subsequent coughing up of the excess mucus and its removal from the lung carries the irritant out, too. Reluctance of the worker to inhale strong irritants minimizes the hazard of these compounds.

PARTICLES

Throughout evolutionary history humans have been exposed to dusty air. The effective system described above for removal of particulate from inhaled air is the response to this environmental problem. However, this system can be overwhelmed, or particles of a significantly toxic nature may be inhaled in the workplace. *Pneumoconiosis* is a general term used to describe the harm done by the inhalation of solids as the result of occupational exposure. Common symptoms include *dyspnea* (shortness of breath), chest pains, fatigue on exertion and *cyanosis* (shortage of oxygen). Accompanying damage to the lung circulatory system can cause strain to the right side of the heart, which is responsible for pumping blood to the lungs. Chemicals also may transfer from particles in the lung into the bloodstream, traveling to other organs and causing damage.

Airborne particulates are classically subdivided into groups based on their source and character. Solid particles, for example, may be dusts if they are produced by agricultural processing or a grinding type of operation, fumes if they arise from high-temperature metal processing (generally metal oxides), and smokes if their source is combustion of organic substances. Dusts are generally at the high end of the size range, perhaps from 1 to 150 micrometers in diameter. Fumes are smaller, running 0.2 to 1 micrometer in diameter, and smokes usually have a diameter of less than 0.3 micrometers. Aerosols may be mists or fogs of water droplets condensed on smaller solid particles, or droplets of organic material such as would result from a spraying operation.

Although toxic hazard primarily relates to the chemical nature of the particulate, the size of the particle is the second most important consideration. Larger particles are more likely to collide with and stick to the mucus lining of the respiratory tract. They are quickly trapped from the inspired air and removed as the mucus is moved out of the lung. The greater penetration by smaller particles lengthens their stay in the lung, increasing the chance of harm being caused. The greatest potential for problems results when particles reach the alveoli. Only the smallest particles are likely to penetrate this far, and for this reason particles a few micrometers or less in diameter are distinguished by the special term *respirable*.

Various problems result from particulates entering the lungs. Dyspnea results from obstruction of the airways. Chemicals in particles may stimulate some contraction of the smooth muscle lining the bronchial tubes. Inhaling dusty air stimulates excess mucus production, narrowing the passages. Finally, the mucus glands and cells themselves may enlarge (hypertrophy) on long-term stimulation.

Particles may produce local effects such as *irritation*. Such irritation can cause fluid to collect in the lungs, a type of *pneumonia*. Some of the particulates that remain in the lung generate *fibrosis*, a scarring of the lung accompanied by a permanent loss of functional capacity. This loss of capacity may not limit capability for physical exertion for a time, and so go unnoticed. Physical activity is limited by the combined capacity of the lungs to provide oxygen to the blood, the heart, and the circulatory system to deliver the oxygen to the tissues, and there is excess capacity of lungs over circulatory system in this process.

SPECIFIC EXAMPLES OF HAZARDOUS PARTICULATES

SILICA

Silicosis is a fibrosis of the lungs caused by inhaling respirable particles (>5 µm) of silica (SiO_2) over a long time period. Phagocytic cells in the alveolus engulf these particles and die, silica being toxic to them. The digestive enzymes released by these cells upon death damage lung tissue, leading to scarring (fibrosis). Victims of silicosis have few symptoms in early stages, but display silicotic nodules in chest X-rays (simple nodular silicosis). In advanced cases, where there is loss of functional lung mass, the victim experiences shortness of breath and coughs sputum. Now X-rays reveal large areas of scarred tissue (conglomerate silicosis). The symptoms develop slowly and may therefore be missed. If the victim is a smoker, the symptoms may be misinterpreted as being the result of the use of tobacco. Infections of the lung are common, and the victim is more likely to develop tuberculosis.

The TLV for dusts containing silica depends on knowing the percentage of silica in the dusts to which the worker is exposed:

$$TLV = \frac{10 \text{ mg/m}^3}{\% \text{ silica}} + 2$$

Silicosis was the first occupational lung disease to be reported. It may result from the inhalation of any uncombined form of silicon dioxide. Grinding with silica abrasives, sandblasting, pottery making or working with sand in a foundry are possible industrial exposures, while sandstone quarrying, "hard rock" mining, or tunneling also provide serious occupational contact. Fibrosis sometimes found in the lungs of coal miners probably results from silica released into the mine air during tunneling and inhaled, rather than from any effect of the coal dust itself.

ASBESTOS

Asbestosis—lung disease caused by inhalation of asbestos fibers—is perhaps the most publicized of the lung fibrosis diseases. What we term asbestos is actually a number of different mineral materials that share the property of being convertible to the familiar fibrous product. They are silicate compounds of calcium, sodium, iron, chromium, nickel, or magnesium. Not all varieties of asbestos present the same degree of hazard. Asbestos strands readily break into short lengths, and enter the air as very small particles, often under 1 μm in length.   The thin fibers lodge in the smaller subdivisions of the air passages, causing irritation and edema. Continued exposure leads to permanent scarring of the lung and a resulting loss of lung capacity and tissue flexibility. Victims often have a cough, and lung infections are very common.

In addition, some types of asbestos are carcinogenic. Bronchial carcinoma (malignant mesothelioma) occurs in association with asbestos exposure. The type of asbestos called crocidolite is a particularly potent causative agent for malignant mesothelioma. Combining the inhalation of asbestos fibers with the smoking of tobacco sharply increases the risk. It is hypothesized that the surface of the fiber, as it lies embedded in the lung, serves to collect and concentrate carcinogens from the tobacco smoke.

Asbestos has found a broad range of applications, largely based on its fibrous character and resistance to high temperatures. These include pipe insulation, floor tile, roofing shingles, cement, clutch pads, brake linings, wall board and plaster, and heat- and fire-resistant garments. It has been used in factories for insulation of steam pipes and boilers, as an air filter medium, in board form for partitions, as fireproofing on beams, and as insulation under the roof to prevent heat loss. Substitutes have been found for asbestos in many applications since the discovery that cancer can result from exposure, so the level of worker exposure has dropped sharply. Although care is taken in new installations not to use asbestos in a hazardous fashion, many older facilities have asbestos present in locations such that workers are at risk of exposure. Particular care must be exercised in removal of old asbestos from such sites, so that the hiring of an expert with the proper equipment and experience is wise.

TALC

Talc is hydrated magnesium silicate, and is usually found in association with deposits of dolomite. It is a component of tile, paper, and paint, and is used as a lubricating powder with rubber products such as tires and gloves, or as talcum powder. It does not cause the degree of fibrosis that we see with asbestos, but occasionally it is a cause of pneumoconiosis.

MICA

The mineral form of mica is called muscovite. It is found as plates, which separate into thin sheets, or as a powdery product. It serves well as a heat or electrical insulator, so it is used in electrical devices and appliances.   The ground material is added to roofing, and it has served as a mold release agent in the rubber industry. It is infrequently responsible for a pneumoconiosis.

FIBERGLASS

Fiberglass or glass wool is simply very long, thin fibers of glass. It is used as insulation and as a reinforcing material in plastics. Tiny fragments of glass wool carried as a dust in the atmosphere can enter the respiratory tract and cause irritation, much as skin contact with glass wool can cause skin irritation.

METALS AND METAL OXIDES

Metals and their compounds may appear in the workplace as the result of grinding, sawing, milling, or otherwise shaping metal parts. Any time metal is melted or formed at high temperature, metal oxide fumes may be released. Several of these fumes are irritating or damaging to the lungs. A set of symptoms very much like influenza is encountered, termed *metal fume fever*. These problems are discussed in more detail in the chapters on metals.

COAL

The mining of coal has caused a great deal of lung damage over the years, producing a characteristic symptomology called *coal miner's pneumoconiosis* (black lung disease). Coal dust deposited in the lungs leads to breathlessness. Coal does not directly cause fibrosis, but other substances in the air of a coal mine, particularly silica, may. The introduction of wet drilling techniques to minimize dust production was a major advance in reducing the magnitude of this problem.

OTHER DUSTS

Another dangerous class of dusts is the natural product particulates. An example is the dust arising from cotton in the course of making cloth. Fibrous material in the air in cotton mills correlates with an ailment called *brown lung disease* in the southern U.S. Other problems have been experienced by farmers

Not all dusts are as damaging as those already presented. When deposition in the lung does not cause inflammation or fibrosis, health problems are significantly reduced. For example, the calcium carbonate (limestone or marble) and calcium sulfate (gypsum) dusts are not as threatening. Similarly, the iron dust and iron oxide fume that welders encounter is not as damaging as many metal oxides. Iron compounds may accumulate in the lungs, a problem termed *siderosis,* but they do not stimulate fibrosis. However, welders need to take precautions because many of the items being welded, as well as welding rods themselves, include more dangerous metals in their composition.

HAZARDS OF GASES

Gases generally cause one of two types of problems for workers. Some are asphyxiants, meaning that ultimately they result in tissues having an inadequate supply of oxygen. Others are irritants, which damage or kill cells lining the nose, throat, and lungs, resulting in discomfort and pain, constriction or spasmatic contraction of breathing passages, or filling of lung passages with fluid.

ASPHYXIANTS

In living organisms the energy needed to perform biological work, processes such as contracting muscle, transmitting nerve impulses and the complex chemical synthesis required for cell repair and growth, is supplied by adenosine triphosphate (ATP). This molecule has three adjacent phosphate groups. When a phosphate-to-phosphate bond is hydrolized (almost always the bond attaching the end phosphate), energy is released to do biological work. By the laws of thermodynamics (these are not in the *Federal Register*), that same amount of energy must be supplied to re-form the phosphate-to-phosphate bond. ATP is reassembled in conjunction with the oxidation of dietary molecules (carbohydrates, fats, and proteins), using the energy released by this oxidation to drive ATP formation.

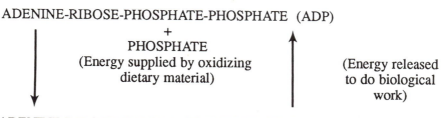

ADENINE-RIBOSE-PHOSPHATE-PHOSPHATE (ADP)

+

PHOSPHATE

(Energy supplied by oxidizing dietary material)

(Energy released to do biological work)

ADENINE-RIBOSE-PHOSPHATE-PHOSPHATE-PHOSPHATE (ATP)

When the supply of oxygen to the body is reduced, these metabolic oxidations slow or cease, ATP supplies dwindle, and vital processes driven by ATP are halted.

The immediate danger from asphyxiation lies in cutting off the supply of oxygen to the brain. Brain cells are more susceptible to the loss of their oxygen supply than are the cells of other tissues. Furthermore, once dead, brain cells are not replaced. Within 5 to 8 minutes of a complete loss of the oxygen supply, serious and possibly irreversible brain damage is taking place. In gradual asphyxiation, the early symptoms are a feeling of euphoria followed by an increase in the respiratory rate. Later there is fatigue and headache, then nausea and vomiting. Finally the victim collapses.

Hypoxia is a general term describing the reduction in oxygen supply to tissues. When it is associated with respiration—the original intake of oxygen into the blood—it is then termed *anoxic hypoxia*. If respiration is normal and the air has sufficient oxygen, but the capacity of blood to transport oxygen is impaired, the problem is termed *anemic hypoxia*. If the blood flow is inadequate due to heart problems or a blocked blood vessel, we have *hypokinetic hypoxia*. Finally, the tissues may not be able to use the oxygen once it is delivered—a problem termed *histotoxic hypoxia*. Asphyxiants are subdivided into simple and chemical subgroups.

Simple Asphyxiants

Simple asphyxiants dilute the oxygen supply to levels that no longer support life. This requires that the diluting gas be present in extremely high concentration. Small concentrations prove harmless, since simple asphyxiants are chemically inert. Nitrogen, helium, argon, and such hydrocarbons as methane belong to this class. These are not detectable by odor—not even methane. The natural gas smell associated with methane is actually trace amounts of a very strong-smelling mercaptan added to facilitate recognition of a leak. Carbon dioxide is a borderline member of this group. It creates problems primarily by displacing oxygen from a working environment. However, as an acid anhydride it exerts a physiological effect on the person inhaling it, for example, stimulating deeper, more rapid breathing. Carbon dioxide is a dense gas, so it may accumulate at higher concentrations at floor level, or at low points in tanks or compartments that workers might enter. Awareness of the hazards of carbon dioxide are important both because of the widespread use of this compound in industry and because it is a combustion byproduct.

Sometimes oxygen-free atmospheres are deliberately produced, as in some special welding techniques. More dangerous is the situation where the worker enters an area unaware that the oxygen supply is dangerously low. This happens in once-sealed compartments in which an oxidative process has been occurring, or in tanks where evaporation of residues of the liquid once stored in the tank has displaced the oxygen.

Table 6.1. Asphyxiants.[a]

Compound	PEL (ppm)	IDLH (ppm)
HCN	10	50
CO	50	1500
H_2S	20 (ceiling)[b]	300
CO_2	5000	50000

[a]PEL = Permissible Exposure Level, time-weighted average, 8-hour day.
IDLH = Immediately Dangerous to Life or Health.
[b]50 ppm; 10-min. peak.

Chemical Asphyxiants—Carbon Monoxide

Chemical asphyxiants (Table 6.1) block the transportation or use of oxygen by some chemical reaction. Important examples include carbon monoxide, hydrogen cyanide, and hydrogen sulfide. Carbon monoxide is an odorless gas. It is found in coal mines, and anywhere combustion is occurring, such as around internal combustion engines and furnaces. An operation of particular concern is the heating of vats or tanks with a gas flame. As the burning gas strikes the cold surface, the gas temperature is lowered below the ignition point of the methane, leading to incomplete combustion and the generation of carbon monoxide.

In the blood, carbon monoxide combines with hemoglobin, the oxygen transport compound. The bond with carbon monoxide is at the same location on hemoglobin as the bond with oxygen, but is more than 200 times as strong. Hemoglobin with carbon monoxide attached is unavailable for oxygen transport, producing a temporary, anemia-like condition. The greater the carbon monoxide exposure, the higher the percentage of hemoglobin tied up in this fashion (Table 6.2). A portion of the hemoglobin of smokers is already incapacitated, since carbon monoxide is found in cigarette smoke, and they are at greater risk.

The Haldane equation, devised in 1912, allows calculation of the degree of saturation of blood with carbon monoxide at various concentrations of CO in the air:

$$\frac{Hb_{CO}}{Hb_{O_2}} = M\left(\frac{p_{CO}}{p_{O_2}}\right)$$

where M is about 245 at normal blood pH, and Hb_{CO} and Hb_{O2} are the relative proportions of hemoglobin binding CO and O_2, respectively.

Table 6.2. Effects of Binding of CO by Hemoglobin.

Percentage of Hemoglobin as CO Complex	Cause/Effect
5	Caused by 1 pack of cigarettes per day
10	Caused by 50 ppm CO in air (the PEL)
15–25	Headaches, nausea
40	Collapse
60	Death

Sample Problem:

Suppose p_{CO} were 10 torr and p_{O_2} were approximately normal (160 torr). What fraction of the hemoglobin is nonfunctional because it is bound to CO?

$$\frac{Hb_{CO}}{Hb_{O_2}} = M\left(\frac{P_{CO}}{P_{O_2}}\right)$$

$$\frac{Hb_{CO}}{Hb_{O_2}} = 245\left(\frac{10 \text{ torr}}{160 \text{ torr}}\right) = 1.6$$

$$\frac{Hb_{CO}}{\text{total Hb}} = \frac{1.6}{1.0 + 1.6} = 0.62 \quad (\text{or } 62\%)$$

Hydrogen Cyanide

Hydrogen cyanide, famous as the lethal gas used in the gas chamber, has a faint bitter almond odor. Salts of this compound are used in some ore extractions, in electroplating, in preparing some plastics monomers, and in a few other industrial processes. In the body, cyanide binds to the compounds that are the prime users of the oxygen in the tissues, blocking the process of energy production in the body. Most often the hazard from hydrogen cyanide arises when a solution containing such salts as sodium or potassium cyanide is accidentally mixed with acid. This converts the cyanide ion in solution into gaseous hydrogen cyanide.

Hydrogen Sulfide

There are no direct uses of hydrogen sulfide, but it does occur as a byproduct in refineries, some plastic and rubber processes, and tanneries. It has a distinctive rotten egg odor at very low concentrations, but at dangerous levels the sense of smell is rapidly blocked. In high concentrations it causes respiratory paralysis.

IRRITANTS

Irritant gases either cause direct damage to the linings of the air passages, or in that moist environment they are converted to molecules that do. They are often referred to as *upper respiratory tract* and *lower respiratory tract irritants,* a distinction based on how far into the lung they are likely to penetrate.

Upper respiratory tract irritants (Table 6.3) are unlikely to cause serious injury if the worker can leave the area, because they are so unpleasant to encounter that workers hold their breath and quickly escape from contact. This tends to limit entry to the nose, throat, and larger passages of the respiratory tract, i.e., the upper respiratory tract. Upper respiratory tract irritants are more water-soluble and are irritants in their present chemical form. A burning sensation and possible spasms of the bronchial tubes are the result of these compounds contacting the lung's surface. Important examples of upper respiratory tract irritants include hydrogen chloride (the gas form of hydrochloric acid), sulfur dioxide, ammonia, and chlorine.

Lower respiratory tract or whole lung irritants (Table 6.4) pose a significantly greater threat. Generally they are less water-soluble, so they reach the lung surface more slowly. The full potential for damage follows a slow reaction with water, converting them into more damaging compounds. For both these reasons the irritation of the lung upon inhalation is not as extreme as with upper respiratory tract irritants, and people are more likely to allow a damaging dose to move through the entire lung. For some of the more dangerous compounds, such as phosgene or nitrogen dioxide, there is a long time lag, even hours, between the inhalation of the gas and the full impact of the compound on the lungs. The irritation causes the lungs to fill with fluid, producing a pneumonia-like condition. In its extreme form a person is drowned in his or her own body fluids. Beyond the immediate crisis, an extreme invasion of the lungs by an irritant can leave the lung permanently scarred, resulting in a reduction of lung capacity. Such paralysis of breathing and fluid in the lungs were the prime cause of death in the tragic escape of methyl isocyanate (actually a vapor) in Bhopal, India, in 1984. Reduced lung capacity is a common complaint among the survivors.

Table 6.3. Upper Respiratory Tract Irritants.[a]

Compound	PEL (ppm)	IDLH (ppm)
HCl	5	100
HBr	3	50
HF	3	20
SO_2	5	100
NH_3	50	100

[a] PEL = Permissible Exposure Level, time-weighted average, 8-hour day.
    IDLH = Immediately Dangerous to Life or Health.

HAZARDS OF VAPORS

Dealing with the evaporation of organic liquids into the workplace atmosphere represents a major regulatory concern. The mechanism by which individual compounds threaten worker health runs across the entire spectrum of toxicology from mild irritation, through systemic damage as the chemical enters the blood, to the causation of cancer. The threat varies with the concentration of the vapor in the workplace atmosphere, and this in turn depends on how the compound is used, ventilation and other safeguards, and the volatility of the liquid that is the source of the vapor. Volatility reflects the rapidity and degree to which the liquid evaporates. Two physical constants measure volatility: the vapor pressure and the boiling point. Vapor pressure measures the amount of a compound that can enter the atmosphere at a given temperature. For example, if the atmospheric pressure were 750 mmHg (slightly below one atmosphere) and the vapor pressure of a compound at the existing temperature were 75 mmHg, that compound would evaporate until 75/750 or 10% of the molecules in the air above the container of liquid were molecules of that liquid. Vapor pressure increases with temperature, and the boiling point is the temperature at which the vapor pressure is as high as the existing atmospheric pressure. Compounds with low boiling points must therefore have high vapor pressures, and would be described as very volatile.

The importance of volatility in industrial toxicology can be demonstrated very clearly by considering again the Bhopal tragedy in India. The compound that caused the deaths and injuries was methyl isocyanate, used to make carbamate insecticides. Toxicologically, it bears a strong family resemblance to other isocyanates, such as toluene diisocyanate, used in the plastics industry. However, methyl isocyanate is very volatile (boiling point, 39° C). With a rise in the temperature of the storage tank, it rapidly converted to vapor, blowing the safety

Table 6.4. Lower Respirator Tract Irritants.[a]

Compound	PEL (ppm)	IDLH (ppm)
Cl_2	1.0	25
Br_2	0.1	10
F_2	0.1	25
NO_2	5.0	50
O_3 (ozone)	0.1	10
$COCl_2$ (phosgene)	0.1	2

[a] PEL = Permissible Exposure Level, time-weighted average, 8-hour day.
IDLH = Immediately Dangerous to Life or Health.

valve and blanketing the town of Bhopal. Had everything been the same, except that the tank had contained toluene diisocyanate (boiling point, 251°C), the amount of vapor escaping would have been very much smaller and the major tragedy would not have occurred.

To list all the compounds whose vapors are hazardous would be a gigantic task, but the recommended exposure maxima of some more common organic solvents are listed in Table 6.5. For any liquid to be used in the plant, referring to the label of the container—or better, to the MSDS—gives warning of any hazard. A danger exists whenever hazardous liquids are used or stored in open containers, are sprayed onto surfaces, evaporate from coatings or from dipped parts, or are used in any other fashion that allows evaporation into the atmosphere. Furthermore, the higher the temperature of the liquid, the greater its vapor pressure, and the more rapidly it will escape into the air. In addition, liquids that are not normally allowed to evaporate become a problem during maintenance procedures, such as valve repair or tank cleaning.

WARNING PROPERTIES OF GASES AND VAPORS

The undetected escape of gases or the evaporation of volatile liquids into the work environment poses a serious health threat. However, some gases or vapors have a strong odor, or are irritants of the respiratory passages. Escape of such compounds will be noticed readily, permitting corrective measures to

Table 6.5. Recommended Maximum Exposures to Solvents.[a]

Solvent	PEL (ppm)	TLV-TWA (ppm)	STEL (ppm)
Acetaldehyde	200	100	150
Acetone	1000	750	1000
Acetonitrile	40	40	60
n-Amyl acetate	100	100	150
Anthracene (coal tar pitch volatiles)	0.2[a]	0.2[a]	—
Benzene	10 (25c)	10	25
n-Butyl acetate	150	150	200
n-Butyl alcohol	100	50	—
Carbon disulfide	20	10	—
Cellosolve	200	—	—
Cellosolve acetate	100	—	—
Chlorobenzene	75	75	—
Cyclohexane	300	300	375
Dibutylphthalate	0.1	1	2
1,2-Dichloroethylene	200	200	250
Diethyl ether	400	400	500
Dimethylformamide	10	10	20
Dioxane	1	—	—
Ethyl acetate	400	400	—
Ethyl alcohol	1000	1000	—
Ethyl chloride	1000	1000	1250
Ethylene glycol	—	50	—
Ethyl formate	100	100	150
Furfural	5	2	10
Furfuryl alcohol	50	10	15
n-Heptane	85	—	—
n-Hexane	100	—	—
2-Hexanone (methyl butyl ketone)	100	5	—
Isobutyl alcohol	100	50	75
Isopropyl acetate	250	250	310
Isopropyl alcohol	400	400	500
Methyl acetate	200	250	—
Methyl alcohol	200	200	250
Methyl cellosolve	25	25	35
Methyl cellosolve acetate	25	25	35
Methyl chloride	100	50	100
Methyl cyclohexa-none	100	50	75
Methylene chloride	500	100	500
Methyl ethyl ketone	200	200	300
Naphtha (coal tar)	100	—	—
Naphthalene	10	10	15
Octane	75	—	—
n-Propyl acetate	200	200	250
n-Propyl alcohol	200	200	250

Table 6.5. Recommended Maximum Exposures to Solvents.[a] (Cont'd)

Solvent	PEL (ppm)	TLV-TWA (ppm)	STEL (ppm)
1,1,2,2,-Tetra-chloroethane	5	1	5
Tetrachloroethylene	100 (200c)	50	200
Toluene	200	100	150
1,1,1-Trichloroethane (methyl chloroform)	350	350	450
1,1,2-Trichloroethane	10	10	20
Trichloroethylene	100	50	150
Xylene (mixed iso-mers)	—	100	150

Source: NIOSH Registry of Toxic Effects of Chemical Substances, 1981–1982, U.S. Department of Health and Human Services. Regulatory Chemicals of Health and Environmental Concern, W. H. Lederer, Van Nostrand Reinhold Company, New York, 1985.
[a]mg/m^3.

be taken before dangerous levels are reached. We say the compound has good *warning properties*. Many compounds are odorless, or will already have accumulated to dangerous levels before an odor is detected. If such compounds are in use in the plant, they should be monitored with special care. A good example is provided by three chlorine-containing gases: hydrogen chloride, chlorine, and phosgene. Hydrogen chloride is a strong acid when dissolved in water, and is a severe irritant to the eyes and respiratory tract. The irritation is so intense that a person exposed to even low concentrations of this gas will struggle to avoid inhaling it, even at levels that would do little harm. Chlorine is not an acid, but reacts with water to produce acid. It is therefore less irritating, so that a person exposed to chlorine might accept dangerous levels into the lungs in an emergency situation. Phosgene is the most dangerous of the three. It reacts only very slowly in the lungs to produce acid, and does not have a strongly repugnant odor. Dangerous levels could easily be inhaled inadvertently, the damage happening some time after inhalation. Hazard should not be equated with repugnant odor or irritation to the lung, but in this case the unpleasant awareness of the compound is a safety feature.

OCCUPATIONAL ASTHMA

Asthma is the inflammatory response of lungs to the inhaling of an allergen—a substance to which the individual is allergic. The difficulty is common, with approximately 3 percent of the population diagnosed as having an asthma problem. As with other allergies, the individual responds rapidly to even very small amounts of the allergen once he or she is sensitized. When a worker displays an allergic response to a chemical inhaled in the workplace, we call it *occupational asthma.* The response may be delayed, making it hard to draw a clear association between the workplace exposure and the symptoms. Skin tests may be needed to identify the allergen.

Response is mediated by antibodies in the lung (often IgE antibodies), and the resulting contraction of the bronchial smooth muscle produces a constriction of the air passages. The victim is short of breath, wheezes, coughs, or generally has difficulty taking in sufficient air, and can be severely distressed. Airborne allergens should be removed from the atmosphere, and sensitized workers should be removed from the vicinity of the allergen. Symptoms of an asthma attack are treated with ß-adrenergic drugs and derivatives of theophylline, which relax the bronchial smooth muscle. Anti-inflammatory steroids may be used in extreme cases.

KEY POINTS

1. Inhaled toxicants may be gases, vapors, particles, or aerosols.

2. The nasal cavity can trap particles, but many times workers are mouth breathers.

3. The pharynx and bronchial tube are mucus-lined. The bends in air passages and the increasing cross-sectional area and consequent slowing of air flow with lung penetration favor particles depositing on the mucus.

4. Mucus in the lungs is continually moved, carrying the trapped particles toward the nasal cavity.

5. Only a thin wall separates air in the alveolus from blood in capillaries. Alveoli have no mucus, but do have phagocytic cells to engulf particles.

6. Polar gases or vapors dissolve in the moist lung lining. Here irritants generate coughing.

7. Nonpolar gases or vapors are not retained long in the lung, but they are more capable of crossing the alveolar wall and entering the blood.

8. Particles include dusts, fumes, and smokes, according to origin.

9. Only particles smaller than a few micrometers penetrate well as far as the alveolus.

10. Irritation causes inflammation, resulting in fluid collecting in the lungs.

11. Some particles generate scarring (fibrosis) in the lungs.

12. Silicosis is caused by long-term inhalation of SiO_2. In advanced cases there is fibrosis and loss of function.

13. Asbestosis is a lung fibrosis problem caused by asbestos fibers. Some varieties of asbestos also cause malignant mesothelioma.

14. Talc and mica occasionally cause fibrosis.

15. Fiberglass is a lung irritant.

16. Many metal and metal oxide particles cause metal fume fever, which has symptoms like influenza.

17. Coal dust deposits reduce lung capacity in a problem called black lung disease.

18. Cotton fibers cause a fibrosis called brown lung disease.

19. Simple asphyxiants are gases that dilute oxygen in the air.

20. Chemical asphyxiants block oxygen either from reaching or from being used in the tissues. These include CO, HCN, and H_2S.

21. Upper respiratory tract irritants are water-soluble and cause immediate irritation on entering the lungs. This limits the willingness of the victim to allow them to go deeply into the lung. They include HCl, SO_2, Cl_2, and NH_3.

22. Lower respiratory tract irritants react slowly with water to produce irritants, causing damage some time after inhalation. These include phosgene and NO_2.

23. Vapors are evaporated liquids. Lower-boiling-point liquids produce vapor in larger quantities.

24. Irritation and odor are warning properties of gases and vapors that make us aware of their presence.

25. Compounds that sensitize the lungs cause occupational asthma.

PROBLEMS

1. In this problem, the objective is to visualize the environment of chemicals entering the respiratory tract and to consider the likelihood of those chemicals being trapped in each region.

   A. Consider a particle entering the nose. Describe changes in the surroundings of the particle as it moves from the nose to the alveolus, and the effect of these surroundings and changes on removal of the particle from the air.

B. What changes in the particle itself can affect the likelihood of its entrapment, including chemical composition, degree of adsorption of water, size, and density?

C. Contrast the movement of a molecule of a nonpolar gas such as methane in terms of the likelihood that it will be stopped.

D. Now contrast the movement of a molecule of a polar and irritating gas such as HCl.

2. Before opening a bottle of CF_3-CH_2-OH you read the warnings on the label and learn it is somewhat acidic and is a respiratory irritant. What effect does a quantity of an irritating gas have in the lungs that increases the likelihood of its removal?

3. You are discussing fibrosis of the lung with a friend, who says "Why should fibrosis increase the burden of the right side of the heart, the side pumping blood to the lungs? Doesn't loss of lung tissue simply mean there is less lung tissue the heart needs to supply with blood?" You reply, "No,…!"

4. You are an industrial hygienist inspecting a plant where these operations are being performed:

(1) Materials and parts are delivered to the plant at an open air loading dock, and (2) are shifted into the plant to sites along the assembly line by gasoline-powered forklift trucks. (3) Relatively low-melting-point aluminum alloys are heated in an open cauldron using gas flames. The molten alloy is poured into molds to form garden tractor transmission housings. (4) The dies are cooled by circulating water, and the housings are clamped in a jig. (5) Holes are drilled for press fit bearings and to insert studs to assemble the components. (6) A grinding wheel grinds one face flat for attachment of the cover. (7) Stud holes are tapped (threaded), and compressed air is used to blow the housing clean. (8) The assembly is dipped in a chlorinated hydrocarbon solvent to degrease the surfaces, blown dry with compressed air, and placed in a basket that is the cathode of an electroplating system. (9) It is passed through a tank of cupric cyanide, sodium hydroxide, and sodium cyanide, where the housing is copper plated. (10) The basket then passes through a water rinse tank and goes into a bath of nickel sulfate, nickel chloride, and boric acid, where the housing is nickel plated, and finally through a second water rinse bath. All baths are electrically heated. (11) The part is again placed in a jig and the outer surface is polished with buffing wheels and a fine abrasive. (12) The surface is covered with a template and spray-painted with the company logo. (13) Heat lamps rapidly dry the paint. (14) Studs are

screwed into the threaded holes with power drivers and bearings are pressed into their housings. (15) Finished parts are packed in Styrofoam pellets in shipping crates for transfer to the main assembly operation in another state.

A.   Identify each step in the operation where you need to be alert for airborne contaminants being generated.

B.   Consider the plating operation. Tanks lose water over time as a result of heating and solution being carried out on parts. Concentrations of plating bath solutions in rinse tanks rise as they are used. You propose that instead of adding water to the plating solution, you add water from the rinse tank, then replenish the rinse tank. How does this reduce waste and what dangerous air contaminant may be prevented by this procedure?

5.   The ACGIH TLV for asbestos is complex, and reveals a great deal about the nature of the problem. It reads as follows:

Amosite, 0.5 fibers > 5 μm/cm^3 TWA
Chrysotile, 2 fibers > 5 μm/cm^3 TWA (95% of asbestos used)
Crocidolite, 0.2 fibers > 5 μm/cm^3 TWA
Others, 2 fibers > 5 μm/cm^3 TWA

A.   For most chemicals, standards describe mg/m^3, ppm, and mg/kg, and depend on measuring the weight of the chemical in the dose volume. How must the dosage be measured here?

B.   Why is it specified that the dose must be of fibers of dimensions less than 5 μm?

C.   Why is the standard for crocidolite so low?

6.   Compare the values for PEL and TLV in Table 6.5. Do they more often agree or disagree? Where they differ, is OSHA or ACGIH more conservative?

BIBLIOGRAPHY

E. J. Calabrese and E. M. Kenyon, *Air Toxics and Risk Assessment,* Lewis Publishers, Chelsea, MI, 1991. Includes a large number of assessments of specific compounds.

D. B. Menzel and M. O. Amdur, "Toxic responses of the respiratory system," in C. D. Klaassen, M. O. Amdur and J. Doull, Eds., *Toxicology: The Basic Science of Poisons,* 3rd ed., Macmillan Publishing Company, New York, 1986.

R. F. Phalen, *Inhalation Studies,* CRC Press, Boca Raton, FL, 1984.

R. P. Smith, "Toxic responses of the blood" in C. D. Klaassen, M. O. Amdur and J. Doull, Eds., *Toxicology: The Basic Science of Poisons,* 3rd ed., Macmillan Publishing Company, New York, 1986.

H. Witschi and P. Netthesheim, *Mechanisms in Respiratory Toxicology,* CRC Press, Boca Raton, FL, 1982.

G. W. Wright, "The pulmonary effects of inhaled inorganic dust," in G. D. Clayton and F. E. Clayton, *Patty's Industrial Hygiene and Toxicology,* Volume I, Part A, 4th ed., John Wiley and Sons, New York, 1991.

CHAPTER 7

PROTECTING THE WORKER, PART I: MONITORING THE PLANT ATMOSPHERE

The totally safe workplace is an unreachable goal, not a reality. However, the efforts of many researchers, government workers, legislators, labor union officers, company managers, and concerned members of the voting public are bearing fruit (Figure I.1). Many dangerous practices of the past have been eliminated. The Right-to-Know laws give the worker access to the information needed for self-protection. Regulatory efforts have moved us toward common recognition of just what constitutes a good working environment, and are a lever to move less concerned managers toward providing those conditions. Heightened awareness of the potential for harm has caused many to be more thoughtful and favorably disposed toward efforts to provide and maintain a safe workplace, and to exert greater effort toward anticipating dangerous situations, rather than to react to a problem only after harm has already occurred. The importance of this progress is made greater by the ever more rapid rate of appearance of new technologies, and the need to change processes more frequently to meet mounting competition in a world marketplace.

These two chapters outline steps necessary to maximize the safety of the workplace with respect to airborne pollutants. Before any health and safety program can be established, it is necessary first to identify the hazards. Once identified, the best measures to protect workers from these hazards must be devised. Finally, a program of continuous monitoring of the workplace for dangerous conditions must be installed to ensure that protective measures are working.

IDENTIFYING HAZARDS

Let us initially assume the role of the compliance officer. Identifying the hazards present for the worker in an existing or planned facility may seem like a monumental task, even when we restrict our consideration to airborne contaminants. A good starting point is the identification of all of the chemicals entering the facility for whatever purpose. With this inventory in hand one should assess the hazards of each chemical. The toxicity and volatility of each chemical can be obtained from a number of sources, the most obvious and readily available of which is the MSDS provided by the supplier of the chemical.

Toxicity and *hazard* are not the same thing. Toxicity data indicates levels of exposure to the chemical that cause harm and the type of harm likely to occur. Hazard includes additionally the *likelihood* of injury occurring, given the manner of use of the compound. It is necessary now to consider how each compound is used in the plant. NIOSH recommends preparation of a flowsheet that lays out the process being performed in a schematic fashion. With the knowledge of where each chemical is used, a walking tour of the plant is taken. At that time, the most likely sites of exposure can be identified.

A large dose of good common sense is important when evaluating plant hazards. A compliance officer is a detective, looking for clues. Candy bar wrappers on the floor indicate workers are eating on the job, and may add chemicals on their hands to the snack. One must see objects not for their function, but for their potential to release chemicals. A forklift is not a device to lift heavy objects, but instead it is a carbon monoxide generator. A valve is not a device to control flow of liquids in pipes, rather it is a potential site for leakage. Particular attention should be paid to unenclosed processes, such as open vats, degreasing tanks, spraying processes, or drying operations. The place where bags of dry chemicals are opened and dumped into a hopper is a high-risk site. In fact, any place where the chemicals are handled or transported deserves special attention. Receiving and shipping departments are places of special concern. Notice how the workers interface with the plant machinery, paying special attention to whether workers are at locations with a high risk of chemical exposure. What is the type and layout of plant ventilation? Sampling of air in high-risk areas is useful to determine if suspected problems are real problems. A summary at this point serves as the foundation for a plan to control hazards to the workers.

Shifting our point of view to that of the plant operator, where the health of the workers is at stake, and where there are legally established limits, noncompliance with which can involve fines and court orders, the control of levels of contaminants in the plant air is a project to be taken seriously. How do we measure levels of dangerous chemicals in factory air on a day-to-day basis?

Once we decide what chemicals are of concern, there is a three-step process: *collecting samples, analyzing samples,* and *deciding if a hazardous situation exists.*

SAMPLING

Two approaches are used for sampling air: the instantaneous or *grab sample,* and the continuous or *integrated sample.*

GRAB SAMPLING

In grab sampling, a single aliquot (sample) of air is captured in some fashion, and is analyzed for the concentration of the contaminant. The method of capturing the sample is usually quite unsophisticated. One may simply open an evacuated bottle in the appropriate location, then seal it immediately. Another approach is to draw a sample of air into an evacuated syringe body, then cap the end of the syringe. In either case, the container is empty at the outset, so there are no previous contents to contaminate the sample. It is necessary to know the temperature and pressure of air being sampled to allow calculation by the gas laws of the exact amount of air that has been collected.

Grab sampling limits the analysis to a relatively small quantity of air, so it is most useful in situations where either the analytical method is very sensitive, or the contaminant is relatively concentrated. Opening the evacuated bottle can be done remotely, so grab sampling can be done where a person might prefer not to go, sampling mine or sewer gas, for example. A larger sample is taken by pumping the air into a plastic bag (Figure 7.1). Such bags are lightweight, are easily stored, and can hold several liters. Any sample taken in the course of a few minutes would be considered a grab sample.

INTEGRATED SAMPLING

Integrated sampling involves pumping a stream of air through a trap, which captures the contaminant. A much larger volume of air is sampled, allowing accumulation of a quantity of the contaminant. This permits the use of less sophisticated methods of analysis, and/or the detection of smaller concentrations of airborne contaminant. Integrated sampling has the disadvantage that it presents values for the average level of the airborne contaminant over the testing period, and does not distinguish a constant moderate level from a low overall level with a series of very high peaks. The indicated moderate level might be in compliance, while the peaks in a relatively continuous low level might exceed ceiling value limits.

Calibration of equipment is important in integrated sampling. The pump that draws the air through the trap must have a constant flow rate. There must be a device to calibrate the pump by measuring that rate of flow. Finally, a trap is needed that effectively and reproducibly removes contaminant from the air. All this must be accomplished in a reasonable period of time. The captured sample must be stable (capable of being stored without significant change until analysis is performed) and in a suitable form for analysis; the system should require a minimum of manipulation in the field; and, if possible, the trap should not use corrosive or hazardous materials.

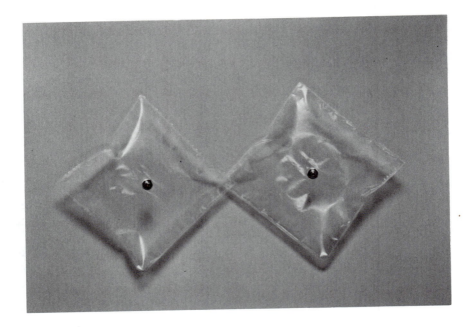

Figure 7.1.   Grab sampling is performed here by pumping a sample of air into these bags. Such bags must be completely evacuated at the start, free of leaks, and constructed of materials that do not release contaminants into the sample.

GENERAL AIR SAMPLING

There are various locations at which we may wish to take an integrated sample of a chemical in the plant air. A general plant air sample is useful to give an overall measure of plant contamination. We are also concerned with escape of chemical at a known or suspected point source, such as an open vat, a spraying operation, or a valve. A variety of stationary devices are available that either collect a sample for later analysis, or give a direct reading of the contamination of the air at that location. Such devices may depend on appearance of a specific absorption of infrared light, change in the transparency of a filter, change in the pressure drop across a filter, scattering of light by airborne particulate, or a variety of other techniques. Devices are available to take samples automatically at timed intervals. It is important to understand the sampling method used by a monitor of this sort to be sure that it is appropriate for the contaminant under study. Installation of any such monitoring equipment should be preceded by careful consideration of chemicals that might be present that would interfere with the analysis.

PERSONAL AIR SAMPLERS

The most important air to sample is air being inhaled by the individual worker. The intake of such a collecting device must be near the face. Unless we wish to attach the worker by a tube to a large stationary device, which would restrict the free movement of the worker and thereby distort the results of the study, the entire apparatus must be small and lightweight enough to be carried about conveniently by the worker. This device, in spite of its small size, must meet the standards for good analysis outlined above.

Such personal air samplers are available, and are in common use. They consist of a small, battery-powered air pump that can be worn on the belt (Figure 7.2), to which a trapping device is attached. A tube pinned to the clothing near the face carries the air to the trapping device (Figure 7.3).

TRAPS FOR COLLECTING CONTAMINANTS

Three types of traps are used on these devices: filters, adsorbent tubes, and liquid traps. The nature of the contaminant determines the best trap to employ. Particulate contaminants are collected on filters, while gases and vapors utilize adsorbent tubes or liquid traps.

Filters

Filters are chosen when the contaminant is a particulate. A good filter strikes a balance between stopping the particles and allowing an adequate rate of air flow. Thus, large filter holes allow good air flow, but fail to trap small particles. Filters are made of a variety of materials, and the composition of filter chosen depends on the method of analysis to be used.

If the total trapped particulate is simply to be weighed, then we require a filter that does not absorb water from the air, altering its weight independently of particulate capture. Polyvinylchloride or polyvinylchloride-acrylonitrile filters are effective for this purpose.

Cellulose esters, such as cellulose acetate, are useful when the contaminant is to be analyzed in solution. To accomplish this, the filter must either be ashed and the residue dissolved, or the filter with the residue on it must be dissolved. Dissolving is accomplished by the use of nitric or acetic acids, sodium hydroxide, or solvents such as acetone or dioxane. This is the procedure if metals are to be measured by atomic absorption.

Figure 7.2. This is a battery-operated portable air pump worn by the worker and used to collect samples from the worker's breathing zone. The rate of flow of the air can be adjusted, and is indicated by the calibrated tube. (Photo courtesy of MSA International, Pittsburgh.)

Where the filter is to be studied microscopically, as would be the case for a study of asbestos in air, polycarbonate filters are superior because of their smooth surface. They can also be dissolved in base or dioxane.

If inertness to the chemicals being trapped is important, glass fibers can be used. They trap particles effectively with a low pressure drop in the flow line, withstand higher temperatures, vary little in weight due to absorption of water from the air, and withstand the use of acids or organic solvents to solubilize trapped material. However, if metals are to be analyzed, especially by atomic absorption, glass may contribute metallic impurities. Teflon® filters are also inert and are nonpolar, so that nonpolar organics are retained.

Figure 7.3.   This worker is wearing a personal air sampling device. Four sampling tubes (traps) are mounted at the air intake level. (Photo courtesy of MSA International, Pittsburgh.)

Paper filters are available in a large range of porosities and weights, and are very inexpensive, but they suffer from some important disadvantages. These include the variability of their ash content if the filter is to be ashed or otherwise broken down during analysis. They change in weight due to the absorption of water from the air sampled, and even change weight without drawing sample through them as the humidity changes, so that accurate determination of the weight of particulate deposited on the paper is impossible.

Respirable Particulate Samples

When the contaminant is in particulate form we must consider the effect of the size of the particles in assessing the toxic threat. Penetration of particles into the lung is a function of particle size: the smaller the particle, the greater the penetration (Chapter 3). A relationship between particle size and percent respirability (ability to reach the alveolus) was devised by the ACGIH and is shown in Table 11.1. Since the larger particles are likely to be carried out of the lung, they do not present the same level of threat as do the smaller particles, which reach the alveolus, where removal is less effective.

When total particulate content of air is measured by noting the increase in weight of a filter after passage of a known air volume, the method does not distinguish small and large particles. Comparing two samples of air with identical weights of particulate contamination per unit volume—one contaminated with particles of all sizes and the other with largely respirable particles—the latter sample might have from tens to thousands of times higher concentrations of the dangerous respirable particles. Measuring of the content only of small particles assesses hazard more accurately. Techniques have been devised to collect the particulate in at least two stages. Commonly, the first is a centrifugal device that spins out the high-inertia large particles, and the second is a filter to collect the respirable dust (Figure 7.4).

Adsorption Tubes

The second class of traps is adsorption tubes—small glass tubes packed with adsorbent and sealed at each end (Figure 7.5). When using these, the ends are broken off and the tube is inserted into the system so that air is drawn through it. The packing is usually activated charcoal or silica gel, although other agents have been used. After the desired volume of air has passed through the tube, it is removed from the apparatus, the ends are capped, and it is sent to a lab for analysis. To be effective, adsorbent should have a high *collection efficiency* (trap a high fraction of the contaminant to which it is exposed), have a high *desorption efficiency* (release a high fraction of the sample at the time of analysis), and have a high *breakthrough capacity* (the amount adsorbed before 5% passes through untrapped and appears on a backup tube).

Measurement of breakthrough capacity is very responsive to changes in the humidity of the air being sampled. Adsorbents have sites on their surface that are particularly effective at binding contaminants due to the presence of charged

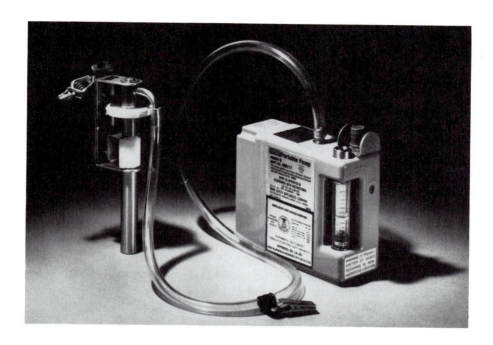

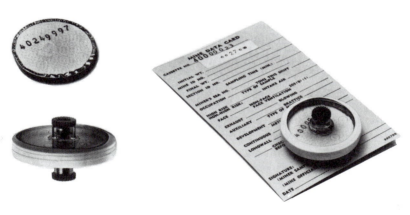

Figure 7.4.   This pump is equipped to measure respirable dust. The air passes first through a cyclone assembly where particles larger than 10 μm are removed. The remaining particles are trapped on the preweighed filter shown below. (Photos courtesy of MSA International, Pittsburgh.)

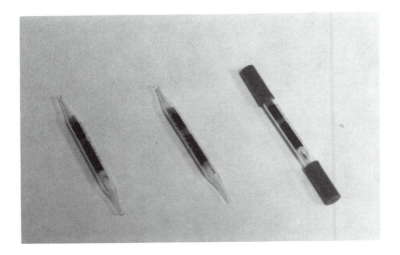

Figure 7.5.   Charcoal adsorption tubes collect samples of organic vapors. The glass ends are broken off and the air sample is drawn through the tube. After collecting the vapor, the ends are capped (right).

groups, and, to a lesser degree, to the shape of the site. Water binds well to these sites, often better than the contaminant itself. As humidity rises, the fraction of sites binding water, and therefore unavailable to bind contaminant, increases. As a direct result, the breakthrough capacity drops. The NIOSH standard requires that *breakthrough capacity should be twice the OSHA recommended maximum level measured at 80% humidity.* Tubes are sometimes designed with a drying agent in front of the adsorbent to remove the humidity from the air. At first this may sound like an ideal solution to the problem of humidity; however, an absorbent used to remove water must not also remove contaminant from the air—a difficult requirement to meet.

Once the sample of contaminant is in the tube, it is necessary to remove it for analysis. The adsorbent binds the gases and vapors to the degree that they have polar character. They can then be removed, either by using a polar solvent, or by driving them off by heating the tube. Once off the tube, one of the techniques for analysis is used to measure the quantity of the contaminant.

Liquid Traps

Bubblers or impingers at one time were the primary technique for entrapment of organic compounds. The air was bubbled through a solvent, which dissolved the molecule, saving it for later analysis. These techniques were improved by adding reagents to the solvent that reacted with the contaminant to form less volatile products. The chemistry going on in the trap sometimes was complex, and on occasion side reactions interfered with the analysis. Liquid traps have largely been replaced by adsorbent tubes today.

ANALYTICAL CHEMISTRY

Before attempting analysis, we need to know what we are analyzing. Those readers not familiar with analytical chemistry must accept that to be given a completely unknown mixture for analysis is to be handed a very difficult task. Gas chromatography and high-performance liquid chromatography (HPLC) can do some remarkable separations of complex mixtures. Even so, there is no guarantee that a single peak displayed during such a separation does not represent more than one component. Furthermore, without a list of probable components and a set of standards for comparison, such a separation only tells us the degree of complexity of the mixture, but says nothing about the composition. Coupling the separation technique with one for identification, for example mass spectrometry, then interfacing the whole with a computer that has a library of identified patterns might do the job for us. However, the cost of such an instrument makes its routine use for spot checks impractical. Fortunately, when analyzing the air of a particular workplace, we normally know what compounds are likely to be present.

ERRORS IN ANALYSIS

How do we assess the quality of the available analytical methods? There are two kinds of error in such measurements. One is termed *determinate error* or *bias,* and describes a problem that causes the system to give an incorrect answer whose error is always on the same side of the correct value, either higher or lower. Suppose we are pumping air through a trap, then removing it from the trap for analysis. Bias results from a flaw in the system, such as incomplete removal of the contaminant from the air, failure to remove the contaminant completely from the trap, or incorrect calibration of the flow rate of the air pump. If the contaminant were not completely trapped, all the readings would be too low, and if the pump were assumed to be moving 2 liters of air per minute through the trap, and it were actually moving 2.5, the calculated value would always be too high.

Bias is detected and compensated for by careful calibration of all parts of the system with standards. Sometimes bias is unavoidably built into the system, as when the recovery from the trap using known samples is always at the level of 95%, which makes all the readings 5% too low. One could then correct by increasing by 5% all the values obtained.

The other problem is *indeterminate* or *random error,* which is the scattering of the values on both sides of the true value. Random error can occur when a lab worker is reading a meter, and is guessing which calibration on the dial is nearest to the needle. Sometimes the value will be too low, and sometimes too high.

Random error is dealt with by running several samples. Reliance on a single sample leaves the question as to whether this is a low value, a high value, or possibly right on the mark. By running several samples, then averaging the results, a much more dependable value is obtained. Furthermore, by dealing statistically with variability in the results, the relative importance of random error in the method can be evaluated. Usually either standard deviation or 95% confidence limit is calculated (Appendix A). The more tightly values cluster around the average value, or the larger the number of samples analyzed, the smaller the value for these error terms.

NIOSH has published standards for the validation of sampling and analytical methods. Measuring a sample that is at the level of the OSHA recommendation for maximum exposure, the value obtained should have a 95% confidence limit of ±25% where there is no bias error, and less if there is a bias correction. Bias should not exceed 10%. The sample should be capable of being stored for one week at room temperature with no more than 10% degradation. Other standards are included in appropriate discussions later in the chapter.

ANALYSIS OF SAMPLES

ONSITE ANALYSIS: STAIN TUBES

It is very useful to be able to spot check air in suspicious locations of a plant, doing an approximate determination of levels of known contaminants in the air during a plant inspection. A variation of the adsorption tube called the *stain tube* is designed to be direct reading, giving an instant assay of the levels of particular chemicals to the inspector without the delay or expense involved in a laboratory analysis.

In a stain tube, a reagent is added to a stationary support (silica gel, polymer, glass, etc.) that reacts with the contaminant of interest to produce a color. The treated adsorbent is packed in a thin glass tube that is sealed at the ends. Stain tubes are used frequently in conjunction with a large syringe device to analyze a grab sample. The ends of the tube are broken off, the tube is attached to the syringe, and a fixed volume of air is drawn through the tube (Figure 7.6). In the earliest tubes, the color produced in the tube was compared to a chart of color standards to determine the concentration of contaminant in the air. Today instead we use a reagent that reacts rapidly with the contaminant and is at a carefully adjusted level in the tube. The contaminant reacts immediately with

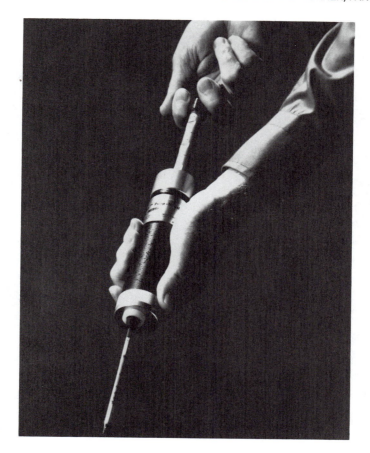

Figure 7.6.   A sample is drawn thorugh the detector tube with a hand pump. For some types of detector tubes the flow rate can affect the reading, so a certain amount of practice with known samples is needed. (Photo courtesy of MSA International, Pittsburgh.)

the reagent at the inlet of the tube, leaving the reagent changed in color. The next contaminant to enter the tube must move farther along to unused reagent, extending the stain. The tube is premarked with calibrations, and the length of stain measures concentration of contaminant in the air (Figure 7.7).

Such a tube can be used in two ways. One can measure the length of stain from a fixed volume of air, or measure the volume of air necessary to produce a standard length of stain. Most tubes sold today are calibrated to read parts per million in the air directly from the markings on the tube when a standard amount of air—typically 10 full strokes of the plunger—is drawn through the tube.

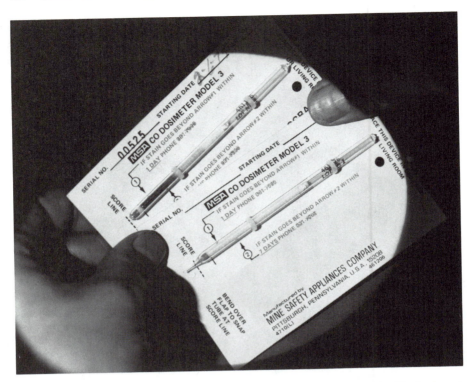

Figure 7.7. Shown here are two detector tubes. The upper is displaying a stain due to exposure to carbon monoxide. (Photo courtesy of MSA International, Pittsburgh.)

Tubes are available for a wide range of compounds (Appendix C). The use of these tubes requires that the operator have some sophistication about the situation under study. The tube is labeled to be used for the analysis of a specific compound. However, on examination, the packings of some tubes are found to be the same for the analysis of several compounds. The company whose list of tubes is presented in Appendix C lists tubes available for 169 different purposes. However, eliminating those entries where the tube is also listed for another compound, and those for which the difference from one tube to another is just a matter of range, we narrow the list to 79 different packings. Many of these may be variations in the treatment of a common packing. As an example of the possible problems, the stain may be the color change in an oxidizing agent such as a chromium salt, and several organic compounds are oxidized by this salt. Clearly there is the possibility of misinterpretation of the results where more than one such oxidizable contaminant is present in the air. Some tubes take this type of problem into account in their design. The portion of the tube where the reaction occurs is preceded by segments that react with likely interfering substances. It is necessary for the person assaying the air sample to be aware both of the contaminants likely to be present in the air under study, and of the method employed in the tube being used.

Two other points must be kept in mind when using stain tubes. First, *these systems are good only for an approximate measure of concentration.* A very low reading assures the inspector that an area is safe, and a high reading alerts the inspector that a problem may exist, indicating the need for careful sampling and laboratory analysis. Second, stain tubes *can deteriorate with time,* and are always dated. Expired tubes should never be used when data is collected for official records. In fact, the expiration date of a tube used should be included as part of the record.

Direct Reading

There are a few examples of instruments that detect concentrations of a gas or vapor directly. For example, a very useful detector is one that sounds an alarm when oxygen levels are too low (Figure 7.8). Workers entering a confined space are therefore immediately warned if a portable air supply is needed.

LABORATORY ANALYSIS OF SAMPLES

The trend over the years has been for laboratory analysis of environmental samples to be done increasingly by instrumental methods. The advantages include greater speed of analysis, possible automation of heavily used procedures (as has happened extensively in the field of clinical analysis), and an elimination of the kind of random error associated with a chemist working at the bench. The need for analyses to be run of various industrial and environmental samples has grown quite rapidly, and a number of individuals have purchased analytical instruments and have started businesses specializing in specific types of analysis.

One must guard against the temptation to assume that because an analytical instrument costs a substantial amount of money, it cannot produce a wrong answer. All instruments should be under the supervision of well trained and informed chemists, and require careful maintenance and calibration. No instrumental method is any better than the standardization techniques used to check its accuracy.

Gas Chromatography

Peter Eller, in his chapter in *Patty's Industrial Hygiene and Toxicology,* summarizes the analytical methods used in the NIOSH/OSHA standards, and the

Figure 7.8.   This battery-powered monitor sounds an alarm if oxygen levels drop below 19.5% or rise above 23%. This is particularly useful to workers entering confined spaces where oxygen levels may have been depleted. (Photo courtesy of Lab Safety Supply, Janesville, WI.)

number of kinds of assays done by each method (Volume IIIA, p. 526). Of 216 analytical methods, 165 utilize gas chromatography. This is not surprising, since the technique is capable of sophisticated separations, and the results can be converted easily to quantitative information. The instrumentation is relatively simple, comparatively inexpensive, and has been heavily used since the 1950s (Figure 7.9). Portable gas chromatographs have been developed that can be used onsite.

A gas chromatograph consists of a packed tube, the *column,* through which an inert carrier gas is flowing, and which is mounted in an *oven* to control temperature (Figures 7.10, 7.11). A constant temperature is necessary for analytical reproducibility. There is a *heated chamber* into which the sample is injected either as a small volume of liquid, which immediately volatilizes, or as

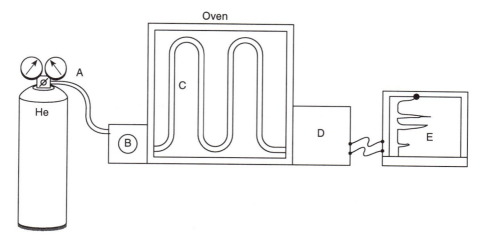

Figure 7.9.   The components of a gas chromatograph are shown in this diagram. The moving phase is an inert gas (helium or nitrogen is used commonly) fed into the instrument at a controlled pressure to assure a constant flow rate (A). Sample, usually a solution, is injected with a very-low-volume syringe (1–10 μL) into the hot injection port (B), where it is vaporized. It is swept through the column (C) by the moving phase gas, where the components of the injected mixture are held back according to their individual attraction for the stationary phase of the column. Separated components are sensed by one of several methods in the detector (D), and a record of their position and relative quantity appears as peaks on the moving paper of a recorder (E).

a compressed gas. *Carrier gas* sweeps the sample onto the column, where compounds are retarded to varying degrees by attraction to the packing of the column. They emerge from the column as a series of separated bands. These pass through a *detector,* which responds to their presence by producing a small electrical current proportional to the amount of the compound. The current is amplified, and information is presented to the operator as a series of peaks above the baseline on chart paper. Under constant conditions of temperature and carrier gas flow rate, the length of time required to move a particular compound through the column (*retention time*) is a constant, and can be used to identify the compound. The area under the peak on the chart paper is proportional to the amount of the compound. Conventional recorders are being replaced with computerized devices that determine and print directly onto the output graph the retention time and relative area of each peak.

Since any column packing and carrier gas can be used on any instrument, the biggest fixed difference from one instrument to another is the design of the detector. Three detectors have had long usage in gas chromatography. The

Figure 7.10. The interior of the oven of a gas chromatograph is shown. The compartment is well insulated and has a heater and blower in back to maintain constant temperature. In gas chromatography the stationary phase is packed into a column and the moving phase (carrier gas) carries the sample through the column toward the detector. Here the stationary phase is a liquid mounted on a solid support and packed in a metal tube. This instrument has two columns mounted in a single oven, each with a different stationary phase to allow flexibility in choice of separation technique used.

simplest and easiest to use is called a *thermal conductivity (TC) detector*. It operates on the basis of the change in resistance of a hot metal wire when the

Figure 7.11.   The view is similar to that in Figure 7.10. Here we have a capillary column: a very long, narrow-diameter glass tube wound onto a rack. The stationary phase is a liquid coated onto the walls of the tubing. Because of the great length of the column, much better separations are achieved by a capillary column.

temperature is changed by the passage of a separated compound across it in the midst of a steady flow of carrier gas. Instruments with TC detectors are stable and easy to run and maintain. However, they are not used much for the analyses of interest in this text because the sensitivity of this method is relatively low.

Flame ionization detectors, on the other hand, are used a great deal in analysis of contaminants. They consist of a very clean hydrogen flame burning between two plates with a large voltage drop between them. The output of the column is added to the fuel for the flame, and when a separated compound enters the flame, it burns and produces a quantity of ions. These ions allow a flow of current between the plates, and this current is amplified. The flame ionization detector is capable of detecting compounds across a wide concentration range, down to quite small amounts. It is the most favored technique, being specified in 144 of the 165 NIOSH/OSHA gas chromatographic methods.

Finally, the *electron capture detector* depends on the ability of molecules containing highly electronegative atoms (fluorine, chlorine, oxygen, etc.) to deflect the electrons being emitted by a source of beta radiation. It is able to detect tiny amounts of molecules containing electronegative atoms. However, it suffers from the disadvantages that it cannot differentiate relative amounts in large samples, it is not sensitive to molecules without electronegative atoms, and it is more trouble and expense to maintain than the other two detectors.

Atomic Absorption Methods

For measuring metals in a sample, the method of choice is atomic absorption spectrometry. Metals in solution are drawn into a flame or stream of hot air. Light of a wavelength that excites electrons of the metal of interest is passed through the stream of rising gas, and a photocell measures the light reaching the other side of the gases (Figure 7.12). Light used to excite the metal atoms is now missing, and from the degree of this absorption of the light, the concentration of metal is calculated. In principle, a separate light source specialized in producing the correct wavelength is used with each metal studied, but lamps are available to be used for more than one metal (Figure 7.13). The flame method, which reads into the microgram range, is the least sensitive of the atomic absorption techniques. The flameless method and the graphite rod technique extend the sensitivity by several orders of magnitude.

Spectrophotometric Methods

Visible and ultraviolet spectrophotometry is useful to assay a few selected compounds. The method also operates on the basis of the measurement of the amount of light used to excite electrons, but the electrons now are in molecules rather than around single atoms. It is necessary that the molecules absorb light in the range covered by the particular instrument. Some molecules do this naturally, while others can be induced to do so by reaction with the appropriate reagent. The method requires only a few milliliters of solution, and can detect milligrams to micrograms of the compound.

Figure 7.12. At the heart of the atomic absorption spectrometer is the burner. The gas supply to this burner is coming from the lower left (A). A wide flame is produced at the top of the burner (B), and the light enters through the window (C) and passes through the length of the flame. The sample is drawn up the fine tube from the flask (D), and is fed into the flame.

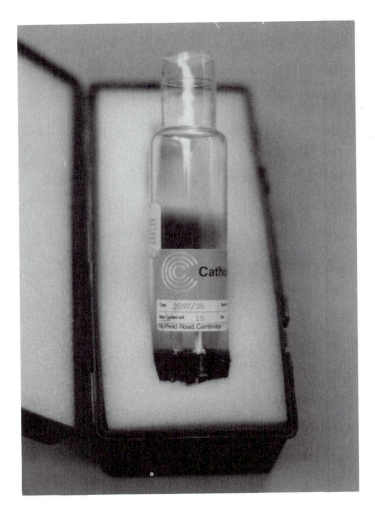

Figure 7.13. Specialized lamps are needed to allow analysis of a variety of metals. Some, like the one shown here, are used for a single element (antimony), while other lamps can be used for more than one metal.

Infrared spectrometry is used for environmental analysis. A gas sample can be read directly, but a long light path must be used because of the relative insensitivity of the method. The lower-energy infrared light is used, and the energy of this light increases the rate of internal vibration in organic molecules. Structures such as carbonyl or alcohol groups each respond to specific wavelengths of infrared light (Figure 7.14). Often a dedicated instrument is used, which reads at a single wavelength, a wavelength absorbed by a group in a

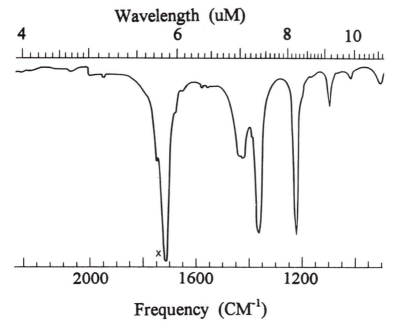

Figure 7.14. A portion of the infrared spectrum of acetone is shown.   The strong absorbance bands each represent a single type of vibration of a specific structure in the molecule. Here, the band to the far left is generated by the carbonyl group. An instrument designed to detect a single wavelength can provide continuous monitoring for compounds in the air absorbing that wavelength.

compound of concern in the plant.   When using infrared spectrometry, it is important to realize that every organic molecule containing a particular structure absorbs light of the same frequency, so, for example, the method would not differentiate between two ketones when the frequency is set to detect the carbonyl group absorption.

Other Methods

Other techniques are used in a few cases for specific assays. X-Ray diffraction is used occasionally, particularly for crystalline silica. "Specific ion" pH meters' electrodes are sensitive to certain ions in solution, are easily used, and have a few applications at present. This method can be expected to expand as new electrodes are developed, particularly since the basic instrument is very inexpensive. High-performance liquid chromatography is a technique that is increasing popular, and will certainly expand its applications in the field. New column packings are being developed, and less complex and expensive instruments are available. It is similar in principle to gas chromatography, but analyzes solutions rather than gases.

Some very sophisticated analyses can result from combining methods. Analysis of individual gas chromatographic peaks by mass spectrometry has the potential of identifying the compound in each peak after an unknown mixture has been separated. Mass spectrometry fragments molecules into charged structures, which are then separated by mass. Each compound produces a distinctive pattern of fragments, which can be stored in a computer. A given analysis can then be compared with items in the computer memory, and so identified. In a similar fashion, infrared spectra can be stored in a computer. A relatively new rapid infrared method, Fourier Transform Infrared (FTIR), can scan the peak from the gas chromatograph, and can identify the molecule from a search of the computer files. These instruments are complex and expensive, and are used largely when the demands go beyond the need for a routine survey, particularly when the identity of the contaminants is unknown.

DECIDING ABOUT THE SAFETY OF THE WORKPLACE

The final stage in our three-part approach to determining the relative safety of the plant air is to make a judgment from the results of analytical work as to whether the level of a contaminant is too high. Here some of the judgment of the industrial hygienist is replaced by government standards. These standards are the listed PEL values of OSHA,[2] of the Mine Safety and Health Administration, or of a state agency that has taken responsibility for the task. Other recommended standards exist that do not have legal status, but can serve as guidelines in the absence of a specific PEL. These include recommendations of NIOSH, the research arm of the federal government in such matters. The ACGIH-TLV values and the American National Standards Institute (ANSI) Z.37 committee report are published values that, in part, were the basis of the PEL values originally set by OSHA. The ACGIH updates their list of values every year, providing for almost all the exposures an industrial worker is likely to experience. The National Academy of Sciences–National Research Council (NAS–NRC) have published short-term public limit (STPL) values.

The management of a plant must be familiar with the regulations and their intent. If this intent is not clear, the appropriate government office should be contacted for clarification. The analytical methods used should be the same as or similar to those used by OSHA in establishing the PEL. It is desirable that laboratory work be done at a laboratory accredited by the American Industrial Hygiene Association (AIHA). Recommendations about the number of samples

---

[2]See 29 CFR 1910.1000.

to be run are provided by NIOSH.[3] Good quality control should be incorporated into the analytical plan, with documentation of the analytical sequence, careful maintenance and calibration of instruments, and cross checking of samples with another laboratory. NIOSH operates the Proficiency Analytical Testing (PAT) program, which evaluates the quality of analytical work done by service laboratories. Careful recordkeeping is essential to meet the requirements of regulatory agencies, and to record the effectiveness of the control systems, and is indispensable in case a chemical involved in the plant operation becomes part of an epidemiological study.

KEY POINTS

1. Grab sampling involves taking a single instantaneous sample.

2. Integrated sampling requires pumping a quantity of air through a trap over an extended time period.

3. Sampling may be done of the general plant atmosphere, of suspected trouble zones, or of the air in the workers' breathing zone.

4. A variety of automatic air samplers may be used for continuous monitoring.

5. Personal air samplers pump air from the workers' breathing zone through a trap.

6. Particulate is collected on a filter. The choice of filter material depends on the analysis proposed.

7. Stain tubes have a color-producing reagent coated onto a solid support. The length of the color zone produced measures the concentration of contaminant.

8. Adsorption tubes adsorb contaminants onto a solid medium.

9. In liquid traps, air is bubbled through a solvent, perhaps with dissolved reagent, to capture the contaminant.

---

[3]N. A. Leidel and K. A. Bush, Department of Health, Education, and Welfare Publication (NIOSH) 75-159, Superintendent of Documents, Washington, D.C.

10. Determinate error (bias) happens when values cluster to one side of the correct values. It is corrected, if it is consistent, by multiplying answers by a correction factor.

11. Indeterminate (random) error results from random variation in some aspect of the analytical procedure, producing a scattering of the answers. Random error is handled by averaging a large number of observations.

12. Gas chromatography, the most used standard occupational analytical method, separates vapors as they are mixed with a gas carrier across a column packing that variably attracts components of the mixture.

13. Metals in particulates are most often vaporized and selectively assayed by light absorption in the gas stream of an atomic absorption spectrometer.

14. Other analytical methods used include visible, UV, and infrared spectrophotometry, liquid chromatography, and mass spectrometers or infrared spectrophotometers coupled to gas chromatographs.

15. Once the concentrations of contaminants are known, the values are compared to prescribed limits.

PROBLEMS

1. Air in a metalworking facility contains large and small particles of a metal oxide. A grinding operation is the source of the large particles, whose average size is 24 µm, and a welding operation is producing a fume whose average particle size is 3.2 µm. Air testing in the shop is performed by weighing a filter, pumping a known volume of air through it, and weighing it again. About 80% of the particles on the filter collected in a sampling of the total room air are of the small size. The density of the oxide is 5.6 g/cm^3.

   A. What is the percentage by weight of small particles?

   A ventilating device is placed at the grinding operation that traps most of the particulate from that source. Filter weigh increases are now only 25% of the previous values on daily testing.

   B. Comment on the relative improvement in safety, assuming the metal oxide is a health hazard.

   C. If the ventilation had been placed by the welding operation, what would have happened to the test results for air contamination?

2. In each of the following scenarios, indicate whether grab sampling or integrated sampling is better, and name the method of chemical analysis that would probably be used on the collected sample.

A. A large underground tank that had stored tetrachloroethylene at a plant has been out of use since the company stopped using that chemical two years ago. It is going to be entered by a worker to check its physical condition with the plan of using the tank to store a different solvent. How will the safety be checked before the worker enters?

B. A welder is assembling custom air ducting for commercial building projects out of galvanized sheet steel. The zinc coating usually has traces of cadmium, which is significantly toxic. The OSHA standard for cadmium (TWA) is 0.1 mg/m^3 with a 0.3 mg/m^3 ceiling, and recommendations by both NIOSH and ACGIH are for a lower standard. Two things are to be assayed: the average daily exposure and the peak level during actual welding.

C. A plant builds small (dorm room) refrigerators, the finished unit without its door is spray painted by automatic equipment and dried by heat lamps as it moves along a line in a ventilated enclosure. Doors are similarly finished on another line. Workers assemble the doors onto the unit immediately afterward. The units are then inspected for quality of finish, and those with flaws go to a station to the side where paint tough-up is performed by hand-held spray guns. We are concerned with overall worker exposure to the paint solvent, particularly by those workers mounting the doors, and by workers doing occasional hand spraying.

3. In each of these analytical procedures, indicate whether the indicated problem produces determinate or indeterminate error. If determinate, will resulting analytical results be too high or too low?

A. Air in a closed coal mine is being sampled prior to reentering the facility. An evacuated bottle of known volume is lowered into the mine and the bottle is remotely opened, then closed again. (1) The pump used to evacuate the bottles is worn, so that the set of bottles to be used was not completely evacuated, all having 0.065 atm of air remaining. (2) Oxygen in the sample is determined by noting the volume drop in the air sample at constant temperature and pressure as the oxygen reacts with an excess of hydroquinone. Volume can be read to ±3%. (3) The combustible gases (here primarily CO and CH$_4$) are estimated by passing air samples through a combustion apparatus and measuring the heat produced. The apparatus is not well enough insulated, so that a percentage of the heat is lost to surroundings.

B.   A personal air sampling pump draws air through a charcoal trap to sample volatile organic material. In the laboratory, a known volume of CS_2 is drawn through the charcoal to extract the trapped organic compounds, and the resulting solution is injected into a gas chromatograph for analysis.   (1) Calibration of the air sampling pump was improperly done, so that the actual volume of air pumped per minute is 7% lower than the calibration indicates. (2) Where concentrations of volatile organics in the air are high, the capacity of the charcoal is slightly exceeded. (3) Samples of 1 µL are injected into the gas chromatograph by a technician by use of a 5-µL syringe. (4) The entrance to the column in the gas chromatograph has particulate deposits, which reduces the flow rate of carrier gas through the column. (5) The technician is opening the sample vials and letting them stand with the CS_2 evaporating for a period of time before injecting the sample into the gas chromatograph.

4.   A plant produces polybutadiene, and at regular intervals at times when the plant is in full operation the level of butadiene in plant air is measured. For one period this data was obtained:

Volume of Air Sampled (liters)	mL of Butadiene
200	1.91
150	1.47
255	2.35
105	0.76
430	5.75

A.   What is the average concentration in mL/L?

B.   What is the standard deviation for this data?

C.   Given the Student t table values in Appendix A, what is the 95% confidence limit?

5.   A spot check is made for solvent in the air near a degreaser using a stain tube at the time parts are being cleaned. The ends of the stain tube are broken, and it is inserted into the syringe type air pump. Instructions say to pump the syringe 10 times, then read the concentration in ppm directly from the tube calibration. The calibration stops at a maximum value of 100 ppm, and the stain reaches 88 ppm after six strokes. From that information, what is the concentration of solvent in the air?

6. You are visiting a plant to do a safety inspection and bring along a case of personal air sampling pumps with the following accessories:

Charcoal adsorption tubes
Cellulose acetate filters
Polyvinylchloride filters
Polycarbonate filters

What would you select to test for each of the following air contaminants at specific plant locations?

Asbestos
Fe_2O_3 fume
Benzene vapor
Total mass of particulates

BIBLIOGRAPHY

ACGIH, *Air Sampling Instruments for Evaluation of Atmospheric Contaminants,* 6th edition, Cincinnati, 1983.

ACGIH, *TLV Values for Chemical Substances and Physical Agents in the Workroom Environment with Intended Changes for 19XX,* American Conference of Governmental Industrial Hygienists, Cincinnati (latest edition, revised annually).

ACGIH, *The Documentation of Threshold Limit Values for Substances in the Workroom Air,* 5th ed., American Conference of Governmental Industrial Hygienists, Cincinnati, 1986.

ACGIH, *Advances in Air Sampling,* Lewis Publishers, Chelsea, MI, 1988.

L. J. Cralley and L. V. Cralley, *Patty's Industrial Hygiene and Toxicology,* Vols. IIIA, IIIB, John Wiley and Sons, N.Y., 1985.

T. Godish, *Air Quality,* 2nd ed., Lewis Publishers, Inc., Chelsea, MI, 1991.

M. Katz, *Methods of Air Sampling and Analysis,* APHA Intersociety Committee, American Public Health Association, Washington, D.C., 1977.

L. J. Keith, *Identification and Analysis of Organic Pollutants in Air,* Ann Arbor Science Publishers, Ann Arbor, MI, 1984.

S. D. Lee, T. Schneider, L. D. Grant and P. J. Verkerk, Eds., *Aerosols,* Lewis Publishers, Chelsea, MI, 1986.

Occupational Safety and Health Act, Public Law 91-596 S2193, 91st Congress, December 29, 1970, Superintendent of Documents, Washington, D.C.

J. B. Olishifski, *Fundamentals of Industrial Hygiene,* 3rd edition, National Safety Council, Chicago, 1988.

C. H. Powell and A. D. Hosey, *The Industrial Environment—Its Evaluation and Control,* 2nd ed., Public Health Service Publication 614, Superintendent of Documents, Washington D.C., 1965.

A. C. Stern, *Air Pollution Vol. III. Measuring, Monitoring and Surveillance of Air Pollution,* 3rd ed., Academic Press, N.Y., 1977.

CHAPTER 8

PROTECTING THE WORKER, PART II: PROVIDING CLEAN AIR

The worker must be provided with clean air. Where exposure problems are revealed, the next step is corrective action. In a plant, we are usually dealing with a point source, and, as those involved with environmental problems will readily indicate, this is a relatively easy problem to solve. Several courses of action are possible.

REDUCING THE LEVEL OF THE CHEMICAL

There are several routes that lead to a reduction of the level of the chemical in the plant air. An obvious first move is to institute an improved program of plant maintenance. Inspections of valves and seals to catch leaks as soon as they occur reduces much needless release of chemicals. Good plant housekeeping serves to prevent the accumulation of dusts or spilled volatiles that can gradually raise levels of toxic substances. Care must be taken in the manner of doing this housekeeping, since a person pushing a broom can be heavily exposed to the dusts being removed. Approved methods are described by OSHA and ACGIH for the removal of spilled chemicals, taking into account volatility and reactivity. Proper disposal of any rags or other absorbents used in such a cleanup is important. This disposal must also consider the possibility of combining chemicals that react with each other. Such chemical incompatibilities are indicated on the warning label of the chemical container and on the MSDS.

REMOVING THE WORKER FROM THE CHEMICAL

Sometimes the safety of the worker can be improved by relocating the job to a spot distant from the source of the chemical. Many chemical processes are readily adapted to remote operations, but often this is not a useful option in a manufacturing plant.

CHANGING THE PROCESS IN THE PLANT

Although it is not always considered, replacement of a chemical in use by a less toxic one is a measure that permanently improves the safety of the plant. Once a process is set up to function in a particular way and is running successfully, there is reluctance to change it. Sometimes the substitute chemical is somewhat more expensive. However, when one studies alternatives, substitution of another chemical could mean a savings when the price of protecting workers adequately against the more toxic chemical is added to the cost sheet.

VENTILATION

We can substitute low-hazard for high-hazard chemicals, or we can redesign processes to prevent the entry of chemicals into the air, but ultimately we must accept that hazardous or otherwise undesirable substances will enter plant air. This air must be removed from the vicinity of the worker, which is accomplished by *ventilation,* the most important tool for bringing a workplace into compliance.

Ventilation must be employed when chemicals present a health, fire, or explosion risk. For certain processes, such as grinding, buffing, and woodworking, local laws may specify ventilation practices. Additional ventilation benefits include improvements in housekeeping of the plant, in employee comfort, and in machine maintenance.

This chapter presents ventilation design features, but not with the goal of having the industrial hygienist or occupational safety specialist design ventilation systems. Initial design is complex, and is the task of a ventilation engineer. However, people concerned with worker safety must understand what comprises a good system and what is needed to assure that a system in place meets the operating parameters for which it was designed.

PARAMETERS OF A VENTILATION SYSTEM

Several terms are used to describe ventilation characteristics. *Air pressure* is the pressure exerted by air molecules randomly colliding with a surface. The reference pressure is atmospheric pressure, the same pressure a weather reporter discusses on the evening news. If we measure air pressure in a duct connected to a fan, it differs from atmospheric pressure. This *static pressure* is above atmospheric pressure (*positive* pressure) when the fan is forcing air into the duct (a blower), and is below atmospheric pressure (*negative* pressure) in the case of an exhaust fan (Figure 8.1). In your home heating system, heated air propelled through the heating ducts to the rooms of the house has a positive pressure, and air being brought back to the furnace through the cold air returns has a negative pressure. Positive pressure moves air from heating ducts into the rooms, and negative pressure causes room air to move into cold air returns.

Air moves through or into ducts at a certain *velocity,* usually measured in meters per minute or feet per minute. Greater velocity means greater friction between the moving air and the duct walls. Overcoming this frictional resistance requires the expenditure of more energy at the blower.

As air moves in the ducts, *velocity pressure* is generated. The moving air particles are now not moving randomly. An object placed in the duct is struck more frequently on the side from which the air is coming than on the side toward which it is going (Figure 8.2). This pressure is what ultimately moves particles through the duct along with the air.

For a given size and mass of particle, there is a velocity of air that has sufficient velocity pressure to keep the particle suspended and moving with the air. The term *capture velocity* is used when the concern is to move particles outside the ventilation system through an opening into a duct, while *transport velocity* describes the velocity needed to keep the particle suspended in the duct until a suitable collector or exit is reached.

Finally there is the air *flow,* the quantity of air moved per unit time through the ducts of the system, which is measured in cubic meters per minute or cubic feet per minute (m^3/min or ft^3/min [cfm]). To increase the air flow in a system of specified capacity, the velocity must be increased. If the air flow is through an exhaust system leaving the building, it is an obvious but sometimes overlooked fact that there must be an equal flow into the building.

GENERAL EXHAUST VENTILATION

Ventilation systems fall into two major classes: *general* and *local* exhaust ventilation. By general exhaust ventilation (GEV), or dilution ventilation, we mean the movement of the entire air mass into, around, and out of the workplace. An indoor facility employs ventilation for heating the plant in cold weather, and for controlling humidity and bringing in cooler air in hot weather. The GEV system may require only these provisions, or it may involve some additional air exhaust facilities operating on a room-wide basis. Chemicals are simply diluted into the workplace air, which is exhausted at a rate that prevents chemical concentrations from rising to a harmful level. GEV is most satisfactory when the contaminants are gaseous and have a low toxic, fire, or explosion risk.

Ventilation engineers sometimes describe the total capacity of a ventilation system in terms of *room changes of air per hour,* i.e., the number of times an

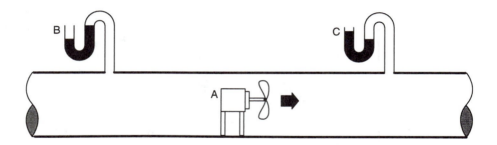

Figure 8.1.   Static pressure can be compared to atmospheric pressure using a water manometer. Such a device consists of a U tube containing water that connects to the duct at one end and is open to the atmosphere at the other. Difference in water height in the U tube measures difference in pressure. When ducts carry air forced through by a fan (A) at the entrance to the duct, static pressure is higher than atmospheric pressure (C), and when air is exhausted from a duct, the static pressure is lower than atmospheric pressure (B).

amount of air equal to the volume of the room is exhausted per hour. Careful design is required to assure that this volume of air flow is adequate, moving air through all parts of the facility and moving it rapidly enough to prevent hazardous materials from rising to harmful levels. In simple terms:

$$E = \frac{Q}{V}$$ Eq. 1

where

E   =   the rate of air exchange in room changes per hour (hr^{-1})
Q   =   volumetric air flow generated by the GEV in m^3/min or an equivalent appropriate unit[1]
V   =   the volume of air in the room in m^3 or equivalent

We make some assumptions: that the generation rate of the contaminant G is constant; entry of the contaminant is by this source only, and exit is by exhausting air only; air in the room is perfectly mixed. Knowing the rate of generation (G, in units of mg/min), the concentration of contaminant in the exhaust air (C, mg/m^3), the rate of air flow (Q), and total mass of contaminant in the room, we

---

[1]Engineers are more likely to use ft^3/min (cfm) than m^3/min.

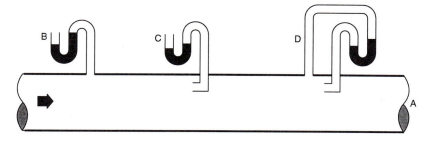

Figure 8.2.   In this duct the fan is at the end marked A and is moving the air to the right. Manometer B measures the difference between atmospheric pressure and static pressure, as did the manometers in Figure 8.1. By directing the inlet of the manometer tube toward the flow of air, manometer C displays the difference between atmospheric pressure and the sum of static pressure and velocity pressure. Since manometer D responds to static pressure at one end and the sum of static pressure and velocity pressure at the other, the reading measures velocity pressure directly.

may now calculate change in amount of contaminant in the room (ΔM) for a period of time (Δt).

$$\Delta M = G\Delta t - QC\Delta t \qquad\qquad \text{Eq. 2}$$

Concentration of contaminant in room air C, the more significant information, is M/V, where V is the volume of the room. Equation 2 can be divided through by V to obtain:

$$\Delta C = \frac{G\Delta t}{V} - \frac{QC\Delta t}{V} \qquad\qquad \text{Eq. 3}$$

The two terms in this equation are the rate of generation of contaminant (first term, $G\Delta t/V$) and the rate of removal (second term, $QC\Delta t/V$). If concentration in the air is rising ($\Delta C > 0$), contaminant is not expelled as fast as it is generated. Concentration does not rise linearly under these conditions, because as concentration rises, the rate of removal also increases. Instead, concentration approaches a *maximum limit concentration* (C_{max}). The equation relating concentration to time [derived in Burgess (1989)] is:

$$C = \frac{G}{Q}\left(1 - \exp\frac{-Qt}{V}\right) \qquad \text{Eq. 4}$$

The system approaches a maximum concentration (t>>0), which is:

$$C_{max} = \frac{G}{Q} \qquad \text{Eq. 5}$$

Not surprisingly, the equation tells us that the maximum concentration reached depends on the rate of generation of contaminant and the rate at which air is exhausted. Once this relationship has been experimentally worked out, we can quickly estimate what a change in either rate of generation of contaminant or rate of removal of air does to the maximum level of the contaminant in the workplace air.

If generation of contaminant ceases, it is useful to predict how rapidly concentration of contaminant is reduced. This is determined as follows:

$$C_2 = C_1 \exp\left(-\frac{Q}{V}(t_2 - t_1)\right) \qquad \text{Eq. 6}$$

Here C_1 is the original concentration and C_2 is the new concentration of contaminant after exhausting air at rate Q for a time period of t_2-t_1. It can be derived (again, Burgess, 1989) that the half-life of the contaminant ($t_{1/2}$), the time required to remove half the contaminant, is estimated from rate of air exhaustion and volume of the room, once generation ceases:

$$t_{1/2} = 0.693\left(\frac{V}{Q}\right) \qquad \text{Eq. 7}$$

With these equations, simple predictions can be made about variations in concentration of a contaminant as conditions change.

These equations were based on assumptions, including that of perfect mixing of air in the room. This would produce ideal transfer of contaminant as it is generated to air leaving the room. In practice, air moving toward the exhaust point mixes with air containing contaminant in a less than ideal fashion, resulting in a buildup of contaminant above predicted values in the vicinity of the source. To include this factor in the equation, Q is multiplied by a constant K:

$$Q_a = KQ_i$$

where

   Q_a is an actual Q
   Q_i is an ideal Q

K has a value greater than one, with less effective mixing generating a larger value. K can also be increased arbitrarily to insert a safety factor into the calculations.

The difficulties of dependence on general ventilation as the only system to maintain a safe environment are several:

1. Exhausting a volume of air equal to the capacity of the building does not assure that all air in the building has been changed once, since air in the center of the pathway between inlets and exhaust vents moves well and is changed more than once, while air in corners and off to the side may be relatively stagnant. The source of a hazardous contaminant may be away from the path of direct and optimal flow of air through the facility.

2. Air carrying a hazardous material may be moved toward the worker or past several other workstations on its way to be exhausted, increasing general exposure to the contaminant.

3. Processes that release larger amounts of material for a short time, and very little between times, demand a high level of air movement during the short time of release. Unfortunately, the time contaminant is being released usually coincides with a worker being at the site. GEV systems do not adapt well to such operations.

4. Seasonal reductions in the ventilation rate may reduce air flow in hazardous areas below desirable levels.

5. A loss of efficiency may happen as a system ages or between times of routine maintenance, as ducts or filters become blocked with dust or fan belts slip.

These problems can be overcome by good design and careful maintenance. The location of inlet and outlet vents is important in preventing stagnant zones in hazardous locations and in minimizing the degree of contact workers away from that location have with the contaminant. Air should move from the cleanest toward the dirtiest locations in the workroom. Standards must be established for normal operation of the system, and a pattern of routine inspection established to assure continued adherence to these standards.

As energy costs rise, interest increases in avoiding the discharge of large volumes of heated air during cold times of the year. This requires either installing heat exchangers that will transfer some heat from outgoing air to incoming air, or installing systems to remove impurities from the air, allowing it to be recirculated. The latter course has obvious monitoring requirements.

Finally, one must always be alert to problems created when there is a *change in process* in a facility. The introduction of new or larger sources of chemicals or particulates may create hazards the previously satisfactory system cannot now handle. A new process may include a substance whose airborne concentrations must be held to lower levels than those previously in use. Even without a change in chemicals or in total level of output into plant air, problems arise if the process change abandons a steady moderate rate of emission in favor of intermittent larger releases. Rearranging the locations of workstations could result in exposure to harmful materials by people previously out of their path. Management concern focuses on how changes affect the cost and efficiency of operating the facility, and it is the job of safety personnel to be alert to how they impact safety.

LOCAL EXHAUST VENTILATION

Local exhaust ventilation (LEV) systems remove air at the point the hazard is generated. This approach has some clear advantages, especially when dealing with particularly hazardous substances or an unusually dirty operation.

1. The hazard can be removed by moving relatively small volumes of air, thus saving energy.

2. The problem of exposing a large segment of the workforce to a locally generated hazard is reduced.

3. If contaminants must be removed from the air before it is discharged from the plant, the task is now restricted to a relatively small air volume.

The employer is rewarded for investment in LEV machinery and ductwork by lower costs for compliance.

DESIGN OF A LOCAL EXHAUST VENTILATION SYSTEM

An understanding of local ventilation system components helps enable the compliance officer to spot flaws and possible bad design features, alerting this person to the need for more detailed analysis of air contamination at particular workstations. A local ventilation system has four components:

1. At the workstation there must be a *hood*, the air intake device designed to capture the hazardous material.

2. Air is then carried through *ducts* away from the site.

3. An *air cleaning device* may be inserted into the ductwork system to reduce the levels of the contaminants in the exhaust air.

4. Finally, there is a *motor driven fan* that drives air from the inlet(s) to the exhaust.

HOODS

The hood can be anything from the open end of a duct placed near the source of contamination to a sophisticated enclosure surrounding the entire operation. Hood design should be appropriate to the contaminant being trapped.   The system fan is moving air through the ducts away from the hood such that air pressure is lower at the hood opening. Room air moving toward this low-pressure zone at the opening carries contaminants with it into the duct.

Room air entering the opening needs to move with enough velocity to carry the contaminants. The term *capture velocity* describes the velocity of airflow needed to move specific contaminants into a hood. Capture velocity varies with the type of contaminant, ranging from approximately 20 to 650 m/min, with large particulate material requiring a greater velocity than vapors or gases.[1] It is hard to determine a reasonable capture velocity value, other than by trial and error.

The distance between the hood opening and the source of contamination is very important. Doing a simplified calculation of the effect of distance from the opening requires use of this equation:

$$v = \frac{kQ}{x^2 + kA} \qquad \text{Eq. 8}$$

where

v = velocity of air flow at distance x
Q = is volume of air passing through the hood opening per minute
k = a constant dependent on the shape of the opening. For square to round openings k is near 0.1
A = the area of the opening

---

[1] Engineers are more likely to use ft/min (fpm).

Volume of air entering the hood per minute is the velocity of air at the opening times the area of the opening.

Using this relationship, we can calculate the effect of distance from the opening. Suppose the velocity of air at a square hood opening is 200 ft/min and the area of the opening is 4 ft². What is the velocity of air flow 2 ft from the opening?

$$v = \frac{kQ}{x^2 + kA}$$

$$= \frac{(0.1)(200 \text{ ft/min})(4 \text{ ft}^2)}{(2 \text{ ft})^2 + (0.1)(4 \text{ ft}^2)}$$

$$= 18 \text{ ft/min}$$

We see that in this example a distance of only 2 ft is enough to reduce the rate of air flow to less than a tenth of its original value.

There are three basic types of hoods: capture hoods (Figures 8.3, 8.4, and 8.5), enclosures, and receiving hoods. A capture hood is placed near the source of contamination and the flow of air is sufficient to carry the contaminant into its opening. When designing a capture hood these points are important:

1. The distance from the hood opening to the source of contamination must be as small as possible. A short distance allows capture with a lower air flow into the hood.

2. The geometry of the source must be considered. For example, a hood opening at one side of a large open vat may be close to one side of the vat, but far from the other side.

3. Movement of air into the hood should be away from the worker.

4. If the contaminant is given an initial velocity in one direction by the process—for example, particulate coming from a grinding wheel—the hood opening should be placed on that side of the process to take advantage of, rather than having to overcome, that velocity.

5. The hood must not interfere with the process or the worker.

6. Cross drafts of air from passing people or vehicles, doors opening and the like can seriously interfere with effective capture of contaminants.

Design features may be added to minimize some of these problems. A round hood opening at the side of a vat draws air more effectively at the center than

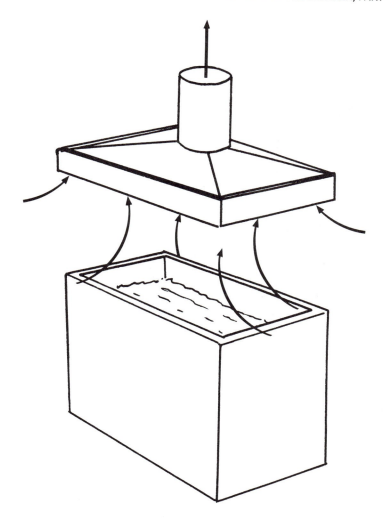

Figure 8.3.   Capture hoods: A *canopy hood* is suspended over the vat or other source of contaminant. Such a design allows access to the workstation from all sides. However, the capture of dense vapors is very inefficient. Light vapors and hot air rise toward the canopy to be captured.

at the edges, so that it is not uniformly effective across the vat surface. Replacing the round hood with a wide slot results in a system that draws relatively uniformly across the entire end of the vat. This modification is not without cost, since a slot generates turbulence in the air inside the duct, increasing its resistance to flow. The effect on Equation 8 is to decrease the value of k, thereby reducing the velocity of the air at a given distance outside the hood opening.

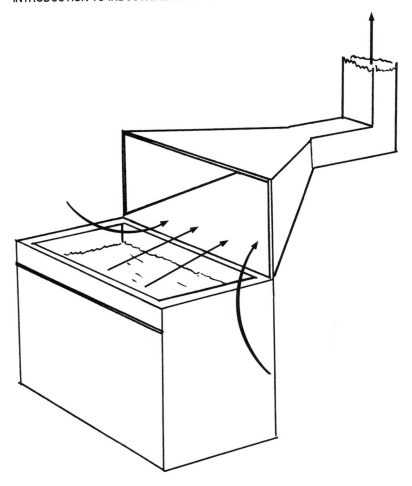

Figure 8.4.   Capture hoods: In this *lateral hood* air moves across the vat into the hood. In such an arrangement air enters the hood from locations other than over the vat. It is important that the air have high enough velocity or that the vat be sufficiently narrow that vapors from the opposite side of the vat enter the hood.

When dealing with a wide vat at which there are restrictions due to access to the vat and location of the workers such that only one side may have a hood opening, air under pressure may be introduced on the opposite side of the vat and directed across the vat to the hood opening. This is termed a *push-pull system*. The velocity of air from the air source must be adjusted to move the contaminants up to but not beyond the hood opening, or else this becomes just another cross draft. This can be done without spreading the contaminant in every direction because a blower can direct air successfully, unlike an air intake

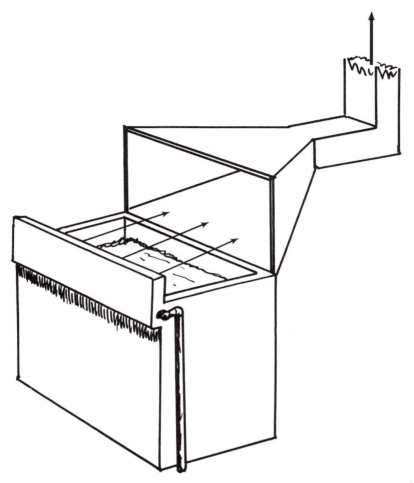

Figure 8.5.   This *push-pull* system differs from the simpler lateral hood in that nozzles direct compressed air across the vat toward the hood. Such a system assures efficient capture of vapors from the side opposite the hood.

which draws from every direction. To illustrate this principle, hold your hand about a foot from your mouth and blow a small volume of air at the hand. Now repeat this, but this time draw in sharply about the same volume of air. Did you feel the moving air in each case?

Simply putting a flange on the opening of the hood increases its effectiveness significantly. This is because air entering a simple opening is partially drawn from behind the opening, and the flange blocks this. The approximate effect of a flange about as wide as the opening itself is to change Equation 8 to the following:

$$v = \frac{kQ}{0.75(x^2 + kA)}$$

Once a flange is added to the hood, the first step has been taken toward building an *enclosure*. Adding sides blocks cross drafts and increases the portion of intake air that is carrying contaminants. Once we have built a box around the source (including an air inlet somewhere), contamination control is achieved with much lower air flows (Q) and the system is less subject to interference in its operation by external factors. This is only possible when it is not necessary to have access to the operation.

Finally, a *receiving hood* takes advantage of velocity given to the contaminant in a specific direction. An air intake hose attached to a saw or grinder so that the particles are directed right into the opening can be very effective. Some local ordinances require such a system to be installed. Another type of receiving hood is a canopy intake placed over a hot process. A canopy hood is a tempting design because it places the hood out of the way of workers and machinery, but it is an ineffective design for a cold process. However, in a hot process the contaminant rises due to thermal upcurrents of air and is conveniently collected. Such an arrangement may move contaminants toward workers at the edge of the vat, and if so should not be used.

DUCTS

Two problems are associated with the design of the duct system: power requirements to move the air, and proper distribution of the air in a multihood system. Beyond the design of the system, it is necessary to have a means of cleaning the system and checking its performance.

The difficulty of moving air through the duct system is called the *resistance* of the system. High resistance must be overcome by more powerful fans, and therefore has a price in energy needed to run the system. Round ducts have a lower resistance than those of other shapes. The higher the velocity of air through the duct, the greater the resistance. To move the same amount of air through a smaller-diameter duct requires that the air move at a higher velocity, so smaller ducts have higher resistance. Now larger fans may be needed. Money saved in initial installation by fabricating smaller ducts may be lost many times over in operating costs. Resistance is also increased by designing sharp bends into the system. Just as it takes more energy to move a Model T Ford down the highway at the same speed as a new Taurus, smooth curves in ductwork lower the energy requirement to move the air (Figure 8.6).

There is a minimum velocity required in the duct when the contaminant is particulate. Heavier particles require greater velocity, with minimum velocities ranging from approximately 600 m/min for fumes from welding or soldering to

Figure 8.6.   Here a small plant has a general exhaust ventilation system. The blower to the right removes air exiting the plant at several points on the ceiling. Notice that the duct avoids sharp bends, allowing air to be moved with less friction. Notice too that as more air enters the main duct, its diameter increases. Done properly, this assures roughly equal flow at each entrance to the duct. The blower on the left serves a local exhaust ventilation installation.

1700 m/min for dust from grinding metals.[3] Too low a velocity results in solids being deposited in the ducts, eventually blocking them. Vapors and gases do not settle out, and an air velocity of around 650 to 1000 m/min is recommended.

The second, and more difficult, problem in duct design is providing proper distribution of air coming from each of several hoods in a complex ventilation system. One possibility is to design the duct diameters so that sufficient air flow is maintained at each hood for the contaminants being captured (Figure 8.6). Another is to place gates—adjustable barriers—in the duct to adjust the flow. A good design based on balancing duct diameters is more efficient, and gates are places to deposit particulates. However, it is easy to change a gated system and rebalance it by adjusting the gates, while a system with fixed duct diameters is very inflexible.

---

[3]Recommendations are given in the ACGIH manual.

FANS

Selection of the correct fan for a system is the function of a ventilation engineer. The safety professional should be concerned that the fan in place is providing correct air flow by checking velocities at hoods and comparing pressures in ducts with design specifications by use of water manometers at access points. Fans should have scheduled maintenance, especially if a belt drive is employed.

Fans are a source of noise. Some fan designs are better in this regard, and in general a larger, slower running fan is quieter.

CLEANERS

A cleaner is used if the exhaust air is hazardous or a nuisance to the general public, or if the air is to be recirculated in some fashion. Cleaner design depends on the nature of the contaminants. Removal of particulate requires a quite different approach from removal of gases or vapors.

Particulate Removal

Most cleaners are designed to remove particulate from exhaust air. Recall the relationship between particle size and the ability of the particle to penetrate deep into the lungs. As a rule, particles under 5 µm are classed as *respirable,* and are therefore more hazardous. Unfortunately, the easiest particle removal methods are effective only against larger particles, and are useful primarily for removal of nuisance dusts.

The efficiency of particulate removal (with the amount usually being expressed in weight terms) is often calculated as:

$$\text{efficiency} = \frac{\text{amount collected}}{\text{amount entering cleaner}}$$

This number can be very deceptive, since respirable particulate may not be removed at all in a system rated as 98% efficient, since the respirable dust particles individually weigh so little.

Removing Large Particulate

The simplest system passes the exhaust air through a large *settling chamber.* Velocity drops sharply in such a chamber, just as water in a fast-moving stream slows as it enters a pond. As velocity drops below the capture velocity of the larger particles, these settle to the cone-shaped bottom of the chamber, from which they can be removed periodically through access at the bottom of the cone. *Baffles* added to the chamber cause air to change direction sharply. These remove some particles as they impact the baffle.

In *centrifugal* or *cyclone collectors* air is swirled around a cone-shaped chamber, moving faster as the diameter becomes smaller (Figure 8.7a). The particles "spin out" against the sides of the chamber and drop to the bottom, much as the bends of the bronchial tubes remove particles from the air in the lungs. Once again, larger particles are removed much more effectively. A *high-efficiency cyclone* has a smaller-diameter cone, spins the air faster and so is more effective at removing smaller particles. Such a cyclone has a much higher resistance and may require installation of a larger-capacity fan.

A *dynamic precipitator* adds a motor-driven impeller to a cyclone to drive the air against the sides of the chamber, which raises efficiency to the level of a high-efficiency cyclone. Respirable particles escape collection even in high-efficiency cyclone and dynamic precipitator designs.

Removing Respirable Particulate

Cloth *filters* are frequently employed as cleaners. A *baghouse* has a large chamber in which air enters at the bottom and has to pass through cloth bags to exit at the top (Figure 8.8). Although respirable particles pass through the pores in new cloth, as the bag is used the larger pores are blocked and respirable particles are removed. Periodically the bags are shaken, so that accumulated particulate is dropped to the bottom of the chamber for removal. This design is restricted to removal of dry particulate, since oily or wet droplets seal openings in the cloth.

Water can be used in a variety of ways to remove both particulate and water-soluble gases or vapors. Water can be sprayed in a chamber or dynamic precipitator, or air can be passed up a scrubber tower in which water is sprayed or flows over baffles or a packing material. By placing used water in a tank, allowing particulate to settle out, water sometimes may be reused.

Electrostatic precipitators utilize a discharge electrode and a grounded collecting electrode (Figure 8.9). Particles are given a charge with the negative discharge electrode, then move toward the positive plate of the collecting electrode. Periodically the accumulated particulate is removed from the plates. Such a system is efficient and adds little resistance to the system. Preliminary cleaning of contaminated air is normally done, perhaps with a settling chamber or cyclone, to reduce the load on the system.

Vapor and Gas Removal

Removal of gases or vapors requires a different approach. Wet scrubbers (above) work unless the contaminant is not water-soluble. Activated charcoal

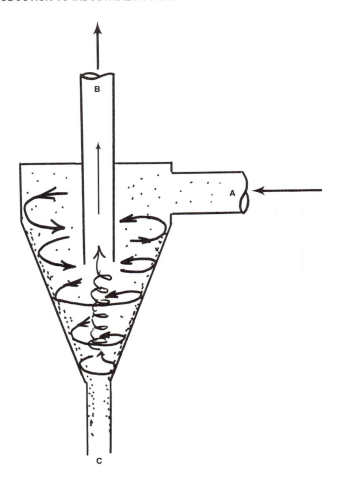

Figure 8.7a. In a *cyclone collector,* particulate-contaminated air is directed into a conical chamber (A), forcing it to spin around the chamber. As it moves down the cone (toward a smaller diameter), the spinning becomes faster. Particles are driven by centrifugal force against the walls of the chamber and deposited there. Faster spin rates deposit smaller particles. Treated air is drawn out of the center (B), and deposited particles are collected at the bottom of the chamber (C) and transferred into a container for disposal.

filters remove most contaminants effectively. Cooling exhaust air may cause the vapor to condense. Passing exhaust air over burners, possibly in the presence of a combustion catalyst, incinerates combustible contaminants.

Figure 8.7b. A cyclone collector on a small industrial building.

USE OF RESPIRATORS

It is desirable to employ an "engineering solution" to problems of airborne contaminants, reducing levels by one of the methods described above. However, there are times when the use of respirators is appropriate. Certainly they should be available for emergencies such as spills, equipment malfunctions, or special maintenance procedures. An adequate emergency preparedness plan should assure that such devices of the correct design are on hand.

There are three classes of face mask type respiratory protection. The first is a *filter mask* worn across the nose and mouth (Figures 8.10, 8.11). In a dusty environment, the air passes through a filter to remove particulate. Such masks are sometimes made of heavy paper and are disposable at the end of the day, or when they become blocked. They may vary in the mesh size of the filter, and therefore in the size of the smallest particle they allow to pass through. It is important to recognize that *a filter mask only traps particulate material.* If the problem is a vapor rather than dust, a filter style of respirator will not serve. Many workers don't understand this, so clear labeling and instruction are important.

To remove a gas or vapor, air must pass through an adsorbent, such as activated charcoal, or a substance that reacts with the gas or vapor. Respirators with this feature have removable cartridges containing the active agent that may be replaced when the adsorbent is spent. Manufacturers provide a number of different cartridges, each color coded. The cartridges used must be correct for the hazardous gas or vapor involved, and the recommendations of the manufac-

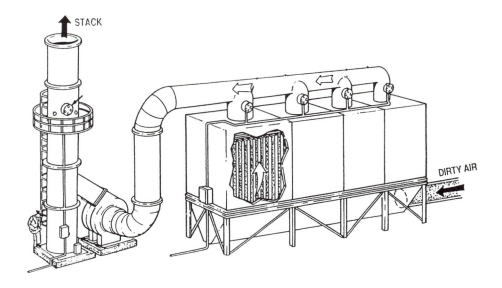

Figure 8.8. In a baghouse, particulate-contaminated air (dirty air) enters the lower end of a chamber. To move to the upper portion of the chamber, air must first pass through hanging cloth filters, which remove much of the particulate. Although it is not shown in this diagram, in most baghouses the filters are periodically shaken, causing accumulated particulate to fall to the bottom of the chamber for collection and removal. Courtesy of Auburn International, Danvers, MA.

turer should be followed. Once again, many workers feel that they are safe when wearing a respirator—even the wrong one—and should be instructed which one is the correct cartridge to employ. Adding a filter outside the cartridge also removes particulate, and may extend the life of the cartridge.

The first requirement of a cartridge respirator is that it fit closely to the face, preventing leakage at the edges. Whatever cartridge is installed, air moving through it encounters resistance. If the respirator leaks at the edges, there will be less resistance to air entering through the leak than by the correct route, and the effectiveness of the device is sharply reduced. The respirator can quickly be tested each time the employee puts it on by a *positive pressure* or a *negative pressure* test. In the positive pressure test the worker blocks the valve through which exhaled air exits, and then exhales. The mask should inflate slightly and

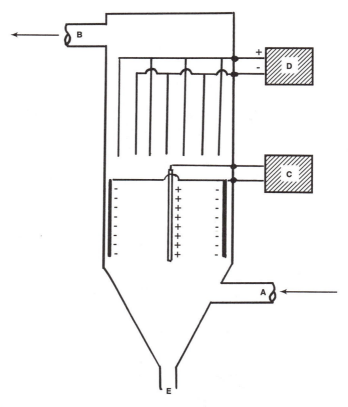

Figure 8.9.   In an *electrostatic precipitator,* particulate contaminated air enters at the lower end of the chamber (A). In order to exit (B), it must pass through an electrical field generated by power supply C, which gives the particles a charge. They then pass by plates given a charge by power supply D, and particles attach to the oppositely charged plates. Collected particulate falls to the bottom where it can be collected for disposal (E).

remain inflated for a few seconds. In a negative pressure test the employee blocks the air intakes on the cartridges, and then inhales. The mask should press against the face. Employees at a workstation who may require a respirator should have one fitted correctly to facial contours. Testing for correct fit could involve exposing the employee wearing the respirator to a harmless chemical with a strong odor. Smelling the odor indicates the presence of a leak.

Where contaminant levels are very high, or treatment by the respirator is inefficient, or oxygen levels are low, a breathing apparatus is used that supplies breathing air (Figure 8.12). A portable or self-contained breathing apparatus (SCBA) includes a tank of air connected to the face mask. Alternatively, the face mask may be connected to an air pump located outside the danger zone.

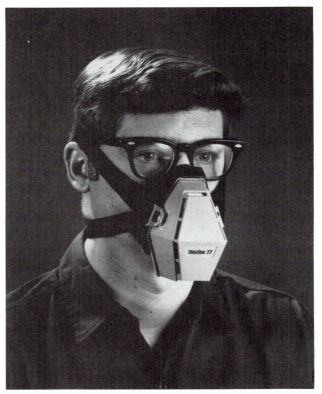

Figure 8.10. Shown here are the parts of a simple filter respirator designed to remove dusts and mists from the inhaled air (above), and the respirator as worn by the worker (below). (Photos courtesy of MSA International, Pittsburgh.)

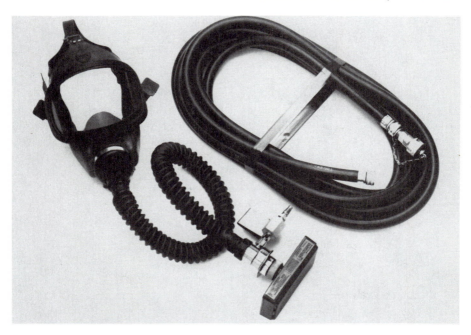

Figure 8.11.   This respirator is for more hazardous atmospheres than the one shown in Figure 8.10. It has a filter adequate for protection against asbestos. For applications involving toxic vapors or gases, it can be connected to an external air supply. (Photo courtesy of MSA International, Pittsburgh.)

Respirator recommendations are included on the MSDS for the chemical. These recommendations are usually based on the level of the contaminant in the air, and in an emergency there is seldom time to do an analysis. It is best then to err on the side of caution.

EVALUATING THE PROGRAM

Once a program to control airborne hazards is in place, evaluation is necessary. Good evaluation has three aspects. First, it is necessary to monitor the atmosphere of the plant to see if the levels of toxic chemicals are safe and in compliance. Second, the condition of the ventilation system should be checked periodically to assure it is meeting design specifications. Third, employees should be trained regarding hazards and the use of such emergency devices as respirators. Finally, good employee health records must be maintained.

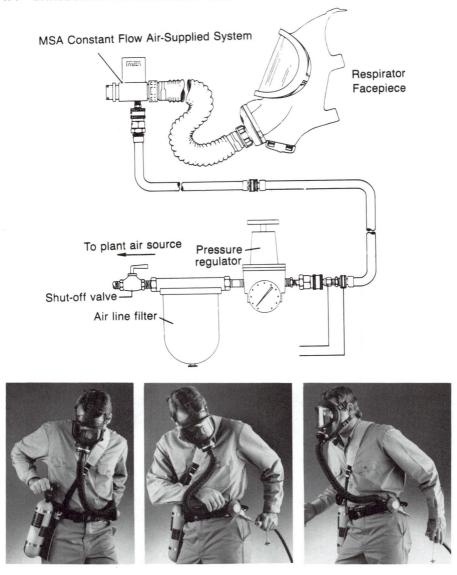

Figure 8.12. Above is a diagram of a system to supply air from an external source. Below is a demonstration of an emergency escape system using a bottle of compressed air (left to right): (1) wearer opens egress cylinder valve; (2) wearer disconnects from the primary air supply if he or she so desires; (3) wearer leaves the dangerous atmosphere. (Photos courtesy of MSA International, Pittsburgh.)

MONITORING THE AIR

Monitoring the air requires several types of measurement (Chapter 7). Concentrations of chemicals in the general plant air should be measured to assure that overall conditions in the plant are satisfactory. Levels in high-hazard parts of the plant should be separately measured. In some cases, it is worthwhile to invest in automatic monitors that record atmospheric levels of chemicals likely to be in high-hazard areas. These could be connected to an alarm system, so that warning is given of unexpectedly high levels of escaping chemical.

When the level of contaminant in plant air is normally high, for example above the action level, more careful continuous monitoring is necessary. The air in the workers' breathing zone can be sampled occasionally by having the workers wear individual monitoring devices. Where there are ceiling limits imposed for the particular chemical, a series of individual readings is necessary to be certain that the ceiling is never exceeded.

TESTING VENTILATION SYSTEMS

It is no more realistic to install a ventilation system and assume it never needs to be touched again than it is to buy a new car and assume it never needs service. At the time of installation the operating parameters of a ventilation system should be measured, first to assure it accomplishes the task for which it was designed, and second to establish baseline values for later comparison to determine if parts of the system need service. Such retesting should be done at periodic intervals.

If changes are made in a plant, such as an expansion or shifting of an operation or the introduction of a new process, the ventilation system should be analyzed to assure it is adequate for the new situation. A given fan moves a given amount of air, hence less air per station as new hoods and ducts are added. Furthermore, a new outlet changes the distribution of air intake, and may require rebalancing the system.

An early concern in analysis of a system is determination of general patterns of air movement. Devices that generate a fine particulate smoke, often of titanium tetrachloride, are useful. Generating smoke near a workstation determines if air is actually moving away from the worker. A hood may draw well at its center, but allow contaminants to escape at the edges of its intended "reach." Air turbulence resulting from hood design may divert contaminants from capture by the hood. Smoke reveals this "spillage" of contaminant. The effects of cross drafts of air generated in the room on hood performance can be assessed. A large amount of smoke released in a room helps estimate how quickly air is changed by a general ventilation system, and may identify regions of stagnant air.

Testing whether a system is operating according to original design parameters involves testing pressures and comparing them to values measured at the time of installation. Atmospheric pressure is expressed in a variety of units: millimeters of mercury, pascals, or pounds per square inch. Static pressure measurements focus on the difference between pressure inside and outside the duct, and are relatively small values compared to total atmospheric pressure. It is common to use a water manometer, in its simplest form a U tube containing water, one end of which is inserted into a hole in the duct. Negative pressure in the duct results in the water level dropping on the atmosphere side of the tube, and the negative pressure is reported as this difference in level in *inches of water*.

Velocity pressure in a duct is measured by a pitot tube. A pitot tube is a variation of the water manometer that has a right-angle bend in the tube inserted into the duct (Figure 8.13). By pointing the tube directly at the air flow, a pressure reading is obtained that is the sum of the static pressure and the velocity pressure. A second tube surrounds the first and is not open in the direction of the air flow, but rather has openings perpendicular to the air flow. This measures static pressure. By connecting one side of the water manometer to each tube, the manometer reads the difference between the total pressure and the static pressure, which is the velocity pressure.

Either a deflecting vane velometer or a rotating vane anemometer are most often used for measuring air velocity at large hood openings. In the deflecting vane velometer, air strikes a vane, which deflects in proportion to the air velocity, and the angle of deflection is read on a scale directly in feet/min (fpm). It can be adapted to different velocity ranges by changing probes, the low end of the low-velocity probe reading 30 fpm. A rotating vane anemometer has a multivaned fan blade that is spun by the moving air, and the speed of rotation is read directly in fpm. A hand-held unit reads down to about 100 fpm. With either of these instruments, error is introduced if air velocity at small hood openings is measured, since the blockage of the hood opening by the instrument distorts the value obtained.

Other devices are also used. The resistance of a heated wire changes with temperature, and if the wire is placed in the air flow, it is cooled in proportion to the rate of flow. Such a heated wire anemometer is particularly useful at low air velocities, reading well below the lowest range of either the deflecting vane velometer or the rotating vane anemometer.

When measuring air velocity at a hood opening, several readings must be taken at different locations in the opening, since the velocity changes from the center to the edges of the opening. In a long, straight duct the centerline velocity is highest, dropping down to zero at the surface of the duct wall. Measurements in a duct should be done 7–8 duct diameters away from a bend or constriction to avoid distortions in the flow from these features.

A basic understanding of the design and parameters of a ventilation system is important to the safety professional. Good ventilation is one of the most important protections for workers when hazardous materials are in use, and systems should be checked to ensure they are performing as is necessary to achieve this end.

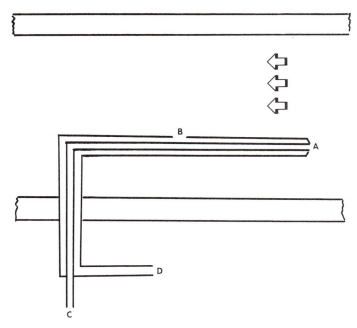

Figure 8.13. A pitot tube allows us to measure both static pressure and velocity pressure in a duct. In this sketch, air is moving from right to left. Opening A in the tube is directed toward the air flow, and a manometer connected to tube C measures the sum of static pressure and velocity pressure. The opening in the outer tube (B) is directed perpendicularly to the air flow, and a manometer connected to tube D measures only static pressure. A manometer connected at one side to tube C and at the other to tube D reads velocity pressure.

EMPLOYEE HEALTH RECORDS

Finally, by law, good employee records of health and exposure must be maintained.[4] If some problem arises involving either an individual worker or the general health of all employees in a sector of the plant, the records are a means of learning the cause of the problem, so the situation will not be repeated. Many times, an epidemiological study of company health records revealed a previously unsuspected hazardous condition. This is particularly true of types of toxicity for which animal surrogate testing is less reliable, for example, in the cases of carcinogens or reproductive toxins.

---

[4]See 29 CFR 1904 and 29 CFR 1910.440.

ENTRY INTO CONFINED SPACES

More than 300 workers die per year as the result of entry into confined spaces. A confined space has limited access ports and inadequate or nonexistent ventilation—a space not intended for continuous human occupation. Examples include storage tanks, silos, bins, utility chambers, and tunnels for pipelines or wiring. Virtually every industrial site has examples of confined spaces. Such areas may need to be entered for maintenance, cleaning, or service of machinery. The most serious hazards are the presence of a dangerous atmosphere and the danger of engulfment of the worker.

CONFINED SPACES WITH BAD ATMOSPHERES

Three classes of problems are found with the air in confined spaces: improper *oxygen* concentrations, *flammable* or *explosive* vapors or gases, and *toxic* vapors or gases. More than one of these problems could be found in a given chamber. A number of portable devices for testing air quality are available, and the air quality must be evaluated before the employee enters the confined space. Blowers attached to flexible ducts may be used to ventilate an area before attempting to enter. This may correct the problem, but if not, the worker must wear appropriate respiratory protection. In any case, emergency respiratory devices such as SCBA must be available if an emergency arises.

A space may have either too little oxygen or too much, but the former is much more common. Oxygen levels are normally 20–21% in the atmosphere. Difficulty breathing is experienced at 14% and mental confusion sets in at 12%. Ten percent leads to unconsciousness and 8% to death. The OSHA requirement is for at least 19.5% oxygen for worker entry without a breathing apparatus. Oxygen can be removed from the atmosphere of a confined space in a number of ways. Oxidative processes such as rusting inside a metal tank use up the oxygen supply. Gases entering or generated in a confined space displace oxygen. For example, a liquid residue remaining in a tank presumed empty can evaporate, and fill the tank with vapor. Here the density of the displacing gas is important. A dense gas concentrates at the bottom of the space, shifting the original air upwards. When sampling the air in a confined space before worker entry, do not assume that a satisfactory reading at one level establishes safety at all levels. High oxygen levels, 23% or greater, create an unusual fire or explosion risk. The OSHA limit upper is 23.5%.

Flammable gases, vapors, or dusts constitute the second class of hazard. Tanks or vats that once stored flammable substances, and are being cleaned or serviced, may well have enough of the substance remaining, particularly residues of a volatile liquid, to create the potential for fire or explosion. The lower flammable limit (LFL) or lower explosive limit (LEL) should be obtained, perhaps from the MSDS, before measuring levels of the material in the space. People associate combustion with liquids, so combustible dusts are less likely to generate concern, but they can constitute a serious threat. Bins containing agricultural products have been known to explode violently on ignition. As a rule of thumb, dust levels sufficient to obscure visibility at 5 feet should be considered dangerous.

Finally, toxic vapors or gases are a concern. Assaying for the known contents or former contents of a confined space is helpful in selecting the test method, but concern should not be limited to these chemicals. Carbon monoxide and hydrogen sulfide are toxic gases commonly found, and a survey for these and other common contaminants is a wise precaution. Remember that many toxic substances have poor warning properties.

PERMIT ENTRY SPACES

OSHA provides a systematic procedure for entering a space defined as a *permit required confined space.* It is therefore a key decision whether or not this procedure is necessary. OSHA defines such a space [29 CFR 1910.146 (b)] as one that:

1. Contains or has a potential to contain a hazardous atmosphere;
2. Contains a material that has the potential for engulfing an entrant;
3. Has an internal configuration such that the entrant could be trapped or asphyxiated by inwardly converging walls or by a floor that slopes downward and tapers to a smaller cross section; or
4. Contains any other recognized serious safety or health hazard.

The employer is required to have a written procedure for employee entry into permit spaces, including permit forms, and must provide training to employees who may either enter or supervise the entry of another into a hazardous confined space. Every such space must be posted with a danger sign indicating that it is a permit space into which unauthorized entry is prohibited. There is a stepwise procedure when such a space is to be entered. A permit must be filled out and signed by an *entry supervisor,* who is responsible for assuring that conditions are acceptable for entry and for terminating the job if they become unsuitable. The space should be entered only by *authorized* (trained) *workers.* There must be an *attendant,* who remains outside to monitor the entrants.[5] The attendant should be in communication both with the entrant and with a *trained rescue team,* should one be needed. Any condition that makes it unsafe to open the space must be eliminated before the space is opened. A barrier must now surround the opening to prevent either workers from accidently entering or falling through the opening, or objects being dropped on employees in the

---

[5]The attendant must *never* attempt to enter the space to rescue workers. There are many examples of rescuers becoming victims, leaving no one outside to summon help.

space. The atmosphere must be tested for (in order) oxygen content, flammable gases or vapors, and toxic compounds in the air. If the air is safe, the employee may enter. If not, forced air ventilation must correct the problem before entry of workers.

Any bin, tank, or silo that contains liquid or flowable solid presents, in addition to questions of the suitability of the air, a problem of engulfment by the liquid or flowing solid. This should be a consideration before the worker enters the space, and a plan for preventing engulfment, and for dealing with it should it happen, should be in place. A worker harness connected to lines that exit the space and perhaps connect to a winch to remove the worker in case of a problem is a wise precaution.

Good preventive practices by industry, and alert response to new findings as they become available, should enable industry to maintain and improve its health and safety record. Although the cost of preventive measures is sometimes high, the cost of not taking those measures is higher yet.

KEY POINTS

1. Good plant maintenance and housekeeping reduce airborne contaminants.

2. Hazard to workers can sometimes be reduced by stationing the operator away from the process.

3. Changing a process to use less-toxic chemicals reduces hazard.

4. The removal of contaminated air is accomplished by ventilation. This is essential if the contaminants pose a health, fire, or explosion risk.

5. Air moving through ducts has a positive pressure in a blower duct and a negative pressure in an exhaust duct. It moves with a velocity resulting from the air flow and the capacity of the system.

6. There is a minimum velocity needed to keep particles suspended in air.

7. A general exhaust system moves the entire air of the workplace using an exhaust fan. System capacity is measured in room changes per hour. Equations are presented to calculate the maximum limit concentration of contaminant, given the rate of generation and system capacity. Rate of clearance when generation ceases is also determined in terms of contaminant half-life.

8. A good general ventilation system should assure that air flow is adequate at the site of contaminant generation, and that contaminant is not carried toward workers. Maintenance is important.

9. Local exhaust ventilation moves air at the point contaminant is generated. Advantages include the requirement to move smaller air volumes, avoidance of spreading contaminant throughout the workplace, and minimization of the volume of air to be treated before being exhausted.

10. Local exhaust systems have hoods, ducts, a fan, and possibly an air cleaning device.

11. Hoods provide a range of degree of enclosure from none to complete. Greater enclosure increases efficiency but impedes access to the workstation.

12. Air velocity drops rapidly with distance outside the hood opening.

13. A good hood system clears air from all parts of the contaminant source, moves it away from the workers, is able to overcome cross drafts, and does not interfere with work at the station.

14. Ducts generate resistance to air flow, requiring more energy expenditure at the fan. Round ducts and larger-area ducts have less resistance. Intake from multiple sources must be balanced.

15. Fans come in a variety of designs that vary in efficiency and noise level. Regular maintenance is important.

16. Cleaners to remove air particles operate with a certain overall efficiency, but most are more efficient for larger particles.

17. Settling chambers lower air velocity and/or provide baffles to cause larger particles to settle.

18. Cyclone collectors spin larger particles out against the system walls. Efficiency is greater in smaller-diameter cyclones or cyclones with dynamic precipitators.

19. A baghouse filters out particles through cloth bags.

20. Electrostatic precipitators give particles an electrostatic charge, then attract them to an oppositely charged plate.

21. Vapors and gases are usually removed either by wet scrubbers or charcoal filters.

22. Respirators, which come in designs specific for particular problems, are useful temporary measures to protect workers.

23. Monitoring air on a regular basis ensures continued compliance.

24. Testing ventilation systems involves comparing velocities and pressures with system norms.

25. Employee health records must be kept as a check that protection is adequate.

26. Special regulations apply to worker entry into confined spaces. An unsuitable atmosphere, the possibility of engulfment by liquid or flowing solid, and entrapment in the space due to its shape are potential problems.

PROBLEMS

1. You are a new industrial hygienist at the original (built in 1924) facility of Acme Foundry and Fabricating in Jackson, Michigan, where machinery beds and transmission housings for heavy industrial machines are produced. In Building A we find at various stations (1) molds being formed of sand, (2) iron being melted in a natural gas–fueled furnace, (3) casting of iron into the molds. Exhaust blowers are mounted in the roof, and air enters everywhere since the furnaces provide an excess of heat even on cold winter days. Gases from the furnace go up a stack. There is (4) an outdoor yard where castings are cooled. In Building B we find in the northeast corner (5) hydrochloric acid baths for cleaning rust from castings stored in the weather and (6) a sand blasting station, and in the northwest corner (7) sites where castings are ground to provide a gasket surfaces and drilled and tapped to insert bearings and studs for assembly. In the southwest corner is (8) an assembly operation where gears, bearings, and shafts from a supplier are installed, and in the southeast corner (9) there are spray painting booths. Two large exhaust fans are mounted on the west wall and air enters along the east wall where it is heated in winter. Building C (10) is a largely enclosed crating, storage, and loading dock area with diesel-powered forklifts and electric engine–driven hoists to load the finished product onto flatbed diesel trucks.

   The company services and modernizes equipment it has built and sold previously. In a separate room of Building C (11) there is a degreasing tank used to clean lubricants from castings from used equipment and a workshop to work on machines that are being modified. Each room has a ceiling-mounted exhaust fan and air inlets from the heating system near floor level, or through the open doors when a truck is being loaded.

   There is a reasonably good rate of general ventilation throughout all buildings of the facility, but there are short periods of high activity at various stations when this is insufficient. Your boss tells you to buy disposable masks where needed for these various workstations to be used at times when air is contaminated.

A.   After inspecting the operation, *list the respiratory hazards* at each operation (by number) in the plant.

B.   When you pick up the safety equipment catalog you find the following:

Catalog No.	Description
35001	O.V.
35002	A.G.
35003	Ammonia and methylamine
35004	Nontoxic dust
35005	Toxic dust/mist
35006	Nuisance odors
35007	Paint spray

Phoning the supplier, you learn that O.V. stands for organic vapors and A.G. stands for acid gases.

1.   If you took one of each of these masks and cut it open, what would you expect to see inside?

2.   What mask should you buy for each station? Does every station need masks? Do masks handle every problem?

3.   When the masks arrive, the boxes are clearly labeled, but the masks themselves are all white face covers with straps. Some have one-way valves for exiting exhaled air, but otherwise they look very similar. All are dispensed from the room at the plant entrance near the showers and lavatory where workers get their coveralls. What steps need to be taken?

C.   By rearranging existing air inlets and exhaust fans, how could ventilation in Building B be improved?

D.   Profits are good and some money is going to be spent in plant modernization. Plant safety people ask for some local ventilation and are told to present a listing of sites that would benefit. What additional information would you need to prioritize the list?

E.   You have enough money in your budget to buy a self-contained breathing apparatus for emergency use. Where is the best place to house it?

F.   What else could be done to improve air quality as plant components and equipment are replaced?

2.   A ventilation system is being designed for an industrial building with 3200 ft^2 of floor space and 12-ft-high ceilings. In this facility premanufactured components of home furnace humidifiers are assembled and packaged in corrugated cardboard boxes. There is no point source of hazardous air contamination, so GEV is considered adequate. It is planned to have a fan blowing outdoor air through a heating and cooling unit, with exhaust vents in the roof.

A.   How will pressure in the building compare to atmospheric pressure?

B.   What airflow must be generated by the fan to provide 3 room changes per hour?

C.   There is a change in the process such that an adhesive dissolved in a chlorinated hydrocarbon replaces some mechanical assembly. The solvent is of low toxicity, but has a strong and unpleasant odor. It is decided to run the system so that the level of the solvent does not exceed 2 mg/m^3 in the air. What would the rate of removal of the solvent be at the maximum concentration?

D.   If solvent is evaporating from the assembly process at 50 mg/min, will the present ventilation system achieve the goal for solvent concentration in the air?

E.   Using the present system, and with the given rate of evaporation, what would the C_{max} be?

F.   At night and over the weekend with the adhesive containers closed, what would the $t_{1/2}$ be?

G.   What engineering change would improve the situation without use of LEV?

3.   Suppose that to capture a particulate coming from a particular process requires a rate of air flow of 20 ft/min.

A. How close must the source be to a roughly square hood of 6 ft^2 and a velocity of air of 190 ft/min?

B. What rate of air flow would be required if the source were 3 ft away from the hood?

4. In the discussion of use of settling chambers to remove large particulate from exhaust air, the addition of baffles to the chamber was described as increasing the efficiency of the system. What price is paid for this addition, ignoring the cost of the baffle installation itself?

5. A large underground storage tank has been used to store fuel oil for furnaces in a plant. It has a single manhole entry point and a narrow aboveground vent. It is leaking, and requires repair. The fuel is drained into other tanks, and a welder is to be sent into the tank to weld the damaged seam.

A. Does this task require a confined space permit? Look in 29 CFR 1910.146 (b) and decide if a "hot permit" is needed.

B. What testing must be done before a worker may enter? In this case special attention must be paid to which hazards this welder could face? What information is needed from MSDS or tables?

C. List personnel required to be available at the time of entry.

D. Suppose ventilation does not succeed in bringing the hydrocarbon level below the lower flammability level. What other alternative is possible?

BIBLIOGRAPHY

29 CFR 1910.94 and 1910.252; Regulations specifying ventilation for specific operations.
ACGIH, *Industrial Ventilation—A Manual of Recommended Practice,* 20th edition, Cincinnati, 1988.
W. A. Burgess, M. J. Ellenbecker and R. D. Treitman, *Ventilation for Control of the Work Environment,* John Wiley and Sons, New York, 1989.
R. P. Garrison, "Ventilation for Contaminant Control," in *The Work Environment,* Volume 1, Lewis Publishers, Inc., Chelsea, Michigan, 1991.
H. J. McDermott, *Handbook of Ventilation for Contaminant Control,* 2nd ed., Butterworth Publishers, Stoneham, MA, 1985.

CHAPTER 9

FIRE AND EXPLOSION

Throughout our lives we hear about fire safety, particularly safety in the home. We are warned not to use space heaters near flammable materials, to maintain our furnaces carefully, and not to store oil-soaked rags in the basement. Schools have fire drills to teach children to escape a burning school building, and every public building has lighted red signs indicating fire exits. Fire has been a menace throughout the history of civilization, and major cities such as Rome, Chicago, London, and Detroit have at some time burned to the ground.

In all industries fire is a concern, and the potential for fire often is increased because certain materials require special precautions and understandings. Actually we have a dual concern, first for the start and spread of a fire, and second for the occurrence of an explosion.

FIRE: GASES AND VAPORS

In order to have a fire, two components are required: fuel and an oxidant. Atmospheric oxygen is the most common oxidant. Thus in a gas furnace the reaction is:

$$CH_4 + 2O_2 \text{ ------> } CO_2 + 2H_2O$$

fuel   oxidant

However, the presence of fuel and oxidant is not sufficient to start combustion. The mixture must be raised to a high enough temperature to start the reaction, usually by a spark or a pilot light. Once started, combustion releases enough heat to keep the mixture temperature above that necessary to initiate the reaction.

It is possible to have a fuel and an oxidant, but in proportions such that the process cannot occur continuously. If the fuel is a small amount of gas or vapor mixed with air and we initiate combustion with a spark, the small amount of heat produced by burning might be insufficient to ignite adjacent fuel, and continuous combustion fails to happen.

Most people have no trouble accepting fire not starting because of inadequate fuel, but many have trouble with the concept that there can be too much fuel. However, too much fuel is really insufficient oxygen, and once again the heat produced when a spark ignites some fuel is not enough to ignite adjacent fuel. If you have trouble with this concept, think of the last time you tried to light charcoal—admittedly a solid fuel—on your grill. You had a great deal of charcoal fuel and you generated heat by burning lighter fluid, but the charcoal was stubborn until *you blew on it,* increasing the oxygen supply. Once it was burning, the hot combustion products rose rapidly upward and new air rushed in to replace them, providing enough oxygen to continue combustion.

For each flammable chemical, we can determine *flammability limits*—concentrations of gas or vapor in air that are at the low or high extremes of ability to sustain continuous combustion. There is a *lower flammability limit* (combustible-lean limit mixture), which is a concentration in air, usually expressed as volume %, below which the fuel/air mixture does not sustain continuous combustion. The *upper flammability limit* (combustible-rich limit mixture) is the highest concentration of fuel in air that burns continuously. Such values should not be treated as absolute physical constants such as melting point or density of a pure substance at 25°C, but rather as guidelines when establishing criteria for safe conditions. For example, flammability limits vary with temperature, pressure, and conditions of combustion. Most importantly, it should be remembered that a closed container whose fuel concentration is above the upper flammability limit will not ignite, but opening the container and allowing more air to enter may produce a flammable mixture.

Another variable arises from the "layering"—or separation by density—of gases or vapors in air. Gases that are less dense than air, such as hydrogen or methane, become more concentrated at upper levels in a container, while dense gases or vapors such as butane or benzene concentrate at the bottom. A dense gas has a molecular weight greater than the 28–32 of nitrogen and oxygen. A container may overall contain a concentration of a gas or vapor fuel lower than the lower flammability limit, but still have the potential to burn at the top or bottom of the container.

EXPLOSION

An explosion occurs when rapid expansion of gases produces high pressures. Fire and explosion may be two sides of the same coin, where both are based on a strongly exothermic chemical reaction, usually combustion in air. Explosions are not limited to combustion, however, and are generally classed as any rapid expansion of gases. For example, an explosion could be the result of failure of a container of gas stored under high pressure.

The difference between a fire and a combustion explosion lies in the rate of generation of gaseous product. A burning fuel is supplied with oxygen as air moves to the fuel, and the gaseous products leave the site. However, when the fuel and oxygen are premixed and on ignition the reaction spreads rapidly throughout the mixture, a large volume of gaseous product is produced from this mixture in a short time period. The pressure produced by rapid expansion of these gases is the explosion. Just as there is a minimum concentration of fuel required to maintain burning (the lower flammability limit), there is also a minimum concentration of an explosive material required for an explosion to occur—the *lower explosive limit*. Pressures due to an explosion can cause structural damage and are life-threatening. An explosion is more likely when the fuel/oxygen mixture is midway between the flammability limits.

PREDICTING HAZARD

In the absence of a table of flammability limits, or when dealing with a compound that is not on the table, we may still be able to estimate or be aware of the hazard of an organic liquid. For example, on the standard diamond-shaped hazard code placed on containers of chemicals, the red upper quadrant warns of combustion hazards. Most organic liquids are flammable, the noteworthy exceptions being chlorinated or otherwise halogenated hydrocarbons.

Industry has been strongly induced to use chlorinated hydrocarbons as solvents because of their inability to burn. The key here is the replacement of carbon-to-hydrogen bonds by more stable carbon-to-halogen bonds. One chlorinated hydrocarbon—carbon tetrachloride—was commonly used as a fire extinguisher fluid before its toxic characteristics were recognized. In its place we sometimes find the chemically similar Freons being used as extinguishers.

Volatility—the tendency of molecules of the liquid to enter the vapor state—is another important liquid property related to fire or explosion hazard. In the vapor state these molecules contribute to the atmospheric pressure in proportion to their numbers. Given time to equilibrate to a maximum value, this contribution to total pressure is called the *vapor pressure* of the liquid. At a given temperature, and given time for eqilibration, the vapor pressure over a pure sample of the liquid is a constant. Higher temperatures increase the proportion of the liquid molecules with enough energy to leave the liquid, and consequently raise the vapor pressure. When the vapor pressure equals the atmospheric pressure above the liquid sample, the liquid boils. Consequently, a liquid with a relatively high vapor pressure has a relatively low boiling point.

If a liquid has high volatility or vapor pressure, and consequently a low boiling point, this is a warning sign for greater fire or explosion risk. If such a liquid is given an opportunity to evaporate, it enters the atmosphere more rapidly and reaches a higher concentration than one of lower volatility. The *flashpoint* of the liquid, the most frequently encountered measure of hazard due to flammability, is the minimum temperature that raises the vapor pressure to the lower flammability limit, thus providing sufficient vapor for combustion on ignition. A flammable liquid therefore should never be stored in an unventilated room or cabinet in which the temperature might exceed the flashpoint. A combination of a leaking container and a spark or other means of ignition could result in a fire or explosion.

Table 9.1. Flammability Data for Common Solvents.

Solvent	Flash Point °C	Flash Point °F	Boiling Point °C	Boiling Point °F	Autoignition Temperature °C	Autoignition Temperature °F
Acetone	−16.7	2	57	134	604	1118
n-Amyl acetate	—	77	149	300	379	714
n-Amyl alcohol	57	134	138	280	327	621
Amyl chloride	3	38	106	223	260	500
Anthracene	121	250	340	644	540	1004
Benzene	−11	12	80	176	580	1076
n-Butyl acetate	39	102	127	260	421	790
n-Butyl alcohol	47	116	117	243	367	693
Butyl cellosolve	61	141	171	340	244	472
Isobutyl alcohol	22	72	107	225	441	825
Carbon disulfide	−22	−8	40	114	125	257
Cellosolve	40	104	135	275	238	460
Cellosolve acetate	51	124	156	313	379	715
Chloro-benzene	32	90	132	270	640	1184
Cyclohexane	3	37	80	176	245	473
Dibutyl-phthalate	157	315	366	690	403	757
1,2-Dichloro-ethylene	17	63	84	183	413	775
Diethyl ether	−41	−40	35	95	186	366
Ethyl acetate	−4	25	77	171	486	907
Ethyl alcohol	14	57	78	173	426	799
Ethyl chloride	−50	−58	12	54	966	—
Ethylene glycol	111	232	197	387	413	775
Ethyl formate	−19	−2	54	130	455	851
Furfural	56	133	161	322	316	600
n-Heptane	—	25	98	208	233	452
n-Hexane	—	7	69	156	247	477
Kerosene	55–73	100–165	151–301	304–574	210	410
Methyl acetate	−13	9	60	140	502	935
Methyl alcohol	0	32	64	147	475	887
Methyl n-butyl ketone	23	73	128	262	533	991
Methyl cellosolve	47	107	124	255	288	551
Methyl cellosolve acetate	56	132	143	289	—	—

Table 9.1. Flammability Data for Common Solvents. (Cont'd)

Solvent	Flash Point °C	Flash Point °F	Boiling Point °C	Boiling Point °F	Autoignition Temperature °C	Autoignition Temperature °F
Methyl cyclohexanone	55	130	163	325	595	1102
Methyl ethyl ketone	−7	19	80	176	516	960
Naphtha V.M.P.	−16.1–6	20–25	100–161	212–320	232	450
Naphthalene	79	174	218	424	527	979
Octane	13	56	125	257	220	428
Paraldehyde	27	81	—	—	283	541
n-Propyl acetate	14	57	102	215	842	1005
Isopropyl acetate	8	46	90	194	460	860
n-Propyl alcohol	22	72	97	207	433	812
Isopropyl alcohol	12	53	83	181	456	852
Toluene	4	40	111	232	552	1026
o-Xylene	30	85	144	291	493	920

Reprinted with permission from: *Fire Protection Handbook,* 17th edition, ©1991 National Fire Protection Association, Quincy, MA 02269.

The *ignition temperature* (autoignition temperature) is the temperature to which the vapor/air mixture must be raised to initiate combustion without a spark or other artificial trigger. Values for these constants for some common solvents are given in Table 9.1.

DUST FIRES AND EXPLOSIONS

When solid particles are suspended in air, and when these particles are capable of reacting with oxygen, there is a possibility of a fire or explosion. Such combustions have been very threatening in coal mines, and the ignition of coal dust has been more carefully studied than many other dust hazards. Fine plant fiber in the air of agricultural storage bins is another common dust fire or explosion hazard.

There are ways in which the dust problem is dealt with in a manner very similar to the approach used with vapors and gases, and other ways in which the approach is quite different. It is possible to define lower and upper flammability limits for dust particles, just as was done with vapors and gases. However, there is more uncertainty involved in making predictions about the degree of dust hazard. In a gas or vapor, the combustible chemicals are present as individual molecules with uniform access to oxygen. This lends gas and vapor hazards a degree of uniformity and predictability. In the case of dusts, the particle size and shape are important variables. Smaller particles have a higher percentage of the fuel molecules at the particle surface and thus exposed to oxygen. Similarly flattened particles present a greater surface area than do spherical particles of the same mass. Where oxygen can contact the fuel molecules more readily, the dust will react more rapidly, increasing the tendency for an explosion to occur.

In general, dusts burn more slowly than vapors or gases because most of the fuel molecules are buried at the start of combustion and only become available as the reaction proceeds. On the other hand, the total amount of fuel present in dusts can be great, so that the total production of heat and gases can be larger than would occur in a gas or vapor burn. This means also that it does not always require a high concentration of dust in the air to create a hazardous situation.

The total fuel available also may increase as a fire or explosion begins. A process producing dust may well result in layers of particles lying on the ground or floor, and the expanding gases at the start of the combustion may add these to the air. For example, in a coal mine there may be only a small amount of dust in the air, but a good supply on the floor of the mine shaft. Stirring this up into the air may produce the conditions for a powerful explosion. In fact, the initial burn may be something other than coal—methane perhaps—but the heat and expanding gases may then generate a coal dust fire.

Finally, the violence of the burn depends on the nature of the reaction occurring. Metal powders oxidize in air, producing metal oxides and actually reducing atmospheric volume by removing oxygen from the air. Expanding gases here are the result of the heat released as the reaction occurs. A carbonaceous dust such as coal or plant fiber also uses up oxygen from the air, but it releases volumes of gaseous combustion products such as CO, CO_2, and H_2O. These add to the gas pressures produced, increasing the destructiveness of the combustion.

EXPLOSIVES

Some chemicals contain both fuel and oxidant, and rapidly produce volumes of gas without need of oxygen. Such *explosives* are a special case, and conditions of proper handling are specific for each chemical. We will not attempt a general discussion of this subject, but strongly recommend that any use or handling be done only after a person has become well informed about the peculiarities of the particular compound.

COMBUSTION PRODUCTS

Hazard from a fire goes beyond the heat and physical destruction. Fire is a chemical reaction, and the reaction products are also a concern. Often the hot gases of the fire are swept upward and away and new air rushes in to replace them. *If the fire is in an enclosed space such that there is no departure of gases from the combustion and no intake of new air, breathing that air may be hazardous.* Combustion products depend on the nature of the fuel, and can present a variety of health problems.

Carbon Dioxide

Most fuels are carbon-based, and complete combustion generates CO_2. Although CO_2 is not highly toxic, as an acid anhydride it shifts blood pH to lower values, which results in a stimulation of deeper and more rapid breathing. At very high levels this effect attains serious proportions. Recognize also that the CO_2 is produced by using up the oxidizer O_2. A fire can seriously deplete air of oxygen and create the possibility of asphyxiation.

Carbon Monoxide

Incomplete combustion of carbon fuel results in CO production. The proportion of CO produced rises sharply as the supply of oxygen to the fire is restricted. Since CO can be oxidized further, it is a fuel. Opening a room in which a fire is smoldering so that new oxygen is introduced can result in explosive combustion of accumulated CO—a serious hazard to firefighters.

Carbon monoxide is also highly toxic (Chapter 6). Levels higher than 0.1 volume % can deliver a fatal dose in a few hours, and above 1 volume % in a few minutes. This is the most common highly toxic combustion product associated with a fire.

Unburned Particulate

Smoke from a fire contains particles of unburned carbon fuel. In the lungs irritation can be extreme enough to cause fluids to form and collect, blocking respiration. This effect is also produced by the other respiratory tract irritants discussed later in this section. Smoke inhalation is the cause of a very high percentage of the fatalities in fires. Smoke also obscures visibility, increasing dangers to those in the vicinity of the fire.

Sulfur Dioxide and Hydrogen Sulfide

Fuels that contain sulfur—most significantly rubber from tires—produce SO_2 on complete combustion. This is an acid anhydride, like CO_2, but it is the anhydride of a stronger acid than is CO_2. As such, it is an upper respiratory tract irritant, so that exposed individuals are very aware of its presence and do everything possible to avoid inhaling it.

Hydrogen sulfide may be present during incomplete sulfur combustion. A strong rotten egg odor is its warning property, but at high concentrations it rapidly overcomes the sense of smell. In the presence of additional oxygen, H_2S burns to form SO_2.

Products of Chlorine Combustion

Although fully chlorinated hydrocarbons are difficult to oxidize, partially chlorinated structures are much more combustible. A variety of compounds containing chlorine result from such combustion.

Hydrogen chloride is a potent upper respiratory tract irritant, so people try hard not to inhale it. As a strong acid, it is capable of serious damage to the lungs if allowed to accumulate.

Phosgene is a much more serious decomposition product of chlorinated hydrocarbons. It is not as irritating, so it is more readily inhaled. Continued exposure to as little as 25 ppm can be lethal.

Hydrogen Cyanide

Combustion of a variety of nitrogenous substances, importantly including silk and wool, several common plastics, and agricultural chemicals, can result in production of HCN. This is a very toxic substance, and even doses in the range of 100 ppm can cause an accumulated blockage of oxygen utilization in the body that is harmful to lethal in under one hour.

Metals

Fires that encompass electrical equipment may volatilize some of the low-melting metals used in solders. Some of these, particularly lead and antimony, represent serious health threats.

FIRE CONTROL

The methods used to extinguish a fire, once started, fall into two classes: (1) deprive the fire of oxygen, and (2) cool the burning material to below the ignition temperature. Most fire extinguishers operate by excluding oxygen from the site of combustion. This may be done with a gas such as CO_2, commonly used in portable fire extinguishers. Such extinguishers spray a cloud of CO_2 particles, which convert to gas and blanket the area, excluding oxygen. Sizable amounts of heat are required to convert the "dry ice" to gas, so the site of combustion is also cooled. Other systems to exclude oxygen use nitrogen gas or Freons.

The chief example of cooling below the ignition temperature is the fireman's traditional method: spraying the flames with a stream of water. Water is not only commonly available, but its high specific heat makes it an effective cooling agent. Steam produced as the water hits a hot surface also serves to displace oxygen. Since it is an electrical conductor, water should not be used on an electrical fire, and there is a danger of scattering and spreading the fire if water were sprayed on burning liquids.

Foams are sprayed on fires to exclude air. The foam is generated in the extinguishers using water and a foaming agent.

Fires have been classified by the National Fire Prevention Association (NFPA) into four groups:

Class A. Fire in ordinary combustible materials such as wood, cloth, paper, and plastic.
Class B. Fire in flammable or combustible liquids
Class C. Fire in electrical equipment carrying power.
Class D. Fire in combustible metals.

Extinguishers are marked with letters according to the class or classes of fire for which they are effective. These letters may be preceded by a number that gives a clue as to the magnitude of fire they are designed to handle: the larger the number, the bigger the fire. *Extinguishers should be inspected and maintained* according to a regular schedule to assure their ability to function.

A porous barrier such as a screen, a *flame arrestor,* is often placed across air vents or access ports to containers of inflammables. If ignition occurs outside the container, the screen cools the gases, preventing combustion from starting inside the container.

Reducing the hazard of working with combustible substances requires recognition and control of sources of ignition. Electrical sparks, flames, hot wires, and static electricity are common. Glowing fragments from friction on a metallic object could be produced during grinding, bearing failure, broken parts dragging from a moving object, and a variety of other sources.

Storing solvents in cabinets that restrict the broadcast of vapors or the spread of fire, should one start, is useful (Figure 9.1). Such cabinets may be vented as a further precaution. Structures or buildings that store combustible liquids or gases may need to be placed far enough away from other operations that any fire or explosion occurring does not threaten other personnel or operations.

KEY POINTS

1. A fire requires fuel and an oxidant in correct proportions, and a temperature high enough to initiate the reaction. Correct fuel mixtures range between lower and upper flammability limits.

Figure 9.1. Special cabinets are available for solvent storage. Vapors are confined and isolated from an ignition source. Should a fire start, the cabinet restricts the air supply, limiting combustion. Such cabinets may also be vented through plant exhaust ventilation. (Photo courtesy of Lab Safety Supply, Janesville, WI.)

2. Explosion is the rapid expansion of gases, generally resulting from a chemical reaction.

3. A liquid that enters the vapor state in high concentration under specified conditions is relatively volatile. A volatile liquid has a high vapor pressure under existing conditions. A liquid with a relatively high vapor pressure has a low boiling point.

4. The flashpoint of a liquid is the temperature at which the vapor pressure produces a concentration of vapor equalling the lower flammability limit.

5. The ignition temperature is sufficiently hot to ignite a fuel–oxidant mixture.

6. Dust can serve as fuel for a fire or an explosion.

7. Explosives contain both fuel and an oxidant, and require no further reactants to produce volumes of gas.

8. Carbon dioxide, carbon monoxide, smoke, sulfur dioxide, hydrogen sulfide, hydrogen cyanide, and chlorinated compounds are potentially toxic combustion products. Metals can be volatilized by combustion or explosion.

9. Fires are extinguished by cooling fuel below the ignition temperature and/or by excluding air.

10. Fires are prevented by eliminating sources for ignition and by correct storage of flammable liquids.

PROBLEMS

1. Regulations regarding fire and explosion prevention are found in CFR 1910.106 Flammable and combustible liquids, 1910.107 Spray finishing, 1910.108 Dip tanks, 1910.109 Explosives, and 1910.110 Storing LPG. Look at this section of the CFR and get a sense of the extent and amount of detail included. Notice that it takes up over 100 pages of the volume.

   A. In CFR 1910.106 (a)(5) we find the definition for purposes of these regulations of *boiling point of a liquid.* Notice that it is defined at one atmosphere. Since boiling point is a physical constant, doesn't this seem to be an unnecessary restriction to place on the definition?

   B. In CFR 1910.106 (a)(14) *flashpoint* is defined. Is the definition provided consistent with the definition in the text? Read how the flashpoint is to be determined. Why is so much detail provided? What does ASTM stand for?

   C. In CFR 1910.106 (a)(18) and (19), how does *flammable liquid* differ from *combustible liquid?* Which presents the higher fire hazard? What is a Class 1B liquid? What is the value of making such distinctions?

   D. You are inspecting a "tank farm"—a storage yard for aboveground tanks of flammable liquids. Of what material should the tanks be constructed? If the tanks are 21 feet in diameter and are not storing LPG, how far apart should they be?

2. A tank contains vapor or gas at an overall concentration below the lower flammable limit. However, the possibility exists that the vapor might accumulate at the top or at the bottom of the tank due to layering in a concentration high enough to ignite. For each of the following compounds, would the hazard exist at the top, at the bottom, or not at all?

   A. $CH_3(CH_2)_6CH_3$

   B. CH_4

   C. CCl_4

   D. Ne

   E. $CH_3-CH_2-CH_2-SH$

   F. CO_2

   G. CH_3-CH_3

3. Considering the possibility of a dust fire or explosion in an agricultural storage bin:

   A. Would an air suspension of small or of large plant fibers be more likely to explode, if the total mass of fiber per m^3 were the same?

   B. How would a fire or explosion differ if the number of particles per m^3 were the same, but in one case the average particle mass were greater.

   C. Would it lower the risk if air were constantly circulating, as opposed to having a sealed bin?

   D. We can determine a lower flammability limit for dust in the same fashion as for a vapor or gas. How would it affect the limit if in two cases the concentration, composition, and size of the particles were the same, but in one they were flattened and in the other they were spherical?

4. You are engaged in the design of a new explosive. What would be the requirements for the chemical?

BIBLIOGRAPHY

P. A. Carson and C. J. Mumford, *The Safe Handling of Chemicals in Industry*, Long Scientific and Technical, Essex, England, 1988. Chapter 4.

J. Grumer, "Fire and explosive hazards of combustible gases, vapors, and dusts," in G. D. Clayton and F. Clayton, *Patty's Industrial Hygiene and Toxicology,* Volume 1 Part B, 4th edition, Wiley and Sons, Inc., New York, 1991.

W. Hammer, *Occupational Safety Management and Engineering,* 3rd ed., Prentice Hall, Engelwood Cliffs, NJ, 1985. Chapters 20 and 21.

H. R. Kavianian and C. A. Wentz, Jr., *Occupational and Environmental Safety Engineering and Management,* Van Nostrand Reinhold, New York, 1990. Chapter 5.

N. I. Sax, *Dangerous Properties of Industrial Materials,* Van Nostrand Reinhold, New York, (latest edition). This is a comprehensive source of information about hazards, including fire and explosion hazards, of chemicals.

Section II

PHYSICAL HAZARDS IN THE WORKPLACE

In this section we deal with threats to worker health and safety that arise from the worker's physical environment. These are a varied lot of problems, lacking the common underlying theme of "problems with chemicals" found in the first section. Many industries employ machines, particularly heavy machines, whose operation involves impact, vibration, and other noise-producing functions. Hearing damage from noise exposure usually comes about gradually and without pain, and may be discounted by the worker as not being a real threat. Radiation damage similarly occurs without the worker's awareness, and may be ignored by the worker if there are no posted warnings, monitoring, shielding, and other precautions. Two different topics are combined in Chapter 12. First there is the prevention of accidents such as tripping, falling, and physical injury from machinery. Secondly, the more recently recognized problems arising from working in improper positions that strain parts of the body, or from performing repeated tasks, are discussed. Finally, heat stress is discussed. Some tasks make exposure to high temperatures unavoidable. It is necessary then to understand the limits to safe exposure to such environments and the signs that overexposure has occurred.

CHAPTER 10

OCCUPATIONAL HEARING LOSS

We all recognize the handicap caused by loss of vision. Most of us have simulated blindness by trying to perform some simple tasks with our eyes shut, and everyone has experienced finding their way through a darkened room. Few ever experience loss of hearing, and are unaware how effectively this isolates us from others. To appreciate this, search around until you find a television program in which the story line is not visually graphic (sex or violent chase scenes—good luck!), but one where a story is being told. First attempt to follow the story line with your eyes shut, getting clues only from the sound-track. Then try watching the screen with the sound off completely. Loss of hearing is a serious handicap.

In a Congressional report,[1] the number of Americans suffering permanent hearing disability is estimated as being between 8.7 and 11.1 million. Approximately a million workers per year experience occupational hearing loss as a result of work conditions, resulting in worker compensation claims in excess of $800 million.[2] Workers can experience a loss of hearing as a result of work conditions. Two causes are distinguished: loss due to a single event, such as an explosion or a blow to the head, and loss due to continuous exposure to relatively high sound levels over a long time period. The "single event damage" results from circumstances that are avoidable, and such worker exposure is accidental. Continuous exposure to sound, on the other hand, is an unavoidable aspect of work in many industries. Any time machinery with moving parts and parts in collision is run, sound is produced. This circumstance is the focus of this chapter as we deal with the ear, the nature of sound and of hearing, how sound is measured, the levels of sound that cause hearing loss, and the means of protecting hearing. OSHA noise exposure regulations arise from this understanding.

---

[1]"Report to the President and Congress on Noise," Administrator to the Environmental Protection Agency, 92nd Congress Document No. 92-63, February, 1982.

[2]Council for Accreditation in Occupational Hearing Conservation.

THE EAR

The ear clearly divides into three anatomical parts (Figure 10.1):

A. *The outer ear.* The ear canal runs about one and one-half inches from the outside to the eardrum. The eardrum is a fibrous membrane sealing the passage at the end of the ear canal. This membrane vibrates in response to sound waves.

B. *The middle ear.* Behind the eardrum is an air-filled chamber. Atmospheric pressure changes cause pressure changes across the eardrum. These are equalized by allowing air to enter or leave the chamber through the eustachian tube, which connects the chamber with the throat. When you are in an aircraft that is descending to land, you refer to the burst of air entering the middle ear chamber to counter the rising outside pressure by saying your ears "popped." Blockage of the eustachian tube, as during a cold, allows a pressure differential to build that reduces the ability of the eardrum to vibrate, reducing hearing acuity. Three tiny bones—the *hammer,* the *anvil,* and the *stirrup*—bridge the chamber and transfer vibration of the eardrum to the end of the fluid-filled compartment of the inner ear. Their lever action amplifies vibrations from the eardrum.

C. *The inner ear.* Two sensory functions are served in the inner ear: equilibrium and hearing. Equilibrium is sensed in the *semicircular canals,* fluid-filled loops containing floating membranous structures. Hearing is served by the snail shell–shaped *cochlea.* The stirrup of the middle ear attaches to a membrane at the opening of the cochlea. The stirrup bone presses fluid into the cochlea, and this fluid then circulates to the center of the spiral and back out by a parallel passage. At the end, another membrane in the inner ear chamber bulges outward to relieve the pressure. Waves of pressure in the fluid cause the walls of the fluid chambers to bulge outward, pressing on a parallel chamber containing hairlike nerve endings. These respond to pressure to produce nerve impulses, which are transferred to the brain. Nerve endings at the wider portion of the cochlea near the membranes separating the inner ear from the middle ear are sensitive to higher frequencies, while those sensitive to lower frequencies are at the center of the cochlea.

THE NATURE OF SOUND

Noise is Unwanted Sound

Sound is the propagation of a pattern of compressions and rarefactions, normally through air, although other matter can serve as the conducting medium. The compressions and rarefactions move at a characteristic rate, traveling

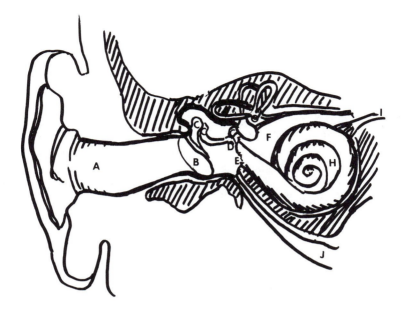

Figure 10.1. Pictured here is a cross section of the ear. Labeled are: A. Outer ear; B. Eardrum; C. Bones of the middle ear; D. Oval window; E. Round window; F. Inner ear; G. Semicircular canals; H. Cochlea; I. Auditory nerve; J. Eustachian tube.

from the source to the ear, which responds by converting the pattern into a signal to the brain. The brain recognizes two properties of this pattern of moving compressions: the intensity or loudness, and the tones, from high-pitched to low.

Sound is visualized as a wave pattern, with peaks at the air compression and valleys at the region of lower pressure that follows the compression. The louder the sound (more intense compressions), the higher are the peaks and the lower the corresponding valleys (the rarefactions). This is termed the *amplitude* of the sound. Some amplitudes are smaller than the ability of the ear to respond, and the sound is not heard. Logically, the greater the amplitude of the sound, the greater its ability to cause damage to hearing.

The brain also distinguishes sounds by the *frequency* with which these compressions and rarefactions stimulate the ear. The x-axis in the wave model is time. Higher-frequency sound has a shorter wavelength, hence more waves are propagated per unit of time. When more compressions reach the ear per second, the brain interprets the sound as being higher in pitch (higher-frequency sound). Musical instruments produce tones of different frequency, and these are blended together to produce the usually pleasing sound of the instrument or instruments. Human speech, the most important sound to us, is a blend of many tones.

FREQUENCY

The measure of frequency is the number of compressions passing by a point—for example, your ear—per second. This number is expressed as *hertz (Hz)*, and has cycles/s (s^{-1}) as units. Everyone is a little different, but *typical young human hearing responds to frequencies from 20 to 20,000 Hz*. Almost everyone has heard—or rather has not heard—a silent dog whistle. This produces a tone above the range of the human ear but not above that of a dog.

Conceptually, sound of a single frequency, called a *pure tone,* is the easiest to understand. It is easy to generate pure tones electronically or with a tuning fork, and we approximate such tones by striking piano keys. Striking a C at the middle of the piano keyboard, then a D or B, produces tones of different frequency, and they have a characteristically different sound.

However, if we move to the right on the keyboard to strike the next C, we hear a tone that is clearly higher in pitch but otherwise sounds the same as the lower C. The relationship between these two different tones, which are said to be an *octave* apart, reflects that the higher C *has exactly twice the frequency* of the lower C. We hear tones at octave intervals as similar tones.

For purposes of analysis, the range of human hearing is divided into octave bands, eight ranges in each of which the highest frequency is twice the lowest frequency. These bands are each further subdivided into three parts, each termed a one-third octave band. Octave bands and one-third octave bands are identified by their center frequency, both are numbered in a standard fashion, and all of this is summarized in Table 10.1. These bands are referred to in the analysis of noise and of hearing damage.

Frequency and wavelength are interdependent measures of the sound wave, as expressed in the equation:

$$\lambda = c/f$$

where

λ = wavelength of sound wave
c = rate of sound propagation
f = frequency of sound (Hz)

Table 10.1. Octave and One-Third Octave Bands.

Octave Band Number	Frequency Range (Hz)	One-third Octave Band Number	Frequency Range (Hz)	Center Frequency (Hz)
sub-octave		14	22-28	25
	22-45	15	28-35	31.5
		16	35-45	40
		17	45-56	50
1	45-89	18	56-71	63
		19	71-89	80
		20	89-112	100
2	89-178	21	112-141	125
		22	141-178	160
		23	178-224	200
3	178-355	24	224-282	250
		25	282-355	315
		26	355-447	400
4	355-708	27	447-563	500
		28	563-708	630
		29	708-892	800
5	708-1413	30	892-1123	1000
		31	1123-1413	1250
		32	1413-1779	1600
6	1413-2819	33	1779-2239	2000
		34	2239-2819	2500
		35	2819-3548	3150
7	2819-5626	36	3548-4467	4000
		37	4467-5626	5000
		38	5626-7080	6300
8		39	7080-8913	8000
		40	8913-11234	10000

The rate of sound propagation varies with the medium and the temperature, but we can approximate it as 1129 ft/s (344 m/s) in air.[3] Under those conditions a sound wave of 13 inches (0.344 meters) wavelength would display a frequency of 1000 Hz. Wavelength is most important as a measure of sound when our objective is to block the sound.

Most sound is not composed of pure tones, but rather is a complex mixture of frequencies. Human speech is complex sound, largely within the frequency range of 500 to 3000 Hz. To interpret speech, we need to hear all the frequencies in use. For example, some consonant sounds have important high-frequency components. Loss of hearing in the high-frequency range makes it difficult to distinguish words that differ only in the terminal consonant (fit, fix, fist).

AMPLITUDE

The amplitude, or loudness, is the sound characteristic of greatest concern in the industrial environment. Therefore, an early requirement is a method to measure sound levels and units with which to express the results. Sound is a form of power, and the *sound power* of a source is expressed in watts. Typical sound power levels of a variety of sources are presented in Table 10.2. However, what we actually hear or measure is fluctuation in air pressure caused by the source, which we call *sound pressure,* and which has pressure units.

One would therefore expect the loudness of sound to be expressed straightforwardly in pressure units. Instead, units are devised that measure *relative* sound pressure, termed *sound pressure level.* "Relative" indicates that there is a reference sound pressure value to which the sound pressure of interest is compared. In principle, the reference value is the quietest audible sound. Since this value is not the same for everyone, it is estimated for convenience to be 2×10^{-5} N/m².[4] A log scale is used to cover conveniently the broad range of values encountered, and the values are made larger by multiplying them by 10. The units of sound pressure level are based on the log of the ratio of the mean square pressure of the source to that of the quietest audible sound, and are

---

[3]The relationship of the speed of sound in air to temperature is as follows:

$C = 49.03(460 + TF)^{1/2}$ ft/s   where T is Fahrenheit temperature

$C = 20.05(T_K)^{1/2}$ m/s   where T_K is absolute temperature

The speed of sound in other media is quite different from the speed in air. In water it is about 4 times as fast and in steel it is about 15 times as fast.

[4]...or 0.0002 dynes/cm², or 0.0002 microbars. The quietest audible sound varies among individuals according to their hearing sensitivity, and varies for everyone by wavelength.

Table 10.2. Sound Power Levels of Commonplace Sources.

Source	Sound Power[a] (Watts)	L_W (dB)
Rocket	100,000,000	200
Air Hammer	1	120
Riveting Machine	0.1	110
Jet at 1000 feet	0.01	100
Shouted Conversation	0.001	90
Average Factory	0.00007	75
Conversation	0.00001	70
Typical Office Sound Level	0.0000001	50
Whisper	0.000000001	30
Quietest Audible Sound	0.000000000001	0

[a]Sound power level for a source of W watts is calculated as follows:

$L_W = 10 \log(W/10^{-12})dB = (10 \log W + 120)dB$

expressed in *decibels* (dB). Sound pressure level (L_p) for a sound of P sound pressure is calculated as follows:

$$L_p = 10 \log (P/P_{ref})^2 \text{ dB}$$

$$= 20 \log (P/P_{ref}) \text{ dB}$$

The acoustic power generated by a sound source is termed the sound power level, or L_W. Once again, the reference level is the quietest audible sound, but the units now are watts, and this level is assumed to be 10^{-12} watts.

$$L_W = 10 \log (W/W_{ref}) \text{ dB}$$

$$= 10 \log (W/10^{-12}) \text{ dB}$$

$$= 10 \log (W) + 120 \text{ dB}$$

SOUND LEVEL METERS

Sound level meters are portable, battery-powered devices used to measure sound pressure levels in the workplace (Figure 10.2). They include a microphone to convert sound to an electrical current, and the current is sent to an amplifier. The amplified electrical signal causes a needle to deflect on a calibrated scale, indicating the sound pressure level. Rather than having the meter cover a typical range of sound levels from 40 to 140 dB, which would make accurate reading difficult, an attenuator is added that allows the meter to be adjusted to read within an appropriate 10 dB range. Thus if the attenuator is set for the 70 to 80 dB range and the needle stops at 6.0, the sound pressure level is 76. Devices are available that mount onto the microphone and produce a standard signal used to calibrate the meter before use.

The human ear is not uniformly sensitive to all frequencies, being most sensitive to sounds of about 4000 Hz, dropping acuity gradually as frequency becomes lower, and sharply as it becomes higher. Most sound level meters include weighting networks, labeled as A, B, and C scales. The A scale varies the response of the meter with frequency, so as to match the response of the human ear. The C scale is flat, so that the sound level meter responds equally well to frequencies across the entire range. The B scale is intermediate. Usually either the A scale is used to estimate potential damage to the ear or the C scale is used when the signal is going to be transferred to another instrument for detailed analysis. An octave band analyzer attached to a sound level meter allows separate analysis of the sound pressure level within each octave band. Such analysis is useful in the design of a control system for a particular sound source.

MEASURING EXPOSURE IN THE WORKPLACE

The sound level meter is used to assess the exposure of a worker to noise at his or her workstation. Sound levels drop with distance from the source, although reflection or absorption of sound by nearby surfaces complicates the sound level/distance relationship. Readings of sound level should therefore be taken as near as possible to the location of the worker's ears. Sound levels vary, so readings should be taken at various times of the day or even on different days. If these readings are taken too close to the sound source, unrealistically high exposure would be indicated, and similarly greater distance infers improperly low exposure levels.

HEARING DAMAGE

Brief exposure to high sound levels can cause a ringing sensation and *temporary threshold shift* (temporary drop in hearing acuity). After a rest period the

Figure 10.2. This portable sound level meter is capable of reading a broad range of sound levels. The cylinder at the top of the instrument is the microphone, and the meter reads the level of sound within the range set by the round control. To the right of the meter is the calibration device, which mounts on the microphone and generates sounds of known level. (Photo courtesy of Lab Safety Supply, Inc., Janesville, WI.)

hearing returns to normal. If exposure to excessive sound is repeated frequently, there is a *permanent threshold shift* (loss in hearing becomes permanent). People vary greatly in their susceptibility to hearing damage by continuous sound exposure, but at higher levels everyone loses some hearing ability. Workers exposed daily to high noise levels display the greatest loss during the first years. If we set the standard for serious hearing loss at a 25-dB or more drop in acuity, over a work lifetime of 40 years 18% exposed to 90 dB and 70% exposed to 115 dB incur such loss. Approximately 75% of the workforce are exposed to sound levels above 85 dB.

An *audiometer* is a device that generates a range of pure tones at known sound pressure levels. Hearing is tested by placing the subject in an environment that is free of interfering sound stimuli, and the hearing threshold is determined for a series of pure tones. The resulting information is displayed as a graph with frequencies along the x-axis and the decibel level of the threshold on the y-axis. Since the decibel scale at each frequency is referenced to the quietest sound *of that frequency* detectable by a person with normal hearing, a person with normal hearing generates a plot that is linear and close to the 0 decibel level. Such a plot is termed an *audiogram*.

The American Academy of Ophthalmology and Otolaryngology has established a standard audiogram presentation, which displays frequencies in octaves across the top from 125 Hz on the left to 8000 Hz on the right. The hearing threshold level in decibels is displayed down the left side, from −10 at the top to 110 at the bottom. One octave on the x-axis has the same length as 20 dB on the y-axis. Hearing thresholds of the left ear are displayed as blue X symbols and at the right ear as red O symbols (Figure 10.3). Such a standard presentation simplifies comparing relative hearing acuity from one audiogram to the next without troublesome conversion calculations.

Many hearing conservation programs routinely measure hearing at 500, 1000, 2000, 3000, 4000, and 6000 Hz. Frequencies from 500 to 2000 Hz are included because these cover the normal range of speech. As a rule, the first sign of hearing damage due to continuous noise exposure is a dip in acuity around 4000 Hz, a "notch" in the audiogram. Including 3000 to 6000 Hz in the analysis permits detection of early damage, even though it may not interfere with the understanding of speech. As hearing loss progresses, this notch deepens and widens. When values for loss greater than 25 dB are obtained on testing, hearing is considered to be impaired.

Appearance of readings indicating loss of acuity on an audiogram does not establish industrial noise as the cause. Many programs add readings at 250 and 8000 Hz. Readings at 8000 Hz are important diagnostically. Observed deafness may be due to some cause other than industrial noise exposure if, rather than a notch at 4000 Hz and some recovery by 8000 Hz, there is an increasing loss of acuity through higher frequencies. A list of other causes of hearing loss includes:

A. Obstruction of the ear canal. This could be due to accumulated ear wax or a foreign object in the canal.
B. Infection. Infection can cause swelling and obstruction of the ear canal. Infections of the middle ear, often secondary to infections elsewhere, cause temporary or permanent hearing impairment.
C. Allergy. Allergic response to some agents may result in ringing in the ears. Continued exposure may lead to permanent damage.
D. Trauma. An extreme sound such as an explosion can do physical damage to the eardrum or the middle ear.
E. Brain damage. Anything that damages the auditory portion of the brain causes deafness. This could include stroke, hemorrhage, or meningitis.
F. Age. Hearing acuity decreases with age, with loss of sensitivity to high frequencies being particularly marked.

FREQUENCY (Hz)

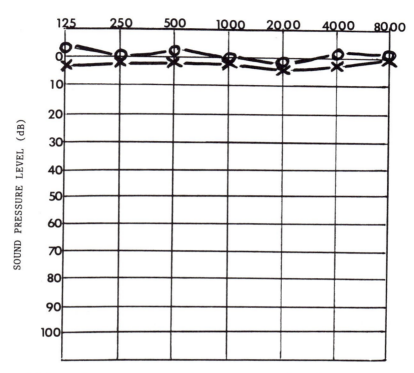

Figure 10.3a.   Audiograms present the response of each ear to a series of pure tones. Zero on the y-axis represents response in individuals with undamaged hearing. This plot presents the audiogram of such a person, with the response of the right ear indicated with circles (normally red) and the left ear with crosses (normally blue).

OSHA REGULATIONS

WORKER PROTECTION STANDARDS

In 1972 an advisory committee to the Department of Labor was appointed to prepare a standard for occupational noise exposure, which submitted recommendations in December, 1973. A standard was then published in the *Federal Register* in 1974 setting the noise exposure standards listed in Table 10.3. The standard for an 8-hour-day TWA is 90 dB. NIOSH has recommended that this

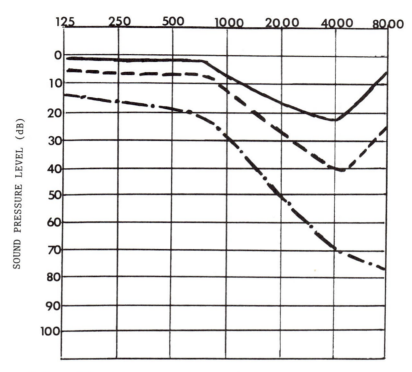

Figure 10.3b.   This plot presents the progressive loss of hearing in one ear of an individual exposed to excessive noise. The top curve is the earlier audiogram, with the lower plots displaying progressive hearing loss. The "valley" at 4000 Hz is typical.

be dropped to 85 dB, and on review the EPA also has recommended that this be lowered to 85 dB.[5]

For purposes of testing compliance, exposures to different sound levels in the course of a workday are calculated as follows. The length and sound intensity of each exposure are logged. At each level the time exposed is totaled (C_1, C_2, ...). Each total is divided by the permissible time of exposure at that level (T_1, T_2, ...) to produce a fraction (C_1/T_1, C_2/T_2, ...). As long as the sum of these fractions is less than one, the site is in compliance.

---

[5]*Federal Register,* October 24, 1974.

FREQUENCY (Hz)

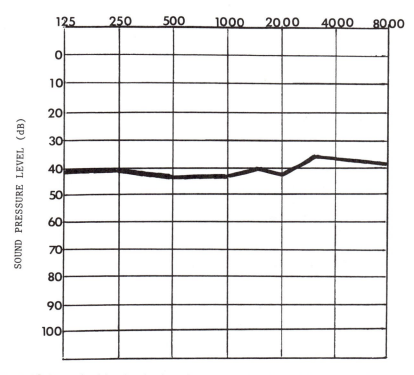

Figure 10.3c.   In this plot the hearing loss is due to middle ear problems rather than noise exposure, producing a characteristically flat curve.

Example:

A worker is exposed for 4 hours to 90 dB, 1 hour to 100 dB and 1 hour to 97 dB. The rest of the time exposure is below regulated levels. Is the site in compliance?

$$\text{Exposure} = C_1/T_1 + C_2/T_2 + C_3/T_3$$

$$= 4/8 + 1/2 + 1/3$$

$$= 1\ 1/3$$

The total exceeds 1, so the site is not in compliance.

HEARING PROTECTION

PERSONAL HEARING PROTECTION

Where it is unavoidable that employees work in areas with a high noise level, personal hearing protection should be worn. There are two types of protection: earplugs and earmuffs. The effectiveness of hearing protection can be estimated by determining the quietest sound audible with the protection off and in place for each of a range of frequencies (Figure 10.4). Earmuffs, if properly fitted to the wearer, are generally more effective attenuators, especially at higher frequencies. No hearing protection completely blocks noise. Sounds can be transmitted through the material of the muff.

Earplugs insert into the ear canal to block the entrance of sound (Figure 10.5). These are generally soft rubber, plastic, or wool material, or are compressible foam cylinders that are squashed and inserted into the ear, where they expand again to fill the ear canal. By filling the canal tightly, they block

Table 10.3. OSHA Regulations — Permissible Noise Levels.

Hours of Exposure per day	Sound level (dB)
8	90
6	92
4	95
3	97
2	100
1.5	102
1	105
0.5	110
0.25	115

Figure 10.4. These earmuffs are worn as personal hearing protection. Notice the noise reduction rating is 20 dB. The foam padding provides a tight fit around the ear, which is important if the device is to be effective. Photo courtesy of Cabot Safety Corporation, Southbridge, MA.

the direct passage of sound through the air. Some earplugs are connected by cords, which minimizes loss if they are removed momentarily for conversation. Earplugs are small, however, and sound may be transmitted through the plug itself. They are particularly useful when exposure to high sound levels is infrequent, since they can be carried conveniently in the shirt pocket, to be slipped into the ears at any time.

Earmuffs fit over the head to place a padded cover completely over the ear, and they then fit tightly against the head. Since sound, particularly high-frequency sound, enters the ears through the leaks in the seals, the better the seal against the head, the more effective the protection. Pads need to be large enough to completely cover the earlobes, and leakage problems are created by the wearing of glasses or eye protective devices. Anything worn over the eyes should be held in place by thin bands around the head, rather than by earpieces.

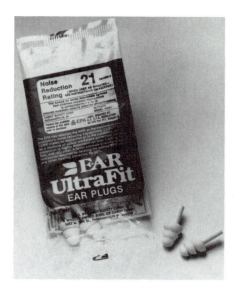 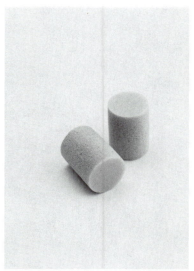

Figure 10.5.   Earplugs are a very portable form of hearing protection that can be carried in the pocket and inserted in noisy areas. In each case the plug is compressed on insertion into the ear canal, and the plug then expands to fill the canal effectively. (Photo courtesy of Cabot Safety Corporation, Southbridge, MA.)

It is important that hearing protection be fitted properly and be as comfortable as possible. Workers generally do not like wearing hearing protection, grow accustomed to high noise levels, and will remove the protection if wearing it bothers them.

ENGINEERING CONTROLS

It is more satisfactory to control the sound level in the workplace than to block the ears of the workers. Workers wearing earplugs or earmuffs are out of communication with one another, and the risk of hearing damage is always present if workers are careless or recalcitrant about wearing protection. Designing engineering controls is a complex field, and such services are best provided by specialized experts. Industrial hygienists are primarily concerned with analysis of the success of such a program, so comments are restricted to a brief discussion of methods employed. References that lead the reader to more detailed presentations are provided.

Controls Related to the Machinery

Machinery that vibrates or otherwise generates sound waves is a noise source. Some benefit may result from improved machine maintenance. Worn, loose-fitting, or unbalanced parts vibrate to produce noise. Improved lubrication eliminates some noise directly, and postpones wear that later increases machine noise levels. Cutting oils lower sound levels of cutting operations and extend tool life.

Some machine designs are noisier than others that perform the same task. For example, hydraulic presses are quieter than mechanical presses. Analysis of the contribution of each machine to the total noise problem focuses attention on the noisiest machines, allowing considerable improvement in sound levels at the time of machine replacement. Ventilating systems should not be forgotten in such an analysis. Some fan blade designs are inherently quieter, and larger but lower-speed fans emit less high-frequency sound. Similarly, some production methods are inherently noisier, and a change in process may result in significant noise reduction. Replacing riveting with welding is an example.

Alteration of the design of existing machines can bring about improvement. A lightweight machine part may vibrate at a high, harmful frequency. Adding weight to the part lowers the frequency of vibration. Vibration is more intense in a flexible part, so stiffening the structure can be helpful.

Often alteration or replacement of machines or processes involves trade-offs of operations cost or ease of maintenance for lower noise levels. In such cases all the alternatives for noise control must be weighed and compared.

Room Design

If we suspend sound sources in the air, they radiate sound energy in all directions. Nearby persons are subjected only to sound propagated in their direction. Furthermore, the farther they are from the source, the weaker the sound pressure they receive. In principle, sound follows the inverse square law, meaning that the fraction of sound reaching an individual is inversely proportional to the square of the distance. A person twice as far from the source should only be exposed to one quarter the sound level.

However, once we place this source in a room, reflection of the sound from the walls and ceiling supplements the direct sound. Walls and ceilings vary in their ability to reflect rather than absorb sound. It is very much like reading a book in a room with a single light bulb. The farther away the bulb is, the less light reaches the book (once again following the inverse square law). However, a larger fraction of the light reaches the book if the walls are painted a highly reflective white. Just as painting the room walls matte black reduces the light reaching the book, replacing surfaces that are highly reflective to sound with acoustic panels reduces reflected sound (Figure 10.6).

Sound is transmitted through solids that the machine contacts. Placing machines on resilient floors, inserting flexible sections into piping or ducts, and changing fastening devices to include resistant pads may isolate and dissipate sound at its source.

Sound Barriers

Enclosures are sometimes the best choice for effective management of a noise problem. One can enclose the machine, enclose the worker, or place a barrier between them. The walls of any of these barriers transmit some amount of sound. Sound strikes the barrier, causing it to vibrate. This vibration, in turn, causes the air on the other side to vibrate. Factors in the effectiveness of sound transmission by the barrier include the mass of the barrier, its area, and its construction materials. Enclosures or walls have doors, and the tightness of the seal around those doors is critical to blocking sound. Acoustic panelling inside an enclosure absorbs sound, reducing the reinforcement of sound by reflection.

Enclosing a machine usually provides a 20- to 30-dB reduction in noise (Figure 10.7). Such enclosures must take account of machine operation and employee safety as well as acoustic effectiveness. Partial enclosures that eliminate the direct path of sound from the source to the worker's ear and are covered with acoustic material can be effective, providing sound exiting from the unenclosed sides of the machine cannot simply reflect from walls or ceiling to the worker. Where a segment of the machine is the prime source of noise, just that structure might be enclosed, providing improved sound levels at less cost.

Enclosing the worker allows other improvements, such as air conditioning to reduce heat stress and air filtration to reduce airborne contaminants (Figure 10.8). It is also often cheaper to build a small worker enclosure than large machinery enclosures that also reduce dissipation of heat from the machine or access for maintenance.

KEY POINTS

1. The ear has three anatomical parts: (1) the ear canal, connecting the outside to the eardrum, (2) the middle ear, containing the bones that transmit eardrum vibrations to the inner ear, and (3) the inner ear, which includes the cochlea, which transforms mechanical vibrations to nerve impulses. The inner ear also includes the semicircular canals, which provide our sense of equilibrium.

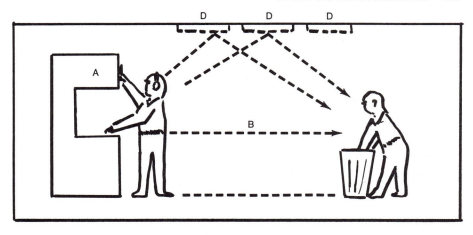

Figure 10.6. Rooms can be altered to lower sound exposure of the worker. Here a machine (A) produces a high noise level. The sound is propagated directly (B) and by reflection (C) to other workers in the room. Adding sound-absorbent panels (D) to the walls or ceiling reduces the exposure of other workers.

2. Sound is a series of air compressions and rarefactions propagated by a vibrating source.

3. Sound is visualized to have a wave form. The frequency or wavelength of the wave determines the tone or pitch, and the amplitude measures the loudness.

4. Frequencies are measured in hertz (Hz, cycles per second, s^{-1}), and undamaged human hearing detects sound from 20 to 20,000 Hz.

5. ctaves are pure tones wherein the frequency of the higher-pitched tone is twice that of the lower-pitched tone. For analysis, the hearing range is divided into eight octave bands or 26 one-third-octave bands numbered from 14 to 40.

6. Human speech falls primarily in the range from 500 to 3000 Hz. Loss of hearing in this range hinders communication.

7. Sound power is measured in watts, and sound pressure, the actual sound level we measure, has pressure units. The quietest audible sound is approximately 10^{-12} watts sound power, or 2×10^{-5} N/m^2 sound pressure.

8. Units of sound level, decibels, are 20 times the log of the ratio of the sound pressure to 2×10^{-5} N/m^2.

Figure 10.7. Noisy operations can be enclosed to protect workers where remote operation of the machinery is possible. The effectiveness of
noise control depends on good engineering practices in the enclosure design. Here, diesel engines of from 1000 to 4000 horse-
power are tested in a Kentucky shop that reconditions locomotive
engines. Unchecked noise levels exceed 120 dBA. The enclosure was designed to control sound levels to no more than 75 dBA
in the control room. (Photo courtesy of Industrial Acoustics Company, New York, NY.)

Figure 10.8. Another approach to protecting workers in a noisy environment is the provision of a quiet room, such as this one in the Chicago Tribune pressroom. During a press run the noise level is 110 dBA, but inside the room it is less than 75 dBA. (Photo courtesy of Industrial Acoustics Company, New York, NY.)

9. Sound level meters read sound pressure directly in decibels.

10. Continuous exposure to high sound levels damages hearing, particularly at the frequency of the source.

11. Hearing is tested with an audiometer, a device that generates pure tones of known loudness.

12. OSHA limits exposure to 90 dB for an 8-hour day, higher levels being permitted for shorter durations.

13. Personal hearing protection includes earplugs and earmuffs.

14. Sound levels in the workplace are reduced by good machine maintenance, substitution of quieter machinery, and the erection of nonreflective panels and sound barriers.

PROBLEMS

1. Explain how each of these structures of the ear contributes to the hearing process:

   A. Eustachian tube

   B. Cochlea

   C. Stirrup

   D. Semicircular canals

2. Visualize the wave pattern used to represent sound.

   A. What do the peaks represent?

   B. What dimension is termed amplitude?

   C. When we talk about the pitch of a sound, this relates to what dimension on the wave pattern?

   D. Which of these dimensions relates to loudness?

   E. What is the relationship between wavelength and frequency?

F.   What is the relationship between wavelength and amplitude?

G.   What are the units of frequency?

H.   If a tone has a frequency of 1000 Hz, what tone is an octave higher? ...two octaves lower?

I.   Under normal conditions, what is the wavelength (ft, m) of a 2000-Hz tone?

3.   A.   What is the approximate range of high-quality human hearing?

B.   What is the approximate range of human speech?

C.   At what frequency does damage to hearing first become evident?

D.   Will typical early damage to hearing result in a noticeable loss in ability to interpret human speech?

E.   Is human speech at the range of greatest acuity in human hearing?

4.   A.   What is sound power and sound pressure, and what are likely units of each?

B.   What are the units of sound pressure level (L_p), and why are they not pressure units?

C.   What is the usual reference sound pressure for sound pressure level, and why was it chosen?

D.   A sound with a sound pressure of 40 N/m^2 has what sound pressure level?

E.   On a typical sound level meter, how would you read the sound pressure level in question D?

5.   An audiogram is a graph.

A.   What are the x- and y-axes?

B.   If a person had perfect hearing, what would their audiogram look like?

C.   With what tool is hearing tested, and how are points obtained for the audiogram?

D.   What is the guideline for describing hearing loss as impairment at any tested frequency, and at what frequencies is impairment most serious?

E.   Two workers employed in a moderately noisy environment are tested for hearing loss. The audiogram of F.B.T. shows a steady drop from mid toward high frequencies, while J.J.H. displays a sharp notch at 4000 Hz. What is a preliminary diagnosis?

6.   A worker is exposed to 90 dB for 2 hours, 100 dB for 1 hour, and 105 dB for 30 minutes. The rest of the time noise levels are well below 85 dB. Is the workplace in compliance?

7.   A small concrete block factory building employs 12 workers in a single room. A worker stands in front of a small mechanical press at one end of the room, feeding sheet metal to the machine and transferring stamped parts to a bin on wheels in which the parts are rolled to workstations in the rest of the room. The other workers process the stampings into finished parts. The stamping machine produces noise levels above OSHA limits, and the operator must wear hearing protection. Suggest at least two engineering controls that might bring levels down sufficiently for the rest of the workers.

BIBLIOGRAPHY

29 CFR 1910.95.

L. L. Beranek, *Noise and Vibration Control,* Institute of Noise Control, Washington, D.C., 1988.

M. Hirschorn, "Compendium of Noise Control Engineering, Part I, Sound and Vibration," July, 1987, pp 16–32; "Part II, Sound and Vibration," February, 1988, pp. 16–28.

B. F. Jaffe and D. W. Bell, "Workplace Noise and Hearing Impairment," in B. S. Levy and D. W. Wegman, *Occupational Health,* Little, Brown and Company, Boston, 1983.

National Safety Council, *Fundamentals of Industrial Hygiene,* 2nd Edition, Chicago, 1979.

J. B. Olishifski, "Occupational Hearing Loss, Noise and Hearing Conservation," in Carl Zenz, *Occupational Medicine, Principles and Practical Applications,* 2nd Edition, Year Book Medical Publishers, Inc., Chicago, 1988.

A. Thumann and R. K. Miller, *Fundamentals of Noise Control Engineering,* Prentice-Hall, Englewood Cliffs, NJ, 1986.

CHAPTER 11

RADIATION

From the standpoint of hazard to workers, we generally class radiation in two groups: nonionizing and ionizing. Nonionizing radiation includes electromagnetic radiation (light), which is no more energetic than ultraviolet light. Of this class, ultraviolet light itself poses the greatest threat, and we shall limit our discussion to it. Ionizing radiation generates ion pairs as it passes through matter. Ion pairs generated in living tissue decompose or transform themselves into more stable structures by a variety of routes, and produce damaging intermediates in this process. Five types of ionizing radiation pose a threat to workers; α, ß, and γ radiation, X-rays, and neutrons. The α, ß, and neutrons are particles, while τ-rays and X-rays are very-high-energy electromagnetic radiation.

NONIONIZING RADIATION—ULTRAVIOLET LIGHT

Light is pure energy, termed electromagnetic energy because of its character, and we most conveniently discuss various types of light by reference to wave characteristics. The energy of light relates to its wavelength: the shorter the wavelength, the higher the energy. In Figure 11.1 we see the electromagnetic spectrum, each wavelength range being identified by its commonplace name. Ultraviolet (UV) light, with wavelengths shorter than 200 nm, is of little concern here because it is rapidly absorbed by air.

Evidence of the ability of UV light to do biological damage is given by the use of UV sources to kill bacteria. In a familiar example, for years barbers have placed their tools under UV lamps when not in use. UV light has little penetrating power, so concerns for workers are limited to skin and eye damage.

DAMAGE TO EYES

Excessive UV exposure causes irritation to the surface of the eye, called *conjunctivitis*. The victim has a sense of a foreign object in the eye, experiences lacrimation (tearing), and may want to avoid exposure to light. The severity

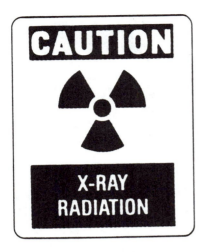 

Figure 11.1. The radiation hazard warning sign is lavender on a yellow field.

of the symptoms is in proportion to the intensity of exposure. Symptoms often clear in a few days.

DAMAGE TO SKIN

Skin exposed to sufficient UV light sunburns. Skin pigmentation is a protection, so workers with dark skin withstand higher doses without symptoms. On chronic exposure there may be damage to skin structural elements, producing a premature aging of the skin. UV light with wavelengths below 325 nm is more damaging. Studies have also established that people with less skin pigment and/or with greater exposure to UV light (generally sunlight) run a higher risk of skin cancer.

WORKERS AT RISK

Obviously, outdoor workers of all sorts are at special risk from sunlight. This is particularly true in summer, especially in warmer climates, as workers take off their shirts, wear shorts, and generally "work on their tans." Lotions are available containing compounds that absorb (block) UV light. They are compounded to provide differing degrees of protection, and should be used by outdoor workers in warm weather.

In addition, there are many people employed at indoor jobs where UV lamps are used for bactericidal purposes: food processors, nurses, barbers, hairdressers, and pharmaceutical company and tobacco company employees. At particular risk are arc welders, since the flash of light produced during welding is an intense UV source. The UV of the welders' arc also converts oxygen in the air to ozone, which can cause respiratory problems.

IONIZING RADIATION

RADIOACTIVE ATOMS

Stability of Atoms

Atoms have nuclei composed of protons and neutrons. These particles are similar in mass, but the proton has a positive charge, while the neutron is uncharged. The identity of an atom is determined by the number of protons in the nucleus. Thus, every atom with six protons is carbon, and every atom with seven is nitrogen. Although every carbon atom has six protons, the number of neutrons need not always be the same. Most carbon atoms have six neutrons, but a small proportion of atoms in all natural collections of carbon have eight neutrons. These variations of carbon would be designated ^{12}C and ^{14}C respectively, the 12 and 14 indicating the total number of nuclear particles. Such variants of an element are termed *isotopes* or *nuclides*.

Stability in an atom requires a correct ratio of neutrons to protons, approximately 1:1 in low-atomic-weight atoms, but with an increasing proportion of neutrons required for stability as the atoms become larger. Although assembly of an atom with a neutron-to-proton ratio that is far distant from the stable ratio is impossible, atoms exist that are near, but not at, a stable ratio. These atoms, termed *radioactive nuclides,* adjust the properties of the nucleus toward greater stability by emitting particles and energy. Such emissions are the ionizing radiation produced by unstable nuclei.

Degree of Instability: Half-Life

Not all unstable nuclei are identically far away from a stable arrangement. Thus, if the stable ratio is 32 neutrons to 27 protons (32/27; ^{59}Co), and a 34/27 (^{61}Co) ratio is wrong enough to be unstable, it is not surprising that a 36/27 (^{63}Co) ratio is farther from a stable ratio and should by some measure be even less stable. This statement is an oversimplification of the rules of stability, but leads to the question: What measures relative instability in radioactive nuclides?

An unstable nuclide has a pathway that enables it to become a different and eventually more stable nuclear structure through nuclear change and emissions.

However, not all atoms in a sample of unstable nuclide embark on that pathway immediately. They remain in the unstable configuration until, at a time which cannot be predicted for an individual atom, alteration occurs. If we observe a quantity of a radioactive nuclide, we see an apparently continuous stream of emissions resulting from this random decaying of individual atoms in the sample. As time passes there is a predictable reduction in levels of emissions because there are decreasing numbers of unstable atoms left to serve as sources.

More unstable nuclides on average wait shorter periods of time before decaying. Relative instability is therefore measured by the rate at which radioactive nuclei decay. If we have two radioactive nuclides of differing instability, a greater percentage of the less stable nuclide decays per unit time. When comparing equal numbers of atoms, the less stable nuclide thus has a higher level of radioactive emissions. As a consequence, a larger proportion of the less stable nuclide decays to become a different nuclide in a given period of time.

As an example, spent fuel removed from a nuclear power plant initially produces high levels of emissions. Radioactive nuclides resulting from the fission process vary greatly in stability. Less stable nuclides disintegrate at a very rapid rate—the primary source of those very high initial levels of emissions. At such high rates of disintegration, these unstable nuclides rapidly disappear, causing the level of radioactive emission to drop sharply. After a period of time, the emissions primarily originate from less unstable nuclides, which are decaying more gradually, and the rate of reduction in levels of emissions slows. Thus, simply storing the spent fuel in a vault at the reactor site for a short time results in a sharp reduction in the levels of emissions (and thus the hazard) of the waste. However, very-long-term storage is then necessary to allow the less unstable nuclides to slowly decay and disappear.

Scientists express the difference in tendency to decay with a unit termed the *radioactive* or *physical half-life* ($T_{1/2}$) of the nuclide. *Half-life is the length of time required for half the atoms in the sample to decay.* If we start with 10 grams of a nuclide and 10 minutes later we have 5 grams—the other 5 grams having become decay product—the half-life of the nuclide is 10 minutes. Another unit to express the same concept is the decay constant (λ) of the nuclide, which is the fraction of the atoms in a sample that decay per unit time. Half-life and decay constant are related as follows:

$$T_{1/2} = 0.693/\lambda$$

Half-life is a constant value for a given nuclide, and in various nuclides these values vary from fractions of a microsecond to millions of years. In our example above of the two unstable n/p ratios, the half-life of ^{61}Co is 1.6 hours, and the half-life of ^{63}Co is 27.5 seconds.

Sample Problem:

It is possible to calculate the level of radioactive emission for a given nuclide at some time in the future if the half-life and the present level of emission are known. The formula is as follows:

$$N_t = N_0 e^{-\lambda t}$$

where

N_0 = present rate of emission
N_t = rate of emission at time t
e = base of natural logs
λ = decay constant
t = time for which N_t is calculated

The present rate of emission of a sample of ^{61}Co is 500 cpm (counts per minute). What will it be in 5 hours?

1. First calculate the decay constant:

$$T_{1/2} = 0.693/\lambda$$

$$\lambda = 0.693/T_{1/2}$$

$$\lambda = 0.693/1.6 \text{ hr}$$

$$\lambda = 0.433 \text{ hr}^{-1}$$

2. Now calculate N_t:

$$N_t = N_0 e^{-\lambda t}$$

[where $-\lambda t = -(0.433 \text{ h}^{-1})(5 \text{ h}) = -2.17$]

$$= (500 \text{ cpm}) e^{(-2.17)}$$

$$= 57 \text{ cpm}$$

Types of Radioactive Emissions

Unstable heavy atoms such as uranium or plutonium may become more stable by a reduction in size, and accomplish this reduction by emitting a fragment of the nucleus called an alpha (α) particle). *An α particle is composed of two protons and two neutrons,* and is the largest and most highly charged of the radioactive particles.

In atoms with too many neutrons for their number of protons, a neutron may spontaneously convert into a proton. An uncharged neutron must lose a negative charge to assume the positive charge of a proton. This is accomplished by emission of an electron, which has a negative charge and only a tiny mass. *An electron emitted from the nucleus is termed a beta (β) particle.* A ß particle has only half the charge and slightly more than a ten-thousandth the mass of an α particle.

In atoms with too many protons for their number of neutrons, a proton may spontaneously convert into a neutron. A proton must lose a positive charge to become neutral. This is accomplished by emission of a *positron,* which is essentially a positively charged electron. Positron emitters are uncommon, and will not be considered further in this discussion.

Changes in the composition of the nucleus are likely to leave the protons and neutrons in an arrangement that is not the most stable possible. As the particles shift into a more stable configuration, thus leaving a higher-energy arrangement to assume one of lower energy, the energy difference is released as *very-high-energy light termed gamma (γ) radiation.* Gamma rays are often emitted slightly after the release of a particle from the nucleus. As light, they have neither mass nor charge. They are very much higher in energy than the UV light discussed earlier.

CHARACTERISTICS OF IONIZING RADIATION

Various types of radiation have different properties, but all ionizing radiation has in common the ability to produce ion pairs when passing through matter. Chiefly, they differ in their penetrating power and in the level of ion pair formation they produce.

ALPHA AND BETA PARTICLES

A charged particle emitted by the nucleus leaves with an amount of energy that is measured by its velocity. As it passes an atom, its charge causes it to remove an electron from the atom, creating an (electron/positive ion) ion pair. This interaction with an electron, whether from attraction by a positively charged particle or repulsion by a negatively charged particle, uses some of the particle energy, causing a reduction in the velocity of the particle. As more ion pairs are produced, the particle is eventually stopped.

The velocity of α particles differs from one nuclide to another, but is constant for any given nuclide. With their +2 charge, α particles are extremely effective at producing ion pairs. In air, an α particle produces from 30,000 to 100,000 ion pairs per cm traveled. With the relatively high density of atoms encountered in tissue, ion pair production is very much higher, so α particles usually penetrate only 50 μm, the thickness of a few cells, before dispersing their energy completely. Alpha radiation from an external source does not even penetrate the already dead keratin cell layer of the skin, and is thus unlikely to do damage. However, if an α emitter gets into the body, the α particle produces a large number of ion pairs at the place it is released, so such an emission can do extensive local damage.

The −1 charge of the ß particle results in the formation of about 200 ion pairs per cm in air. At a given point in tissue the passage of a ß particle causes less local damage than does an α particle. However, this lower rate of dispersal of particle energy results in much longer path lengths, on the order of a few cm of tissue. Exposure to an external ß emitter is therefore hazardous.

GAMMA AND X-RAYS

Both gamma and X-rays are pure energy, lacking both mass and charge. The energy of such emissions is measured by the frequency or wavelength,[1] with greater energy associated with higher frequency (shorter wavelength). The greater the energy of the light, the greater its penetrating power.

X-Ray tubes produce a range of X-ray energies, depending on which of the electrons of the atom are excited, up to some maximum value. Higher-maximum-energy X-rays result when higher voltage is applied to the X-ray tube. Often the lower-energy, thus less-penetrating or "softer," X-rays are removed by a metal screen, allowing only the high-energy rays to be used.

The frequency of γ rays depends on the radioactive atom that is the source. It is possible to identify the nuclides present by analyzing the frequencies of the γ rays coming from a complex source. In general, γ rays are higher in energy than are X-rays.

Since high-energy light is without charge, its mechanism to generate ion pairs is different from that of charged particles. Light strikes the electron layers of an atom, and the energy of the light is absorbed by an electron. The electron now has too much energy to remain attracted by the nucleus, and leaves, generating an (electron/positive ion) ion pair. The production of ion pairs per cm of air is lower for γ or X-rays when compared to α or ß particles.

---

[1]Frequency is related to wavelength. The product of the two is the speed of light $(3.0 \times 10^8$ m/s in a vacuum).

NEUTRONS

Neutrons are emitted with a variety of energies, and the energy of the neutron determines its penetrating power. Low-energy neutrons pass through perhaps a centimeter of tissue, while high-energy neutrons may pass through several centimeters.

Since a neutron is uncharged, it does not produce ion pairs directly when passing through air or tissue in the manner of α or ß particles. By collision with other atoms a variety of changes may be generated. A nucleus may be broken into charged fragments or may have a proton ejected. Such charged pieces are now potent ion pair generators. As neutrons lose energy by such collisions, they become low enough in energy to be "captured" or incorporated into a nucleus, usually of a hydrogen or nitrogen atom in tissue. This nucleus is made unstable by the capture, and may decay radioactively after a short time. The ionizing effects of a neutron are termed *secondary emissions,* since they are due to the changes brought about by the neutron rather than by the neutron itself.

SHIELDING

Protection from a source of radiation is provided by *shielding*—the placing of matter around the source. As radiation interacts with the matter, it disperses its energy and is stopped. Lead is associated with shielding, and it is an excellent material to use for this purpose. However, its only special characteristic is that it is a very dense form of matter. Any substance can act as a shield, as long as enough matter surrounds the source to stop the emissions. Lead is often used because a radiation from a source may be blocked without resorting to ungainly thicknesses of shielding.

The amount of matter necessary to shield a source depends on the penetrating power of the emissions. Relatively little shielding is needed for α particles, while γ rays require a great deal. One method to express the penetrating power of a particular source of radiation is to measure the thickness of a layer of a particular material needed to block half the emissions, termed the *half thickness* in that substance.

EXPOSURE TO IONIZING RADIATION

We are constantly exposed to radioactive emissions because some of the atoms around us—or even that are part of us—are radioactive nuclides. For example, the ^{14}C described above as being part of all carbon samples is a ß emitter. In addition, radioactive nuclides are sometimes used in the workplace. As a result, in some workers, natural exposure to radioactive emissions may be supplemented by exposure in the workplace.

As a result, in some workers, natural exposure to radioactive emissions may be supplemented by exposure in the workplace.

Radioactive Nuclides in the Workplace

Because they act chemically exactly as do the stable nuclides, radioactive nuclides are used as *tracers*. By detecting their radiation we can follow compounds in which radioactive nuclides are included without disturbing the chemical changes being undergone by those compounds. This is useful in research laboratories, and hospitals use radioactive tracers diagnostically. For example, the efficiency of concentration of radioactive iodine by the thyroid gland measures the activity of that gland. Tracer studies in medicine and research use small amounts of radioactive material.

Powerful γ radiation sources may be used to kill cells. This is a useful means of killing bacteria and fungi in packaged food to preserve it. Such sources are also used in hospitals to destroy inoperable tumors. High-energy sources must be treated with great care to avoid unwanted human exposure to the radiation.

X-Rays

When a sample of metal is bombarded with high-voltage electricity, electrons surrounding the nucleus of the metal atoms acquire high energy. They lose this energy again in a short time in the form of high-energy light called X-rays. X-Rays are similar to γ rays, but fall in a somewhat lower range of energy. Such high-energy light can penetrate matter to a marked degree—the characteristic of X-rays that is most often utilized. Examples include medical diagnosis and the checking of luggage in airports. Although small doses bombard us from outer space and small amounts are produced by television or computer screens, our exposure to X-rays is primarily from devices deliberately designed to generate them.

Neutrons

When a fission reaction occurs in an atomic power plant, neutrons are released. This is the only likely source of exposure to these particles, unless a person is in a research laboratory with a particle accelerator or deliberate neutron source such as ^{252}Cf.

MEASUREMENT

In order to work with ionizing radiation quantitatively, it is necessary to have devices to detect the radiation and a standard set of units used to express levels of radiation. Researchers who first detected ionizing radiation used photographic film, which is exposed by the radiation. We have since developed more sophisticated devices, but as you will see, we still make use of film to estimate personal exposure to radiation.

MEASURING DEVICES

Counting devices in use are based on two general principles: measurement of the degree of ionization produced in a sample of gas, and the amount of light generated when certain liquids or crystals are irradiated. Personal dosimeters use additional detection techniques.

Geiger-Müller Counters

The Geiger-Müller meter (G-M counter) is a very common and much used device for radiation measurement that is based on ionization of gases. Portable battery powered G-M counters can conveniently be taken into the workplace to measure exposure at the workstation, contamination of clothing, or contamination of the skin of the worker (Figure 11.2).

In this instrument a tube is filled with an easily ionized gas. Electrons produced by ionizing radiation are attracted to a positively charged wire, and on striking it produce a pulse of electricity. The strength of the signal is multiplied by ionization trails as the electrons accelerate toward the wire. Such a pulse is recorded as a "count," and the instrument accumulates a total of counts (Figure 11.3).

A G-M counter measures ß, γ and X-rays. Alpha particles cannot penetrate the window and so are ignored by the tube. By changing the design of the detector tube so that the sample is placed in a chamber with easily ionized gas flowing through, the window is eliminated (a windowless counter) and α particles can be counted. It is also possible to put sufficient shielding over the window of the counter to block ß particles, making the instrument sensitive only to high energy electromagnetic radiation.

Scintillation Counters

Scintillation counters utilize the fact that certain liquids or crystals emit light when struck by radioactive emissions. The amount of light, measured using a photomultiplier, indicates the level of radioactive emissions. These are primarily research tools, used in laboratories.

Figure 11.2.   Shown here is a Geiger counter used to scan work areas for spilled radioactive isotopes.

Dosimeters

A dosimeter is a device worn by workers who may be exposed to radiation, One type of dosimeter, called a thermoluminescence detector (TLD), contains a chip of lithium fluoride. Radiation elevates electrons in the atoms of the chip to higher-energy positions that are stable enough to remain there for an extended time. The dosimeter is read by placing it in a device that heats the chip, causing the electrons to return to their "ground state" or normal low-energy position. As they do so they emit light, which is read by the measuring device.

A film badge is a convenient and low-cost dosimeter (Figure 11.4). It contains a small piece of photographic film, which is "exposed" by radiation just as it is exposed by light. After wearing the badge for a specified time period the film is developed. The degree of exposure of the film estimates the dosage of radiation absorbed by the worker.

A pocket dosimeter to measure X-ray and γ ray exposure looks like a pen and is worn clipped in the pocket (Figure 11.5). It contains a quartz fiber in a chamber, and the fiber carries an electrostatic charge that moves it to a zero point on a scale. Radiation generates ion pairs in the chamber, which discharge the fiber, causing it to move on the scale. The chief advantage of this device

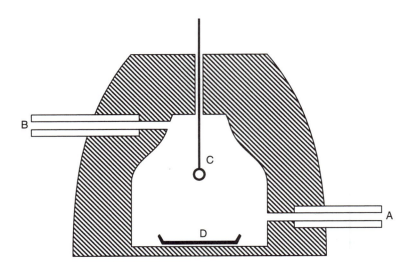

Figure 11.3. Here is one type of Geiger counter detector. Ionizing radiation passes through a readily ionized gas, causing an electron cascade to the anode. This is detected as an electric pulse in the external circuit, and is recorded as a single count. Gas enters the chamber at A and exits at B.

is its ability to be read by the worker at the time of exposure so as to estimate exposure as it happens, rather than having to send the device away to be read.

UNITS OF RADIOACTIVITY

There are two types of measurement that interest those who work with radioactivity. Units of emission describe the rate of disintegration of a radioactive nuclide, and are the usual data collected in experiments involving use of such nuclides. Industrial hygienists are more interested in units of exposure, which describe the dosage an individual has received.

Units of Emission

Essentially what we wish to know when radioactivity levels are measured is how many radioactive nuclides have disintegrated in a unit of time. The basic SI (Système International) unit is therefore the becquerel (Bq), which is the amount of nuclide that produces one disintegration per second. The Bq is relatively new, and older data as well as many of the newer measurements instead use the curie (Ci). One Ci, originally the radiation produced by a gram of radium, is defined as 3.7×10^{10} disintegrations per second—a very large

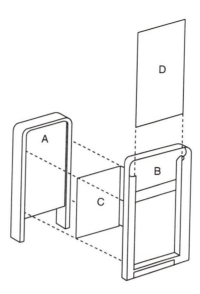

Figure 11.4. A film badge dosimeter is a very simple device. A and B are the
back and front respectively of the housing.   The film (C) is
snapped in between them, and the personal I.D. of the wearer (D)
slides into the front.

measuring unit. We therefore normally use small fractions of the basic unit
such as millicuries (mCi, 10^{-3} Ci) or microcuries (μCi, 10^{-6} Ci).

When one points a GM counter at a radioactive sample and reads numbers
of emissions passing through the tube, this is termed counts per minute (cpm)
or counts per second (cps). This value is not the measure of Bq or Ci for that
sample. For that to be true, all emissions from the sample would need to have
been beamed toward the counter tube. Emissions occur randomly in all directions, so even placing the counter right against the sample will only detect about
half the disintegrations—the other half having been directed away from the
window. However, if counts are taken in a fashion that is geometrically reproducible with respect to the sample, valuable information about relative rates of
disintegration are obtained.

Units of Exposure

The roentgen (R) is based on the energy absorbed by air from a radioactive
source, and one R is 83 erg/g. This unit has been used principally to measure
X-ray or γ ray exposure, often expressed as rate of exposure in milliroentgen per

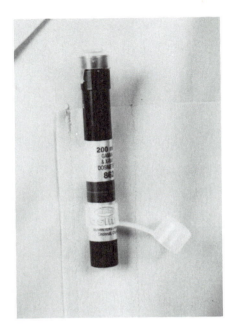

Figure 11.5.   This pocket dosimeter (left) detects exposure to X-rays and γ rays. It is recharged in the stand shown (right). (Photos courtesy of Lab Safety Supply, Janesville, WI.)

hour (mR/h). However, the energy absorbed by more dense matter such as tissue is greater by 10–20%. Because of the difficulty of translating R into tissue exposure, this traditional unit is being used less nowadays.

The rad is the most common unit in use, and is equal to an absorbed dose of 100 ergs per gram (erg/g) or 0.01 joules per gram (J/g) of whatever material is absorbing the dose. Differences in the ability of material to absorb the energy are defined out of this unit, but actually measuring the energy absorbed by a specific sample of material is difficult. *Rate of exposure* is often expressed as millirads per hour. A dose of 100 rad equals the SI unit, one gray (Gy). (Expressing exposure in Gy is not yet commonplace in the U.S.)

Different types of radiation create different levels of damage. This is indicated above, for example, in terms of the number of ion pairs produced per cm of dry air. This means that a rad of α particle exposure produces more damage to tissue than a rad of X-rays. The relative hazard of each type of radiation is expressed as the *quality factor* (QF) of the radiation. X-rays, γ rays, and ß particles have QF values of 1, while α particle values are about 10.

When damage is caused by taking the radioactive nuclides into the body, the chemical nature of the nuclide leads to its preferential deposition in certain parts of the body. This fact is recognized when talking about dosage absorbed by a given organ by indicating the *distribution factor* (DF) for that organ.

The *rem* is a unit that measures the dose *equivalent*—the likely damage to specific tissue rather than just the energy absorbed. We calculate rems by multiplying the dosage in rads by the quality factor for the type of radiation, and sometimes also by the distribution factor for the organ that is the target:

$$rem = rad \times QF \times DF$$

(The SI unit for dose equivalent is the sievert (Sv), which equals 100 rem, and, like the Gy, is not in common use in the U.S.)

EXPOSURE STANDARDS

For individuals in unrestricted areas the maximum permitted whole body dose is 0.1 rem per calendar year.[2] Special and detailed permitted exposure levels are described for individuals in restricted areas in the Code of Federal Regulations.[3]

Caution Signs

In order to alert workers to a radiation hazard, OSHA prescribes a standard warning sign. The symbol is the shape of a three-bladed propeller, and the colors are purple on a yellow background. Such words as, "CAUTION — RADIATION AREA" or "CAUTION — HIGH RADIATION AREA" should be included (Figure 11.1).

Special Circumstances in Radiation Exposure

A variety of special circumstances influences the degree of damage caused by exposure to ionizing radiation. These include the dose rate, tissue or part of the body exposed, and the age of the person receiving the dose.

Standards for exposure to radiation indicate a maximum total exposure in a given period of time. It matters whether the dose received is distributed broadly

---

[2]See 10 CFR 20.1301 (a) — Standards for Protection Against Radiation.

[3]10 CFR 20.1301 (c); 29 CFR 1910.96 (b).

across the indicated time (low dose rate) or entirely in a short period of exposure (high dose rate). If a small amount of damage occurs in a cell, the cell can repair the damage as it occurs. Since we are continuously exposed to radiation from our environment, damage from radiation is a natural occurrence and we have repair mechanisms. However, if damage occurs faster than the cell can repair it, the cell dies from accumulated damage. *A given amount of radiation delivered at a high dose rate produces greater damage than the same amount delivered at a low dose rate.*

Which cells in the body absorb the radiation is also important. We can differentiate cells as specialized and unspecialized. For example, the various types of cells in the blood have specific roles for which their special structural features adapt them: red blood cells carry oxygen, white cells engulf invading bacteria, and so on. These are specialized cells. They arise by a process of maturation, acquiring these specialized characteristics in a stepwise fashion, from very unspecialized cells in the bone marrow. We see another example of this in the epidermis of the skin, where cells at the base of the keratin layer divide to produce cells that migrate outward toward the surface of the skin and gradually become the fiber-filled cells of the keratin layer as they migrate.

Specialized cells are less susceptible to radiation damage than are the rapidly dividing unspecialized cells, sometimes called stem cells, from which they arise. Stem cells in the bone marrow, in the epithelial lining of the intestines, and in the ovaries and testes are all exceptionally sensitive to radiation.

Finally, the age of the person dosed with radiation is important. Younger individuals with a more rapid rate of cell growth are more readily damaged, with the greatest damage from a given dose level occurring in a fetus. In recognition of this, the practice of using X-rays to determine the position of an infant before delivery has been minimized, and recommended maximum exposure for pregnant women is lower.

Radiation from the Environment

The limitations in occupational exposure to ionizing radiation are *in addition to* the background levels of exposure that everyone experiences. Largely unavoidable radiation exposure comes from the environment, health sciences exposure, and a few other sources.

Radiation strikes the earth from space—an exposure that is partially blocked by the atmosphere. Our altitude above sea level is therefore a factor, exposure being about three times higher at 10,000 feet than at sea level, where we benefit from the maximum blocking by the atmosphere. Average exposures range from about 75 mrem/year in a high-altitude state like Wyoming to about 38 mrem/year in Florida. Average exposure in the U.S. is 44 mrem/year. Obviously, air travel raises this exposure. In addition, we are exposed to the emissions from radioactive nuclides in our surroundings in rocks, soil, and bricks, estimated to range from 15 to 140 mrem/year according to location, and averaging about 40 mrem/year in the U.S. We have internal radioactive nuclides such as ^{40}K which add another roughly 18 mrem/year, bringing the total to around 100 mrem/year from all natural sources for the average individual.

Health care involves X-rays for many people and exposure to diagnostic radiopharmaceuticals for a few. An estimate from 1970 places the average exposure from medical sources at 73 mrem/year, but a lessened dependence on diagnosis by X-ray techniques has probably lowered this figure more recently.

A few miscellaneous sources add a little to our nonoccupational radiation exposure. Television and computer screens emit soft X-rays, adding perhaps 2 mrem/year. In total, our nonoccupational exposures average about 175 mrem/year per person, with individual totals varying widely according to location and experiences.

BIOLOGICAL EFFECTS OF RADIATION

Radiation disrupts chemical structures to produce free radicals and ion pairs. This is the *direct effect* of the radiation. These unstable products then undergo further change, generally involving interaction with the molecules around them, in order to return to being stable chemical structures. The sum of these interactions and changes is the *indirect effects*. Direct effects are virtually instantaneous with the passage of the emission, and the highly energetic and reactive direct products cause the indirect effects in the next tiny fraction of a second. There is no time to intervene after exposure to radiation to block these events. Treatment must involve dealing with the consequences of the exposure.

The biological effect of radiation depends on the changes in living tissue produced by these reactive species. Radiation is not selective in its chemical targets, and direct damage occurs to the entire range of molecules in the path of the emission. However, the majority of the molecules in tissue are water, so if we understand what happens to water on irradiation, we understand a large percentage of the events occurring, and by example we understand the nature of the others. Important reactions resulting from irradiation of water are given in Table 11.1. From these typical reactions occurring in an irradiated cell, we see that:

1. Free radicals are generated.
2. Hydrogen peroxide is generated.
3. The amount of hydrogen peroxide generated is increased by increased oxygen concentrations in the tissue.

Once generated, free radicals initiate a chain of destructive events. The unpaired electron is inherently unstable, and stabilizes by removing an atom from a surrounding molecule along with one of the two electrons that bonded that atom to another, as in the second step of the interaction of a hydrogen free radical with oxygen. The original free radical is now stable, but the molecule with which it collided is now a free radical. This sequence of collisions leaves

Table 11.1. Reactions Resulting from Ion Pair Formation in Water Due to Radiation.[a]

$$\text{H}_2\text{O} \xrightarrow[\hspace{2cm}]{\text{energy (34 eV)}} \text{H}_2\text{O}^+ + \text{e}^{-1} \quad \text{(Ion Pair Formation)}$$

$$\text{H}_2\text{O} + \text{e}^{-1} \xrightarrow{\hspace{1.5cm}} \text{H}_2\text{O}^- \qquad \text{(Electron Capture)}$$

$$\text{H}_2\text{O}^+ \xrightarrow{\hspace{1.5cm}} \text{H}^+ + \text{OH·} \quad \text{(Free Radical Formation)}$$

$$\text{H}_2\text{O}^- \xrightarrow{\hspace{1.5cm}} \text{H·} + \text{OH}^-$$

$$\text{H·} + \text{O}_2 \xrightarrow{\hspace{1.5cm}} \text{HO}_2\text{·} \qquad \text{(Free Radical Interactions)}$$

$$\text{HO}_2\text{·} + \text{H}_2\text{O} \xrightarrow{\hspace{1.5cm}} \text{H}_2\text{O}_2 + \text{OH·}$$

$$\text{H·} + \text{H·} \xrightarrow{\hspace{1.5cm}} \text{H}_2 \qquad \text{(Loss of Free Radicals by Collision)}$$

$$\text{H·} + \text{OH·} \xrightarrow{\hspace{1.5cm}} \text{H}_2\text{O}$$

$$\text{OH·} + \text{OH·} \xrightarrow{\hspace{1.5cm}} \text{H}_2\text{O}_2$$

[a]The first step (Ion Pair Formation) is the direct result of the radiation; the rest of the steps result from this initial reaction. As a result of electron capture in water, there are both positively and negatively charged water molecules, which form free radicals as they react to form more stable structures. Free radicals stabilize by extracting atoms from other molecules, but the source molecule in turn is left as a free radical. This "chain reaction" of the free radicals causes miscellaneous damage to important biological molecules in the cell. This terminates by chance collision of two free radicals to form stable molecules.

the molecules of the tissue chemically altered, and to the extent that important biological species are involved, the cell functions are impaired.

Hydrogen peroxide is a strong oxidizing agent, and destroys biologically functional molecules. Tissues have an enzyme that converts hydrogen peroxide to harmless products, but it can only prevent damage if it encounters the hydrogen peroxide before a sensitive functional species does.

IMMEDIATE EFFECTS OF LARGE DOSES

Regulations are designed to hold total radiation exposure (occupational and otherwise) to under one rem per year for the typical person. There are no immediate symptoms related to exposures at these levels. Here we consider the health effects of dosages that go far beyond regulated levels—exposures that only would occur in the case of a serious accident. The symptoms appear in a short time and result from extensive direct tissue damage by the radiation. Table 11.2 summarizes the severity of symptoms due to increasing levels of exposure.

Below 100 rad, there is little chance that the exposure will be fatal in the short term. However, symptoms are produced that reflect the sensitivity of the less specialized cells. Rapidly dividing cells that produce sperm cells are damaged, resulting in a drop in the sperm count. If the dose is confined to a small area of the body, there is loss of hair as the cells at the base of hair follicles die, and the skin itself displays erythema, as in a burn.

Above 100 rad, effects on the cells that produce the epithelial lining of the gastrointestinal tract appear. Within a few hours a loss of appetite (anorexia) is displayed, then nausea at higher doses. By 200 rad vomiting is likely, and by 250 diarrhea is probable.

Above 500 rad, damage to the blood cell–producing cells in the bone marrow becomes a serious, potentially lethal problem. The altered pattern of production of white blood cells is obvious in 1 to 2 days, and if the victim survives, this pattern may not return to normal for an extended time period. These effects are dose-dependent, becoming more serious with greater exposure. Cataracts may result over a longer time period from these exposure levels.

Above 1000 rad, damage to the epithelium of the G.I. tract becomes life-threatening. Lining destruction leads to extensive internal bleeding, and death often follows in a few days. Beyond 5000 rad the destruction of the nervous system becomes the immediate problem, and at higher levels the rapid and massive cell death is immediately fatal.

LONG-RANGE EFFECTS OF MODERATE EXPOSURE

Levels of radiation described in the previous section only occur in extreme cases involving accidents, and they are not the basis of concern in setting worker exposure standards. A small proportion of the damage to cells resulting from the generation of free radicals or hydrogen peroxide in the cells is to the DNA molecules. A small percentage of this damage alters the DNA so as to convert the cell to a malignant pattern. Thus, in a dose-dependent fashion, the frequency of cancer is increased by radiation exposure. In terms of time after exposure, leukemia is the earliest type of cancer to appear. Although the latent period is longer, other types of cancer result from radiation damage and in the long term are more common.

Cataracts, sterility, and premature aging are other long-range results of exposure to radiation. Cataracts result because lens cells in the eye are not replaced when destroyed by exposure. Without these cells to maintain the lens, it gradually loses transparency. Sterility in males does not necessarily require complete elimination of viable sperm, only a reduction in numbers. This may

Table 11.2. Effects of High Whole Body Radiation Dosage.

Dose Level (rads)	Health Effect
1	No health effects detected in the short term.
10	Developmental effects in very young embryos.
10^2	White blood cell count decreases.
10^3	Damage to the G.I. tract causes vomiting, diarrhea, nausea. Depressed blood cell production. Death: 1–2 weeks.
10^4	Nervous system damage. Coma and death: 1–2 days.
10^6	Massive cell death. Death: immediate.

occur in the short term as a result of low exposure to radiation, but then the sperm count may rise to normal again. Higher exposure can lead to a permanent drop in sperm production. In females, since the ovaries contain the lifetime supply of egg cells, the response pattern does not include the restoration of fertility after a waiting period following lower levels of exposure.

KEY POINTS

1. Damaging radiation is divided into nonionizing and ionizing radiation.

2. Nonionizing radiation includes UV light from the sun or artificial sources that can damage the eyes and skin.

3. Atoms with a ration of protons to neutrons that is distant from the ideal adjust the ratio by radioactive emissions.

4. The degree of instability is reflected in the rate of radioactive disintegration, measured as the half-life, or time required for half the nuclides in such a sample to disintegrate.

5. Radioactive emissions include charged particles (α: 2 protons and 2 neutrons, β: an electron, positron: a positively charged electron); high-energy light (γ and X-rays); and neutrons.

6. Matter is ionized by the passage of these emissions. Charged particles draw or drive electrons from atoms, high-energy light excites the electrons so that they leave the atom, and neutrons drive charged

particles out of the nucleus, and then draw or drive electrons from atoms.

7. Radioactive emissions can be blocked by placing matter in their path (shielding). Charged particles are more easily blocked than high-energy light.

8. Levels of radioactive emissions can be measured by G-M counters and scintillation counters.

9. Levels of exposure are determined by dosimeters worn by the worker.

10. Units of emission include the becquerel (Bq = 1 dps) and the curie (Ci = 3.7×10^{10} dps).

11. Units of energy absorbed include the roentgen (83 erg/g in air), the rad (100 erg/g in tissue), and the rem (rad × quality factor of radiation × distribution factor in tissues).

12. The exposure ceiling for workers in unprotected areas is 0.5 rem/yr. In protected areas the ceiling is higher, and follows a formula.

13. It is more damaging to receive a dose of radiation in a short time period than the same dose over a long time period.

14. The average American receives 175 mrem/yr from a combination of radiation from space, environmental radioisotopes, internal radio-isotopes, medical treatment, and TV or computer screens.

15. The biological effect of low radiation doses depends on the formation of free radicals in body fluids. The most common final product is H_2O_2.

16. Large doses of radiation can be lethal. Over 100 rad destroys cells lining the G.I. tract. Over 500 rad depletes the supply of blood cells. Over 5000 rad destroys the nervous system.

17. The most serious long-range effect of radiation exposure is an increase in the occurrence of cancer.

PROBLEMS

1. Using a periodic chart, answer the following:

   A. How many protons are in the nucleus of the elements V, Ag, and K?

B.   What particles are found in the nucleus of ^{15}N, ^{23}Na and ^{235}U?

2.   Write balanced nuclear equations of this form:

$^{14}C \longrightarrow {}^{14}N + {}^{0}\beta$

for the nuclear reactions indicated, using these symbols:

alpha $^{4}\alpha$; beta $^{0}\beta$; neutron ^{1}n; proton ^{1}p; positron $^{0}\beta$

$^{26}Na \longrightarrow$ beta

$^{212}Rn \longrightarrow$ alpha

$^{111}Sb \longrightarrow$ positron

$^{16}O + {}^{1}n \longrightarrow$ proton

3.   Refer to the definitions of units for measuring exposure to radiation in 29 CFR 1910.96 (a).

A.   What dosages are equivalent to one rem?

B.   What does this assume is the quality factor of ß, γ, X-radiation, neutrons, high-energy protons, and heavy particles of high penetrating power?

4.   How does the standard for maximum permitted whole-body dose for workers in unrestricted areas compare to the average dose from naturally occurring sources?

5.   Find the section in CFR that lists the maximum permitted dosage for workers in restricted areas.

A.   What is the relationship between whole body exposure permitted here and permitted for workers in an unrestricted area?

B.   Read Table G.18. The permitted dose to hands, forearms, feet, and ankles is greater than to whole body or to a series of body regions. Why, and why is each specific body region so designated?

C.   Dosage can go beyond levels of Table G.18. What is the ultimate limit?

D.   J.W.J. is a 45-year-old worker with an accumulated occupational dose of 110 rem. Must she be isolated from radiation exposure this year?

    E.   S.D.W. is a 17-year-old worker. What are his exposure limits in a restricted area?

6.   A.   The half-life of ^{29}Al, a ß emitter, is 6.6 minutes. What is its decay constant?

    B.   A sample of ^{24}Na, a ß emitter, is counted with a G-M counter, and reads 2550 cpm. If the $t_{1/2}$ is 15.03 hr, what will the emission level be in 2 hr?

BIBLIOGRAPHY

C. H. Hobbs and R. O. McClellan, "Toxic effects of radiation and radioactive materials" in C. D. Klaassen, M. O. Amdur and J. Doull, *Toxicology: The Basic Science of Poisons,* 3rd ed., Macmillan Publishing, New York, 1986.

A. C. Upton, "Ionizing radiation" in B. S. Levy and D. H. Wegman, *Occupational Health: Recognizing and Preventing Work-Related Disease,* 2nd ed. Little, Brown and Company, Boston, 1988.

G. L. Voelz, "Ionizing radiation" in C. Zenz, *Occupational Medicine,* 2nd ed., Year Book Medical Publishers, Chicago, 1988.

C. Zenz and A. L. Knight, "Ultraviolet exposures" in C. Zenz, *Occupational Medicine,* 2nd ed., Year Book Medical Publishers, Chicago, 1988.

CHAPTER 12

WORKSTATION DESIGN

In this chapter we deal with the layout of the workplace, and how this impacts the employee from the standpoint of possible physical injury. Two topics are included in this brief survey: protection from accidental injury, and design of the task to prevent physical strain and the resulting bodily damage. An accident is a single event causing harm to the worker, and is sometimes termed *overt trauma*. A slip and fall accident is a common example. Injury that is the result of a long series of events overstressing some part of the body is termed *cumulative trauma,* or *repetitive motion disorder,* for example, lower back pain resulting from repeated lifting of heavy loads for a long time period. *Ergonomics* or *human factors engineering* is a field of investigation focused on the design of tasks to best fit human anatomy and physiology, and requires study both of the capabilities of the human body and engineering aspects of the task. One goal of ergonomics is the elimination of cumulative trauma.

Some hazards discussed here are preventable by simple regulatory control, for example, requirements to place guards on moving belts, and handrails around elevated workstations. Others, particularly in the area of cumulative trauma, are not as simple. To deal in detail with all aspects of these topics is beyond the scope of an introductory volume of this sort. The reader is directed to references that have greater depth of discussion and detailed information.

PREVENTION OF ACCIDENTS

Throughout life we learn self-protection skills: fear of heights, avoidance of obviously hot objects, leaving the path of a moving object. This sort of acquired common sense must be involved in creating a safe workplace. This is supplemented by experience with typical work situations and records that reveal what aspects of a job lead to accidents. Such records reveal that common problems involve slip and fall situations, problems in handling and shifting objects, contact with moving parts of machinery, and hazards from high-energy electrical installations.

To avoid having to rediscover common hazards in each workplace, standards are set by OSHA and presented in 29 CFR 1910. Statements range from general principles to specific regulations. For example:

Subpart D: Walking-Working Surfaces

General Principles

1910.22 General requirements (a) *Housekeeping* (1) All places of employment, passageways, storerooms, and service rooms shall be kept clean and orderly and in a sanitary condition.

Very Specific Regulations

1910.23 Guarding floor and wall openings and holes (e) Railing, toe boards, and cover specifications (1) A standard railing shall consist of a top rail, intermediate rail, and posts, and shall have a vertical height of 42 inches nominal from upper surface of top rail to floor, platform, runway, or ramp level. The top rail shall be smooth surfaced throughout the length of the railing. The intermediate rail shall be approximately halfway between the top rail and the floor, platform, runway, or ramp. The ends of the rails shall not overhang the terminal posts except where such overhang does not constitute a projection hazard.

Slip, trip, and fall accidents generate 16.4% of all injuries in the workplace. Prevention of these accidents is the subject of Subpart D, from which the above quotations are drawn, and Subpart F (Powered Platforms, Manlifts, and Vehicle-Mounted Work Platforms). Floor openings, stairs, ladders, scaffolds, and a variety of powered platforms are covered in detail in these sections.

Studies show that an unexpected change in the surface encountered while walking or climbing raises the likelihood of an accident. This could mean a change from a surface "gripped" well by a person's shoes (high coefficient of friction) to a more slippery surface (low coefficient of friction). Coefficient of friction values range from 0 (very slippery) to 1. Values of 0.7 or higher, for example a brushed concrete surface, are very safe (Figure 12.1).

Good housekeeping—keeping the floor free of objects that can cause one to trip or substances that decrease the coefficient of friction (water, small round particles)—reduce hazard. While walking, a person's heel may skim within 1/4 inch of the floor, so that even an irregularity lower than 1/2 inch can cause a person to trip. 49 CFR 31528 provides standards for floor surfaces designed to prevent accidents.

Any time materials are moved in a plant, objects (sometimes quite heavy objects) are set into motion, with risk to workers. Aspects of the design,

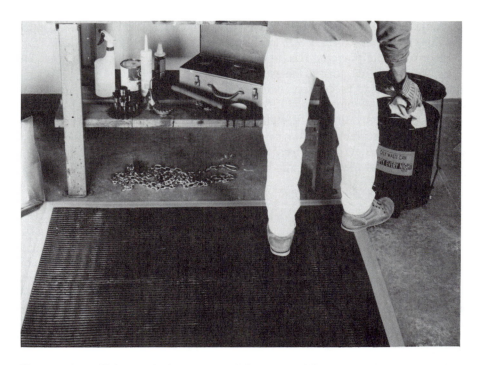

Figure 12.1. This mat helps prevent falls by providing the worker with a high-friction surface. Abrasive granules are bonded to the surface, and openings allow spills to drain through the mat. Such a resilient mat also reduces fatigue for workers who stand all day, and insulates the workers' feet from a cold floor. (Photo courtesy of Lab Safety Supply Janesville, WI.)

maintenance, and operation of machinery to do this moving are presented in Subpart N (Materials Handling and Storage).

Moving parts on machinery can cause serious injury either directly to the body or by snaring clothing or hair. It was recognized early on that in a safe workplace, guards are in place over moving belts, cutters, shears, saws, and other such devices to prevent worker contact. Regulations are presented in Subpart O (Machinery and Machine Guarding).

Subpart S (Electrical) deals with design, safety-related work practices, safety-related maintenance, and safety requirements for special equipment. Specifications that include extensive design detail are given for equipment operating at 50 volts or greater.

Workers should be instructed about the hazards of operations in their work area as a part of compliance with the *Right-to-Know* regulations. Signs posted to inform workers are classed as follows: *danger* warnings (red, black, and white) inform about immediate danger requiring special precautions, *caution* warnings (black and yellow) inform about potential hazards or caution against unsafe practices, and *safety instructions* (green and white) (Figure 12.2).

Identifying the potential hazard at a workstation may lead to recommendations for protective gear to be worn by the worker. In 29 CFR 1910.135, standards for head protection (hard hats) intended to protect workers "from falling and flying objects and from limited electrical shock and burn" are presented (actually the ANSI standard Z89.1 of 1969). Safety glasses are appropriate for certain jobs, may be required, and must meet the standards in 29 CFR 1910.133. Safety shoes have metal toe covers that protect feet from falling objects, and must meet standards in 29 CFR 1910.136 (ANSI Z41.1 of 1967).

VIBRATION

Vibration and sound (Chapter 10) have much in common. Both are described as wave phenomena, and the frequency and amplitude are significant parameters. Vibration operates at much lower frequencies than sound, most vibrations of interest being in the range of 1 to 300 Hz (cycles/second). Just as sounds are generally more complex than pure tones, vibration may include a complex spectrum of frequencies.

Vibration can affect the comfort and efficiency of an employee, and, in extreme cases, can damage the health. Vibration is usually transmitted to the body through the floor to a standing worker, through the seat to a seated worker, or through a hand tool in use. Most problems arise from the use of vibrating hand tools such as air hammers, chain saws, and riveting machines, or by riding in vehicles such as tractors or forklift trucks on rough surfaces.

All objects have a natural (resonant) frequency such that if they are exposed to a particular frequency of vibration, they also begin to vibrate (resonate). The entire body resonates somewhere between 4 and 8 Hz, the range which therefore generates the greatest discomfort. Body organs are not rigidly assembled into a single resonating structure, but have their own individual frequencies. As a result, vibration of a particular frequency may generate discomfort or pain in a specific organ. For example, a particular worker may feel abdominal discomfort in response to a particular vibrational frequency.

Workers who use hand tools for long periods, particularly those that generate vibration of one hundred to a few hundred Hz, may suffer damage to nerves and blood vessels in their hands. These individuals then experience numbness, pain, and loss of control in their hands. The fingers appear pale, and the problem was termed "white finger disease," or *Raynaud's disease*. If exposure to the vibration is allowed to continue, the problem becomes irreversible (Figure 12.3).

Workers experience problems reading instruments or control systems when either they or the machine are vibrating. Adjusting to the movement is fatiguing, is more difficult as the frequency increases, and can result in inability to perform a task.

Figure 12.2. The Right-to-Know regulations require that workers be informed of hazards. When signs are used for this purpose, a color code is used to distinguish notices according to their urgency. DANGER warnings are red (The word danger) and black (the lettering) on a white field. CAUTION signs have black letters on a yellow field.

ERGONOMICS TO PREVENT CUMULATIVE TRAUMA

Overt traumas present a simple picture since the injury occurs immediately due to a failure of protection. Poisoning by chemicals, cell damage by radiation, or hearing damage by high noise levels all allow quantitative prediction of the result of an unsafe condition. Cumulative traumas are harder to predict and avoid, and the industrial hygienist is more likely here to recognize the problem after and because injury has occurred. Experience has established that certain parts of the body are particularly susceptible to injury, including the fingers, wrists, elbows, shoulders, and back. The vertebral discs of the lower back can be compressed by heavy lifting, eventually producing lower back pain.

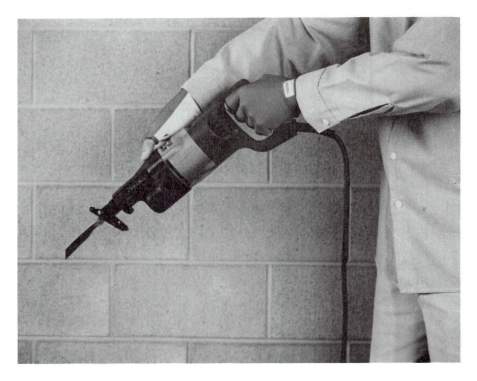

Figure 12.3. This worker is protected from the vibration of this hand-held power tool by vibration-absorbing gloves. (Photo courtesy of Lab Safety Supply, Janesville, WI.)

Knowing the contributing factors helps anticipate problems: jobs requiring *repetitive movements,* an awkward *position,* and the need to use *force.* Additional contributing factors include pressure on some part of the body, such as the finger or thumb, to run a tool. Spending long time periods in the same position, i.e., seated at a desk, may result in a continuous small insult to some body part, which eventually leads to a disorder. However, it is hard to assess the level of risk of injury due to sitting in an ill-fitting desk chair for long stretches during the workday in a fashion parallel to accurately measuring amounts of benzene in the air and assessing the degree of risk at particular concentrations.

Progress is being made in anticipating workstation layouts likely to produce cumulative trauma. It is useful to consider the relaxed or neutral positions for parts of the body, then design the job to keep those parts as close as possible to those positions. Arms are below shoulder level, wrists are straight, elbows are close to the body, the head faces straight forward, and the back is vertical with a natural curve. Armstrong and Lifshitz prepared a checklist of job factors to predict upper body cumulative trauma. (This checklist appears with the answer to Problem 5.) Using this checklist in different job situations, they obtained high correlations between low scores and the occurrence of cumulative trauma. Computer comparison of the type of trauma with individual responses on the survey suggests corrective action to prevent the trauma.

Sometimes repetitive movements or awkward positions are hard to eliminate, particularly in assembly type operations. In such instances, other preventive practices may be employed. Workers can be shifted from one workstation to another to change the patterns of movement. Employees can be trained to do simple stretching exercises before or during the job that relieve stress and break harmful patterns. Sometimes a job makes demands on certain muscles that can be strengthened by exercise so as to prevent trauma.

Computers may be used in a more sophisticated fashion in workstation design. Inputting detailed presentations of athletes competing allows coaches to pinpoint changes in style to improve performance. In the same fashion, the movements involved in performing a task at a workstation can be fed into a computer programmed to detect the potential for cumulative trauma.

EXAMPLES OF SPECIFIC PROBLEMS

Hands, Wrists, and Elbows

One of the facilities that gives humans an advantage over other species is the ability to grasp and manipulate objects in a sophisticated fashion. Complex machinery is built into our hands so that we can flex and bend our fingers, and also bend and twist our wrists. If muscles to do this were in the hand itself, it would be an unwieldy structure. Instead, most muscles are in the forearm, and are connected to the bones of the hand by tough flexible cords called tendons, which attach at one end to a muscle and at the other to a bone, so that when the muscle contracts, the tendon moves the bone. Take a moment to feel the tendons in one hand as you bend and flex your fingers. Appreciate the variety of movements you can accomplish with your fingers. Now feel your wrist, and recognize that it is primarily a bony structure. Tendons and the median nerve reach the hand through a passage in the wrist bones about the diameter of a small coin called the *carpal tunnel*.

When your hand and arm are in a straight line, the tendons pass smoothly through the carpal tunnel. When the wrist is bent, however, tendons rub on the walls of the tunnel and on one another. Continuous abrasion eventually leads to *tenosynovitis*—inflammation of the tendons. The swelling tendon can press on the nerve, leading to loss of sensation, weakening of muscles, and eventual loss of grip and other hand functions. These problems are termed *carpal tunnel syndrome*.

To avoid such problems, work functions should be designed so the worker performs operations with a straight wrist. Often hand tools can be redesigned to accommodate this "bending the tool, not the wrist" (Saunders and McCormick). For vertical work surfaces, a hand tool with a pistol grip keeps the wrist straight, while work on horizontal surfaces is best done with a tool that is in line with the work and is grasped like the support pole on a bus or subway car. Braces may be used that hold the wrist straight to prevent the start or worsening of carpal tunnel syndrome (Figure 12.4).

Repetitive finger action results in a type of tenosynovitis commonly called *trigger finger*. Use of the finger becomes restricted such that the person can bend the finger, but has to pull it in order to straighten it again. In general, controls should not be operated in a repetitive fashion by the index finger, although this seems to most people to be the most natural operation. Less trouble is experienced when thumb controls are used because, unlike the fingers, the muscles and tendons of the thumb are in the palm of the hand rather than the forearm, so tendons operating the thumb do not pass through the carpal tunnel.

In the forearm we have two bones, the *ulna* and the *radius,* which meet at the elbow with the *humerus,* the bone of the upper arm. Rotating the wrist—a common motion in tasks such as driving a screw—involves the rotation of the radius in the elbow. Inflammation due to friction in the elbow causes *epicondylitis* or "tennis elbow."

Finally, operations should minimize exertion of pressure applied with fingers or palms. This compresses blood vessels and nerves in the hand, leading to *ischemia* (loss of circulation) or numbness in the hand. Tools should be designed to spread the compression load as broadly as possible, in order to minimize pressure at a single point.

Lower Back

Lower back pain is a common worker complaint resulting from the handling of objects: lifting, carrying, pushing, and pulling. Twisting, reaching, and bending increase difficulties. Of these, lifting is the most common cause of lower back pain. Compression of or damage to vertebral discs is often the cause of pain. Workers may be trained in appropriate ways to perform tasks that reduce back strain. The most effective approach is to redesign tasks to reduce back involvement.

CONCLUSION

This discussion has only looked briefly at some major topics in occupational safety. Accidents are most effectively prevented by appropriate barriers, labeling of hazards, warning signs, and worker education. Cumulative trauma is more complex and is evolving in new directions as more is learned about the relationship of the structure of the human body and the performance of tasks.

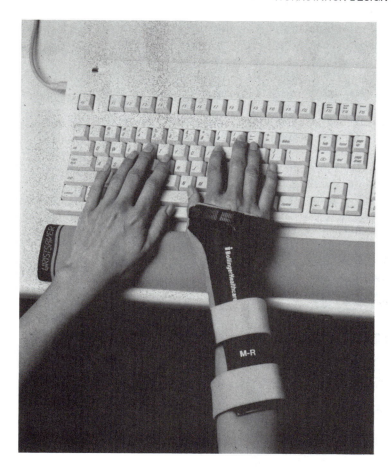

Figure 12.4. The brace holds the worker's wrist straight, preventing the type of tenosynovitus where tendons abrade in the passage through the wrist, a problem called carpal tunnel syndrome. (Photo courtesy of Lab Safety Supply, Janesville, WI.)

Much can be accomplished by arranging the task so that the worker is in an optimal posture and is minimizing strain on the body as the task is performed. Redesign of controls and hand tools not only prevents injury but also increases efficiency.

There has been an emphasis in this chapter on factory problems, which has left completely unmentioned the many other important job classifications, particularly with respect to cumulative trauma. Workers operating keyboards in front of visual display terminals are perhaps the most rapidly increasing segment of the workforce. Extensive ergonomic studies focus on this group of employees. References provide interested readers with a much broader coverage of the material.

KEY POINTS

1. An accident is a single event causing harm—an overt trauma. Long-term overstressing of the body, causing harm, is cumulative trauma.

2. Standards to prevent accidents are spelled out in 29 CFR 1910.

3. The frequency of slip and fall accidents is reduced by providing proper railings and barriers, avoiding unexpected changes of the floor surface, and practicing good housekeeping.

4. Materials handling generates risks of workers being impacted by machinery used for moving.

5. Moving parts on machinery require guards.

6. Electrical equipment operating at over 50 volts requires special safety design.

7. Hazards should be labeled with warning signs.

8. Vibration affects employee comfort and can cause harm.

9. Repeated physical operations, awkward positions, heavy lifting, and long periods spent in the same position can lead to injury.

10. Awkward and/or repeated wrist bending in performance of a task can cause problems with tendons and nerves running through the carpal tunnel. This is termed tenosynovitis or carpal tunnel syndrome.

11. Repeated finger action results in a special tenosynovitis called trigger finger.

12. Repeated twisting movements of the forearm produce epicondylitis.

13. Pressure on fingers or palms cuts off blood circulation.

14. Correct lifting and limits on the weight of objects lifted help prevent lower back pain.

PROBLEMS

1. Ladders are frequently involved in accidents, so it is not surprising that ladders are closely regulated. Even so, you may be surprised at the degree of detail of the regulation. Turn to 29 CFR 1910.21 (c).

    A. What is the specific subject matter of this section?

    B. How many kinds of ladder are separately defined?

    C. What is wane?

    D. What is shake?

    E. Is section 1910.21 (c) all, or in part, a set of regulations?

    F. Compare 1910.21 (c) and (d). Why is this done?

    G. Now look at section 1910.25. What is this section?

    H. OSHA is sometimes charged with regulating too closely, creating overly detailed rules that make compliance and enforcement difficult. Suppose you were going to work as a janitor in a plant, and would be replacing fluorescent tubes while standing on a high wooden ladder. Which of these sections in 1910.25 would you feel should be eliminated in the interest of simplification?

    I. Next time you are in a hardware store or a lumber yard, look for OSHA stickers on the ladders. What does this sticker tell you?

2. Define each of these terms:

    A. overt trauma

    B. Raynaud's disease

    C. tenosynovitis

    D. carpal tunnel syndrome

    E. trigger finger

    F. epicondylitis

    G. ischemia

3. You may at some time have looked up at workers washing windows on a tall building and wondered how safe such a job can be.

    A. What are the hazards?

Turn to section 29 CFR 1910.66.

    B. How many pages are devoted to this topic?

    C. What is the kind of "rope" that can be used on such a platform and what safety margin is required regarding its load bearing ability?

    D. Suppose a platform has two suspension wires, each rated at 5000 lb, and the working load of the platform is 2000 lb. Is the system in compliance?

    E. How often must the wire ropes be inspected, and how would an inspector know if the ropes were in compliance?

    F. Look at Figures 1, 2, and 3. What are the systems illustrated designed to prevent?

    G. What are the regulations regarding design and operation of a platform in the wind?

4. Decide whether each of these signs should be labeled DANGER or CAUTION, or should be presented as safety instructions, and indicate the correct color code in each case.

THIS EQUIPMENT STARTS AND STOPS AUTOMATI- CALLY	EYE WASH FOUNTAIN	WEAR HEARING PROTECTION IN THIS AREA	OXYGEN IN USE–NO SMOKING OR OPEN FLAME

5. A worker is employed to custom cut 19th century style wood trim for the restoration of old houses. The job requires running a 12-lb sabre saw around curved templates while the wood is in a jig at chest height. The metal handle of the saw is parallel to the wood surface during cutting. The worker turns the saw on by squeezing a trigger with the forefinger, release of that pressure automatically stopping the saw. At the end of the cut the saw must be lifted clear of the wood surface. Referring to Table 1, indi-

cate aspects of this job that are likely to produce cumulative trauma, indicate the type of worker difficulty that could result, and suggest improvements in the layout of the job to reduce strain.

CHECKLIST FOR ANALYSIS OF UPPER EXTREMITY CUMULATIVE TRAUMA DISORDER RISK FACTORS

1. Physical Stress:

1.1. Can the job be done without contact of fingers or wrist with sharp edges?
1.2. Is the tool operating without vibration?
1.3. Are the worker's hands exposed to temperatures greater than 70°F?
1.4. Can the job be done without using gloves?

2. Force:

2.1. Does the job require less than 10 pounds of force?
2.2. Can the job be done without using a finger pinch grip?

3. Posture:

3.1. Can the job be done without flexion or extension of the wrist?
3.2. Can the tool be used without flexion or extension of the wrist?
3.3. Can the job be done without deviating the wrist side to side (ulnar or radial deviation)?
3.4. Can the tool be used without ulnar or radial deviation of the wrist?
3.5. Can the worker be seated while performing the job?
3.6. Can the job be done without "clothes wringing" motion?

4. Workstation Hardware:

4.1. Can the orientation of the work surface be adjusted?
4.2. Can the height of the work surface be adjusted?
4.3. Can the location of the tool be adjusted?

5. Repetitiveness:

5.1. Is the cycle time above 30 seconds?

6. Tool Design:

6.1. Can the thumb and finger slightly overlap around a closed grip?
6.2. Is the span of the handle between 5 and 8 cm?
6.3. Is the handle of the tool made from material other than metal?

6.4. Is the weight of the tool below 10 lb?
6.5. Is the tool suspended?

From: Lifshitz, Y. and T. Armstrong: A Design Checklist for Control and Prediction of Cumulative Trauma Disorders in Hand Intensive Manual Jobs, *Proceedings of the 30th Annual Meeting of Human Factors,* pp 837–841 (1986).

BIBLIOGRAPHY

T. J. Armstrong and Y. Lifshitz, "Evaluation and Design of Jobs for Control of Cumulative Trauma Disorders," in ACGIH Industrial Hygiene Series, *Ergonomic Interventions to Prevent Musculoskeletal Injuries in Industry,* Lewis Publishers, Chelsea, MI, 1987.

D. B. Chaffin, "Biomechanics of Manual Materials Handling and Low Back Pain" in C. Zenz, *Occupational Medicine: Principles and Practical Applications,* 2nd ed., Year Book Medical Publishers, Inc, Chicago, 1988.

O. B. Dickerson and W. E. Baker, "Practical Ergonomics and Work with Video Display Terminals" in C. Zenz, *Occupational Medicine: Principles and Practical Applications,* 2nd ed., Year Book Medical Publishers, Inc, Chicago, 1988.

B. S. Levy and D. H. Wegman, *Occupations Health,* Little, Brown and Company, Boston, 1983. Chapter 9.

W. Marletta, "Trip, Slip and Fall Prevention," Chapter 12 in D. J. Hansen, *The Work Environment: Volume 1,* Lewis Publishers, Chelsea, MI, 1991.

S. Meagher, "Hand Tools: Cumulative Trauma Disorders Caused by Improper Use of Design Elements," in W. Karwowski, *Trends in Ergonomics/Human Factors III,* Elsevier, North Holland; New York, 1986.

S. W. Meagher, "Design of Hand Tools for Control of Cumulative Trauma Disorders," in ACGIH Industrial Hygiene Series, *Ergonomic Interventions to Prevent Musculoskeletal Injuries in Industry,* Lewis Publishers, Chelsea, MI, 1987.

D. J. Osborne, *Ergonomics at Work,* 2nd ed., John Wiley and Sons, New York, 1987.

M. S. Sanders and E. J. McCormick, *Human Factors in Engineering and Design,* 6th ed., McGraw-Hill Book Co., New York, 1987. Chapter 11 deals with hands and hand tools.

S. H. Snook, "Comparison of Different Approaches for Prevention of Lower Back Pain," in ACGIH Industrial Hygiene Series, *Ergonomic Interventions to Prevent Musculoskeletal Injuries in Industry,* Lewis Publishers, Chelsea, MI, 1987.

W. Taylor and D. E. Wasserman, "Occupational Vibration" in C. Zenz, *Occupational Medicine: Principles and Practical Applications,* 2nd ed., Year Book Medical Publishers, Inc, Chicago, 1988.

CHAPTER 13

HEAT STRESS

Harm termed *heat stress* results from too great an increase in body temperature. Some jobs involve working at high temperatures, and so have the potential to cause such harm. Indoor examples include laundries and processing of metals, glass, chemicals, or food. A number of outdoor jobs, including roadwork, agriculture, and construction, involve potential heat stress in hot weather.

BODY TEMPERATURE CONTROL

Humans, like all mammals and birds, have physiological mechanisms designed to maintain a constant body temperature. Such a system has advantages, allowing high physical activity in spite of decreasing ambient temperatures, but the price we pay is the need to operate within a narrow range of body temperature. When our surroundings become hotter or colder, we must have corresponding regulatory mechanisms to control our internal temperatures. We first experience discomfort, then illness, in hot surroundings when these mechanisms are inadequate to prevent a rise in body temperature.

We do not have a uniform body temperature, but rather have a warm core surrounded by a cooler shell. When we measure body temperature to assess our state of health, we measure *core temperature* by placing the thermometer in the mouth, ear, or rectum. Even these temperatures are not the same, with rectal temperatures running slightly higher than oral temperatures. Everyone is aware of the standard 98.6°F or 37°C "normal" temperature.[1] In fact, *normal* can range from just above 36°C to just below 38°C in a resting person, and climb to above 39°C during hard physical exertion. When the core temperature drops to 35°C, we are in a state of *hypothermia,* and by 27°C death occurs. A core temperature as high as 40°C is described as *hyperthermia,* and death occurs by 42°C. Skin temperatures are much more variable, normally being around

---

[1]Brush up on the conversion of °F to °C [°C = 5/9(°F −32)]. You may be used to using °F, but references in this field all use °C.

33–34°C, but dropping as low as 22–27°C in cold weather and rising close to core temperatures during heat stress.

HEAT GENERATION AND DISTRIBUTION IN THE HUMAN BODY

Heat is produced in the warm core, and is transferred by blood circulation to the skin to be lost, much as a car transfers excess heat generated in the combustion processes into surrounding water, then circulates that hot water to the radiator to transfer the heat to air. Two body processes release heat as the result of chemical reactions that overall are exothermic: the metabolism of nutrients, and muscle contraction. Nutrients can be from the diet, or they can be from reserves such as either body fat or glycogen from liver or muscle. Understand that nutrient metabolism operates continuously, even while we are sleeping or when effort expended is no greater than pushing the remote button to change channels. However, when physical effort is greater, rates of metabolism rise to supply the fuel for increased muscle contraction. *Physical work sharply increases generation of internal heat, and heat produced is proportional to effort expended.*

BODY TEMPERATURE REGULATION

Core body temperature is regulated by processes controlled by the brain, primarily by the small segment at the lower surface of the brain called the hypothalamus. Temperature sensory input is interpreted in the hypothalamus, and it directs two compensatory actions to counter a core temperature rise. First, smooth muscles lining blood vessels in the skin are relaxed, causing the vessels to dilate and carry a greater volume of blood through the skin. In a light-skinned person this is seen as a reddening of the skin or flushed appearance during exercise. The loss of heat to surrounding air is thus increased. Second, sweat glands are stimulated to coat the skin with sweat (a very dilute salt solution). The evaporation of this water requires heat that is supplied by the skin, cooling the skin. Evaporation of a gram of sweat requires about 0.58 kcal of heat. The net loss of about a liter of water through sweating is tolerable, but a much greater loss produces discomfort, increased heart rate and thirst. Water should be taken frequently during physical work in a hot environment to compensate for loss as sweat, but simple response to thirst usually leads to inadequate water intake. Water should be taken in amounts that exceed that needed to quench thirst.

Salt is also lost as the worker sweats, approximately 3–5 g/L. Given that the worker may produce more than 4 L/day of sweat, and that typical daily intake of salt is 10–15 g/day, supplementation by using more salt on food or by taking salt tablets may be necessary in more extreme cases. When the worker does not replace the lost salt, the kidneys retain less water, since the mechanism in the kidney for retaining water involves moving it back into the blood along with recaptured salt. This leads to dehydration. Salt deficiency is diagnosed by observing that urine has abnormally low levels of chloride ion.

The effectiveness of sweating as a method of increased heat loss is increased as: (1) more skin surface is exposed to air, (2) the humidity of the air is lower, facilitating evaporation, and (3) the air is moving so that a layer of water-saturated air does not build up at the surface of the skin.

Both methods of losing body heat depend on the surface area of the body. Increased skin area provides more surface at which blood vessels are losing heat to the air, and more surface from which sweat is evaporating. A slender person has a greater area of skin per kilogram of body mass than does a stocky or obese individual, and is less likely to suffer problems of heat stress, all else being equal.

Heart Rate

Working under conditions of heat stress leads to a higher heart rate. In fact, heart rate can be used as a measure of heat stress. One cause is the need to move additional blood to the surface of the skin as dermal vessels dilate. The additional use of nutrients and oxygen by muscles in a person physically working requires that the heart increase blood output to the muscles. The actual rate increase varies with the fitness of the person's cardiovascular system. A person in good physical condition requires a smaller increase in heart rate to respond effectively to a given rise in core temperature. A rate of 180 to 200 per minute is the maximum rate sustainable in most adults, and that rate can only be supported for a few minutes.

OTHER FACTORS IN RESPONSE TO HEAT

Acclimatization

People *acclimatize* to working in a high-temperature environment. The first day working under high-temperature circumstances, a person may show signs of stress. These lessen with each succeeding day, and disappear within two weeks. Once a worker leaves the high-temperature work environment, acclimatization is lost.

Primarily, acclimatization is seen as an adjustment in the production of sweat. A larger volume of sweat is produced in the acclimatized worker, as much as 8 L of sweat in an 8-hr shift. Another aspect of acclimatization is that the sweat contains only 1–2 g/L of salt, reducing the loss of sodium ion by this route.

Prescription Drugs

There are a number of prescription drugs that can interfere with adaptation to work in a hot environment. Diuretics and antihypertensives may interfere with the volume of blood in circulation or with the heart response to stress, and could increase the hazards of heat stress. A worker taking prescription drugs should consult with a physician about the implications of the particular drug to heat adaptation.

Alcohol and Social Drug Consumption

Binge type alcohol consumption results in the production of quantities of dilute urine. The individual is now much more susceptible to dehydration working in a hot environment, and therefore to heat stroke. As a CNS depressant, alcohol interferes with heat adaptation. Other social drugs can similarly have detrimental effects on adaptation, even to the point of increasing the chances for heat stroke and death.

Age

Older workers have greater problems with work in a hot environment. Capacity to work under stress is reduced by age due to factors such as a decrease in cardiovascular efficiency. Older workers do not begin to sweat as readily as younger workers, and produce a lower volume of sweat when they do. This results in a lessened ability to lower core temperatures. The older worker has a higher core temperature doing the same task as a younger worker, and requires longer to recover during a rest period.

Physical Conditioning

Two workers of the same age, but differing in physical conditioning, also differ in their ability to withstand heat stress. Improvements in the circulatory system result from physical conditioning. A fit person moves blood to the skin more effectively and experiences less strain on the heart doing this in a hot environment.

SUMMARIZING HEAT FLOW

People quantitatively studying heat gain and loss by workers express each aspect of heat flow with a symbol, and place these symbols into equations that facilitate quantitation of each aspect. For a person in equilibrium, that is, whose body temperature is not changing, the sum of additions and subtractions to body heat is zero[2]:

$$M \pm R \pm C - E = 0$$

where

M   is heat production due to metabolism
R   is radiative heat transfer
C   is convective heat transfer
E   is heat lost due to evaporation

A person working as an industrial hygienist is unlikely to use this equation for calculations in the field. However, it is useful conceptually to see the components spelled out in this fashion and to get a sense of the contribution of each.

Metabolic heat production (M) is the heat energy released by the oxidation of dietary material to produce energy storage intermediates (largely the famous ATP molecule), and the use of energy storage intermediates to drive muscle contraction and other body processes. Heat production is hard to measure directly, but can be estimated easily by measuring oxygen intake.[3] Heat production is roughly 5 kcal/L of oxygen. Oxygen intake varies from 0.3 L/min in someone resting to 2–4 L/min in a hard-working fit person.

As a warm body, a person loses heat to cooler surroundings by radiative and convective heat transfer. Radiative heat transfer (R) is the transfer of heat energy through space according to the equation:

$$R = K_R \Delta T$$

where

K_R is a constant
ΔT is the temperature difference between the person and the surrounding objects

---

[2]Another term that could be added to the equation would represent heat conducted between the body and objects it contacts, but this is usually very small and can be ignored.
[3]Using oxygen consumption to estimate metabolic heat production is sometimes termed "indirect calorimetry."

This is a two-way street, however, since hot surrounding objects transfer heat to the person. Similarly, convective heat transfer is the transfer of heat to or from air contacting the skin. The equation is very similar:

$$K = K_C \Delta T$$

where

K_C is a constant
ΔT is the temperature difference between the person and the surrounding air

Once again, the direction of transfer is normally from skin to air, but in a very hot environment it would transfer in the other direction. The constant is modified here by surface area of the body, degree of contact of skin with outside air, and rate of air movement.

Heat loss due to evaporation of sweat depends not on temperature differences, but on the relative humidity (water content) of the surrounding air. Dry air accepts water evaporating from the skin more rapidly, producing a higher rate of heat loss. The higher the skin temperature, the more rapid the evaporation, and again the more rapid the heat loss.

ILLNESSES DUE TO HEAT

It is hard to classify a particular heat-related illness without medical diagnosis. Rises in body temperature and heart rate are common symptoms. The safest approach to emergency treatment in what appears to be a serious situation is to assume the worst case—heat stroke—until adequate analysis by an expert has been performed. We present here classes of heat problems.

HEAT CRAMPS

A worker may experience severe muscle *cramps* working in a hot environment. This is generally described as being due to loss of fluids and salt as a result of excessive sweating, leading to inadequate circulation in muscle tissue. Specific causes of the muscle contractions are a subject of disagreement, but rest and intake of fluids containing a small amount of salt are often recommended as treatment. Careful attention to adequate fluid intake during exertion in hot work environments helps prevent heat cramps.

HEAT EXHAUSTION

Heat exhaustion may also be called *heat prostration*. The shift of blood flow to the skin to lower body temperature creates a demand for greater blood volume. This, combined with a loss of fluids as a result of sweating, leads to circulatory collapse. The victim feels fatigue and weakness before the collapse, followed either by feeling faint or by actually fainting. The victim should lie down, perhaps with the head low, and take slightly salty fluids.

HEAT STROKE

Heat stroke is the result of body temperature rising to very high levels, in the range of 40–41°C (105–106°F). Serious tissue damage occurs at these temperatures, especially to liver, kidneys, and the brain. Systems for temperature regulation may not be working properly as a result of high brain temperatures, so in spite of high body temperature, sweating may have ceased. The person has a headache, is fatigued, and feels dizzy. The pulse rate is rapid, the victim becomes disoriented and quickly becomes unconscious. Convulsions may occur. It is important to bring the body temperature down, but immersion in ice or cold water, as used to be recommended, is no longer considered wise. Cooling skin to that degree causes blood vessels to contract, reducing transport of core heat to the skin. Wetting the skin and moving air to increase the cooling rate by evaporation is useful. This is a very serious problem, with a high risk of death or brain damage, and the victim should receive medical care as quickly as possible.

EVALUATION OF WORKING CONDITIONS

Over the years a number of units have been devised to measure the potential of a work environment to generate heat stress. All have one failing or another. Those that include all relevant factors are cumbersome to use, while those that omit some factor obviously suffer for the omission. The Wet Bulb Globe Temperature index (WBGT) is the method that has been adopted in all proposed heat stress standards, so we shall discuss only that technique. WBGT was first devised by the military, and has the important advantage that it is measured rapidly and easily. Three temperature measurements are involved: *dry bulb temperature* (T_A), *natural wet bulb temperature* (T_{NWB}), and *globe temperature* (T_G).

Dry bulb temperature is simply the measurement of air temperature with an ordinary thermometer. A wet bulb thermometer has a wick that carries water to the bulb of the thermometer. The evaporation of this water cools the bulb, so the wet bulb temperature is lower than the dry bulb temperature. If the humidity of the air is high, evaporation is slowed and the amount of cooling is decreased, just as would be the case with sweat evaporating from the skin. In this case the difference between wet bulb and dry bulb temperatures is smaller. A *natural* wet bulb thermometer is exposed to the existing air movement, and moving air expedites evaporation from the bulb, just as it would from the skin.

Globe temperature is measured by inserting the bulb of the thermometer inside a thin copper sphere that has been painted nonreflecting black on the outside. The bulb picks up heat radiated from surroundings of the globe and from the sun, so globe temperature is likely to be higher than dry bulb temperature.

Each of these three temperatures is multiplied by a fraction, and the sum of these is the WBGT. Outdoor WBGT is calculated as follows:

$$WBGT = 0.7T_{NWB} + 0.2T_G + 0.1T_A$$

The heavy emphasis given to T_{NWB} emphasizes the importance of evaporative cooling to body temperature control. Indoor WBGT is calculated in this fashion:

$$WBGT = 0.7T_{NWB} + 0.3T_G$$

Notice that now the T_A is omitted from the calculation.

Other measures of potential heat stress include the measurement of air velocity. This is an awkward measurement to perform in the field and was omitted as impractical in WBGT. However, the effect of air velocity is inferred by using the T_{NWB} in the calculation.

EXPOSURE STANDARDS

At this time OSHA has not issued standards for protection of workers from heat stress parallel to those for exposure to chemicals or noise. There is a statement in the original OSH Act language of 1970 that provides general coverage charging each employer to protect workers by providing:

"...employment free from recognized hazards causing or likely to cause physical harm."

Recommendations have been sent to OSHA both by NIOSH and by a Standards Advisory Committee on Heat Stress, but at the time of writing of this text a standard has not been written.

ACGIH prepared a set of standards that were published in 1974 (Table 13.1). These are threshold limit values (TLVs) based on the assumption that an acclimatized worker, fully clothed, whose core temperature is at or below 38°C (100.4°F) is not experiencing heat stress. However, one cannot constantly

Table 13.1. ACGIH Permissible Heat Exposure TLVs.

Work Regimen	Work Load (kcal/hr)		
	Light (≤200)	Moderate (≤350)	Heavy (≤500)
	TLV (°C/°F WBGT)		
Continuous	30.0/86	26.7/80	25.0/77
75% Work	30.6/87	28.0/82	25.9/79
50% Work	31.4/89	29.4/85	27.9/82
25% Work	32.2/90	31.1/88	30.0/86

Source: "1993–1994 Threshold Limit Values for Chemical Substances and Physical Agents and Biological Exposure Indices" Reprinted with permission of the ACGIH, Cincinnati, OH.

measure core temperatures of workers, so the standards describe instead the conditions in the workplace, including the WBGT index, the severity of the workload, and the percentage of time engaged in continuous work (Table 13.1). Standards assume workers wear clothing that does not trap air, preventing evaporation of sweat, and that conditions of the rest area are approximately the conditions of the work area.

NIOSH recommended a set of standards[4] in 1972. These standards placed a ceiling limit WBGT temperature for continuous heavy work by an acclimatized worker for over 1 hr of 26° C. Beyond that limit, one of a number of work practices should be employed to assure that core temperature does not exceed 38°C. Due to extensive objection to this proposal, a committee was established, the Standards Advisory Committee on Heat Stress, which modified the proposal as shown in Table 13.2, and submitted the revision in 1974. In 1986 NIOSH updated their proposal,[5] defining recommended alert limits (RALs) for healthy nonacclimatized workers and recommended exposure limits (RELs) for healthy acclimatized workers. NIOSH also defined for each group a ceiling limit (CL) above which workers must be provided with and must correctly use protective clothing and equipment. The qualifier "healthy" refers to a variety of criteria relating to the worker's age, health, degree of body fat, and other physical characteristics. The proposed regulations would require heat measurements to

---

[4]Criteria for a Recommended Standard...Occupational Exposure to Hot Environments, U.S. Department of Health, Education and Welfare, NIOSH, HSM-72-10269, 1972.

[5]Criteria for a Recommended Standard...Occupational Exposure to Hot Environments, Revised Criteria 1986, U.S. Department of Health, Education and Welfare, NIOSH, April, 1986.

Table 13.2. Standards Advisory Committee Recommendations.

Workload	Low Air Velocity ≤1.5 m/s	High Air Velocity >1.5 m/s
	Threshold WBGT Values (°C/°F)	
Light (≤200 kcal/hr)	30/86	32/90
Moderate (<300 kcal/hr)	28/82	31/87
Heavy (>300 kcal/hr)	26/79	29/84

be taken at least hourly during the hottest parts of the workday, during the hottest part of the year and during heat waves.

The International Organization for Standardization (ISO) published standards in 1982 that are based on WBGT measurements and the premise that worker core temperature should not exceed 38°C.[6] Their recommendations closely resemble those of the ACGIH.

KEY POINTS

1. The body has a warm core temperature and a cooler skin temperature. Humans maintain core temperature between 36°C and 38°C. Higher temperatures are termed hyperthermia and lower temperatures are termed hypothermia. Reaching extreme body temperatures is fatal.

2. Excess metabolic heat generated in the core is transferred to the skin to be lost.

3. Core temperature is regulated by the hypothalamus, which expands skin blood vessels to carry more warm blood to the skin and stimulates sweating to cool the skin by evaporation to dissipate excess heat.

4. Sweating leads to a loss of water and salt. Water must be replaced by frequent fluid intake, and in extreme cases salt supplements are needed.

---

[6]"Hot Environments—Estimation of Heat Stress on Working Man Based on the WBGT Index (Wet Bulb Globe Temperature)," ISO 7243-1982, 1982.

5. During heat stress, heart rate increases due to greater vascular volume as skin vessels expand. Muscular activity calls for greater blood flow to muscles.

6. Workers acclimatize to a hot environment, eventually producing a greater volume of more dilute sweat.

7. Older workers control body temperature by sweating less effectively.

8. Use of prescription drugs may—and use of alcohol and other social drugs does—reduce workers' ability to handle heat stress.

9. Thinner and more fit individuals handle heat stress more easily.

10. The body heat balance is the sum of changes in metabolic heat production, radiative heat gain or loss, convective heat gain or loss, and loss by evaporation.

11. Heat cramps are muscle cramps caused by inadequate delivery of blood to muscles during heat stress.

12. Heat exhaustion results from circulatory inadequacy as the combination of greater demands for blood flow and loss of fluids results from heat stress.

13. Heat stroke, often fatal, results when the brain response to heat stress ceases as core temperatures rise to very high levels.

14. Wet bulb globe temperature (WBGT) index, used to evaluate working conditions, is based on dry bulb, normal wet bulb, and globe temperatures. These are weighted and added according to formulae that differ for indoor and outdoor work.

15. No OSHA standards have been written to regulate workplace heat stress. ACGIH has written standards, and NIOSH and the Advisory Committee on Heat Stress have proposed standards, all based on WBGT values.

PROBLEMS

1. A small, windowless cinder block toolshed with a corrugated metal roof nailed to beams laid on the cinder block is constructed on the grounds of the equipment yard of a large construction company in South Carolina with

the original intention that yard employees would spend only brief periods of time inside the building. As the business expanded, the building was changed to a place to house inventory and maintenance records, and workers began to spend longer times in the building. With age, the roof became blackened and rusty. In summer weather, temperatures rise to uncomfortable levels in the building.

A. Does placing an oscillating fan in the room lower room temperature? Explain in terms of the process of losing heat from the body why the situation is improved by such a fan.

B. If instead an exhaust fan is installed in the wall, how specifically does this help? Would air inlets be of help, and if so where would they best be located?

C. The metal roof is contributing to the heat problem. What terms in the equation for heat balance in the body are affected by the roof? What could be done to the roof to reduce the heat problem in the building?

2. Three workers man a station in a plastics recycling facility at which an open vat is heated by gas burners. Washed shredded scrap plastic milk bottles are delivered in an overhead hopper and dumped into the vat to be melted. The vat contents then are stirred by impellers driven from above. Excess water from the washing process boils off the vat. One side of the vat has a walkway with a pipe safety railing and devices to control the hopper, stirrer, and valve to transfer molten contents to a blow molding machine. The mix is sampled and analyzed, and pigments and plasticizers are added by dumping bags of appropriate dry solid chemicals into the vat from the walkway. The molten plastic product is then piped to a blow molding machine to be formed into squeezable red bottles to contain a detergent product. *Worker 1* controls the hopper and stirrer, and is stationed most of the time on the walkway. *Worker 2* carries bags of chemicals up steps, opens the bags by pulling the tear tab across one end, and dumps the contents down a chute into the vat. Worker 2 must wear a face mask, safety glasses, and a plastic coverall because one of the chemicals is a skin irritant. *Worker 3* removes samples from this and other vats from the walkway with a dipper and takes them to the nearby lab station where decisions about amounts of chemicals to add are made. Worker 3 wears a protective face mask at the time samples are removed. The gas burners are controlled automatically by temperature sensors in the vat. Analyze the heat stress potential of these three jobs.

3. Four male construction workers are working outdoors in hot, slightly breezy weather shoveling fine gravel into trenches. Worker 1 wears no

shirt or hat, and his skin is constantly wet with sweat. Worker 2, also shirtless and hatless, is dripping sweat onto the ground. Worker 3 has on an open weave hat and a cotton tee shirt which is remaining wet with sweat. Worker 4 wears a leather cowboy style hat and a blue denim jacket over a sweat-soaked tee shirt. Contrast the effectiveness of body cooling of these four workers.

4.   A.   In a very hot, dry indoor environment such that the air temperature is above body temperature, is it better to expose as much skin as the law allows, or is clothing an advantage?

   B.   In such a workplace, outside air can be drawn in through a water spray, and the air is cooled by the heat removed to evaporate the water (evaporative cooling). T_A is lowered. Will T_G be improved, made worse, or be unaffected by this process? Will T_{NWB} be improved, made worse, or be unaffected by this process?

   C.   Below are data taken before and after installation of an evaporative cooling system.[7]   Calculate WBGT for each set of data.   Do the changes in individual values agree with your predictions above?

	Before Installation	After Installation
$T_A(°C)$	42.8	29.4
$T_G(°C)$	46.1	35.0
$T_{NWB}(°C)$	24.2	24.4

5.   In an aluminum plant, one job involves skimming dross (impurities floating on the molten metal) using a ladle. The worker is receiving a great deal of radiant heat from the molten metal. A sheet of aluminum with a window and holes to reach the ladle is installed between the worker and the molten metal. Once the shield is installed, which of the three temperatures should show the greatest change?

Below are data from such an installation[7]. Calculate WBGT for each set of data.

---

[7]Data is from: R. S. Brief and R. G. Confer, *Med. Bull.*, (1973), **33**, 229; discussed in J. E. Mutchler, "Heat stress: its effects, measurement and control," in G. D. Clayton and F. E. Clayton, *Patty's Industrial Hygiene and Toxicology*, Volume 1A, 4th ed., John Wiley and Sons, New York, 1991.

	Before Installation	After Installation
$T_A(°C)$	47.8	43.3
$T_G(°C)$	71.7	43.3
$T_{NWB}(°C)$	36.3	29.8

BIBLIOGRAPHY

A. J. Kielblock and P. C. Schutte, "Physical work and heat stress," in C. Zenz, *Occupational Medicine: Principles and Practical Applications,* 2nd ed., Year Book Medical Publishers, Inc. Chicago, 1988.

J. E. Mutchler, "Heat stress: its effects, measurement and control," in G. D. Clayton and F. E. Clayton, *Patty's Industrial Hygiene and Toxicology,* Volume 1A, 4th ed., John Wiley and Sons, New York, 1991.

SECTION III

SPECIFIC HAZARDS

In this section of the text, we describe a few important industries. In the course of outlining the procedures and practices of these fields, specific hazards are described and/or methods of protecting workers. This is not meant to be an encyclopedia of industry, but rather a sampling to introduce important industries and show the application of the principles developed earlier in the book.

CHAPTER 14

METALS, PART I

GENERAL PRINCIPLES

Throughout the history of civilization, metals have been at the heart of technological advance. Early civilizations exploited copper and bronze for tools and weapons. Then the iron age replaced a copper and bronze–based society. For a century, the tonnage of steel a nation produced annually was the prime measure of its industrial strength. In recent years we have seen new materials rise in importance. Aluminum replaced wood and canvas in aircraft construction. Now, high-performance aircraft have passed through the period of aluminum construction into one of titanium structures. We are amazed to learn that the annual volume of plastics produced in the U.S. exceeds that of steel, and to read of experiments with plastic automobile engines.

Health problems due to production and use of familiar metals have generally led to regulations designed to prevent their recurrence. In this rush of progress we must learn to deal with a host of new problems. New materials, new uses for familiar materials, and new technologies for preparing and forming metals may bring with them a variety of new hazards. An awareness of the types of problems that have occurred helps us to anticipate the kinds of difficulties to expect. In this chapter we consider some ways to deal with familiar problems, hoping this serves as a guide to dealing with new problems as they arise.

ORE MINING, PROCESSING, SMELTING, AND REFINING

Searching for ore deposits has historically been a search for riches. Early in the American west, prospectors gambled years of their lives and endured hardship and danger in the hope of finding a rich ore lode. Today, exploitation of a newly discovered ore deposit has become a more controversial issue, with developers and environmentalists often in opposition. For the most part, the richest deposits have already been utilized, and new technology has been devised to allow us to make use of "lower-grade" deposits. We have even returned to the waste piles of earlier operations to find exploitable sources. On the negative side, using less-concentrated sources means that a greater volume of the Earth's surface must be disturbed to obtain the same amount of raw material. Increased recycling of already extracted metals becomes more important in such a scenario, and increased amounts of scrap are being used as the raw material—a development generating its own set of hazards.

MINING

Underground Mining

Exploitation of ore finds since the earliest times of civilization often involved traditional mines, with shafts and tunnels drilled out by miners following the seam of ore. Hazards include dust-filled air, possible collapse of shafts, use of explosives, noise, high temperatures, flooding, and accidental explosions (especially in coal mines). Silicosis has been a serious health threat to "hard rock miners"—miners drilling in silica rock. In ancient times the health of miners was not a concern, since this work was done by slaves. Since the time of the industrial revolution there has been an effort to improve the safety and quality of life of miners. Wet drilling methods (drilling under a stream of water to prevent dust in the air), wetting down broken rock and ore before transport, placing bags of water in the hole with explosives to reduce dust production, the use of protective breathing devices when necessary, and monitors to test air quality are examples of measures employed to improve miners' health and safety.

Ventilation is very important in a mine to remove dusts from drilling or blasting, exhausts of internal combustion engines, gaseous end products of welding or blasting, and other contaminants. Combustion and blasting products include CO_2, CO, SO_2, and nitrogen oxides. Methane is a problem in coal mines, and in recent years we have recognized problems with the radioactive gas radon in some underground mines. Ventilation is also needed to supply oxygen to workers, since they are largely cut off from the atmosphere in a mine shaft. Where underground temperatures are high, ventilation is a means of cooling the work environment. With long, dead end shafts and limited access to fresh air, underground mines present unusual challenges for the design of good ventilation systems.

Mining presents serious problems of noise exposure. These arise particularly from blasting and drilling, but also from some ventilating equipment and machinery used to move ore out of the mine.

Surface Mining

Where ores lie at or close to the surface, open pit mines (surface mines, strip mines, open-cut mines, quarries, placer mines) are employed. Huge power shovels load ore directly onto trucks or railroad cars. Dust-filled air and noise are still hazards, but the job is very much safer than below-ground mining. Dust exposure is less serious in the open air, where it disperses rapidly, and in surface mining relatively few workers are close to the dust sources. Wet drilling may be used to reduce worker exposure, and in large drilling rigs the worker may be in a ventilated control cab. Broken rock and ore may be wetted down before loading and hauling, and once again the operators of the equipment may be in ventilated cabs. Dependence on water spraying to control dust may require that protective practices change in regions with cold winters as temperatures drop. Road surfaces are always temporary in surface mines, and may be serious sources of dust if not coated or wetted down appropriately. Under circumstances where workers leave the ventilated cabs occasionally, training is important to inform workers about the hazards of airborne particulate and how to minimize exposure. Silica dust is still a special threat, and air sampling is important to assure exposures are within recommended limits.

There is a tendency to be more concerned about noise in indoor situations, perhaps because of problems with the reflection of sound from indoor surfaces, but the drilling and earth moving equipment in surface mines produces high noise levels, and blasting is a special problem. The cabs from which workers control the equipment can be designed or retrofitted to control sound. Sound levels should be monitored and appropriate hearing protection provided where necessary, taking into consideration the time spent by the operator outside the cab.

Open pit mines, particularly for iron or copper, are some of the most massive examples of man altering the face of the Earth. These huge holes in the ground are one of the most negative aspects of this sort of endeavor. Historically, once the ore or coal had been removed, the sites were abandoned, and remained as nonproductive land and scars on the landscape. Efforts to restore old mine sites and laws to prevent present day operations from becoming the scars of the future have improved this situation, but one doesn't need to be a dedicated environmentalist to have serious reservations about allowing new surface mining operations in areas of natural beauty or near where one lives.

Mine Safety Regulations

In recent years three laws have defined the federal role in mine safety. The federal Metal and Nonmetal Mine Safety Act[1] of 1966 required annual inspections of underground mines, empowered the inspectors to issue notices of violation and to close mines, required worker training, mandated that injuries be reported, and established procedures for generating new health and safety

---

[1]Public Law 89-577 (1966).

standards. The federal Coal Mine Inspection Act[2] of 1969 extended coverage to surface mines. Benefits were assured to workers with black lung disease, and plans for research to identify health and safety hazards and control them were addressed. The powers of enforcement of federal mine inspectors were increased. Before the passage of the federal Mine Safety and Health Act[3] of 1977, responsibility for federal regulation of mining health and safety was in the Bureau of Mines in the Department of the Interior. The new law shifted it to the Department of Labor along with OSHA, setting up a parallel enforcement agency known as the Mine Safety and Health Administration (MSHA). Thereafter NIOSH was charged with responsibilities for researching questions of mine safety and health, and for recommending new standards. This legislation completely replaced the Federal Metal and Nonmetal Mine Safety Act of 1966 and extended to all miners the provisions directed at coal miners in the 1969 law. The actual health standards are found in 30 CFR Chapter 1. As with other standards, they are a mixture of general requirements and very specific statements concerning particular serious concerns. As a result of new technology and enforcement of government regulations, great progress has been made toward improving conditions for mine workers, but mining, particularly underground mining, is still a relatively hazardous occupation.

ORE DRESSING

Typically, ores contain only a small percentage of the desired metal, so *benefaction,* processing to remove a proportion of the clay, dirt, or rock that does not contain the desired metal, is necessary. Ores must be broken down into a powder. This is termed *comminution,* and begins with breaking up the large fragments from the mine using large crushers. The resulting small fragments are transferred to a rod or ball mill to be ground to powder. As these operations proceed, *sizing,* the separation of powdered product from the larger fragments, is done. This involves the use either of *screens* or *classifiers.* In classifiers, smaller particles are separated from larger ones by their ability to be suspended in a stream of either water or air.

Then the metal-rich components must be concentrated by separation from nonmineral-bearing dirt and ground rock. Techniques for this include *froth flotation,* which separates ores from impurities by chemically modifying the surface of the mineral-bearing particles to reduce their attraction to water, then generating foams that selectively attach the mineral particles. The chemicals used are sometimes specific for the particular mineral. pH adjustment is important, and is accomplished by use of bases such as sodium hydroxide,

---

[2]Public Law 91-173 (1969).

[3]Public Law 95-164 (1977).

sodium carbonate, or lime, and by acids such as sulfuric or sulfurous acid. *Depressant* reagents may be added to prevent the flotation of an unwanted mineral, while allowing another to be collected. These include sodium silicate, sodium fluoride, sodium cyanide, lime, chromates, phosphates, and polymers such as starch. Some of these are also used to clean mineral surfaces after separation. Foams are produced using organic detergent-type compounds. Obviously the harmful properties of these chemicals must be considered and worker exposure monitored. Worker training about chemical hazards is important. Finally, the separated solids are usually "dewatered" in a tank in which particles settle and are drawn off the bottom, while the clarified water layer is removed from the top. Wastes are often pumped as water suspensions to disposal sites.

Minerals may also be separated from waste by differences in density. This may employ air or water as the fluid medium. The chief dangers are from dusts, especially where the ore includes such hazardous components as toxic metal impurities or silica rock. Powdered minerals may be mixed with binders and rolled into balls, which are baked into hard pellets in a furnace. This form will be delivered to the smelter.

The mills and crushers described above are major sources of noise, as are the ventilating systems, pumps, conveyors, motors, and shakers involved. Noise levels should be measured at all workstations and hearing protection required as needed. Ventilation may serve to control heat as well as dust if furnaces are involved.

Piles of waste from below-ground mines or from crushing operations can represent a health and environmental hazard as rain percolates through and extracts acids or traces of potentially toxic materials such as metal salts.

REFINING

Obtaining a free metal pure enough to be useful from upgraded ore is the process termed *refining*. Three refining methods are employed: smelting, electrolytic purification, and hydrometallurgy.

Smelting

In a smelter the ore is converted to free metal by a heat and reduction process. The largest-volume such process, the smelting of iron, is presented in detail as an example. Iron smelting is done in a furnace over 100 feet tall called a blast furnace, which is capable of producing as much as 4000 tons of iron per day. The raw materials are iron ore (an iron oxide), crushed limestone ($CaCO_3$), and coke (a carbon material made by roasting the volatile materials out of coal in a coke furnace). Blast furnaces are usually located on waterfronts in regions close to sources of these three components so that the cost of transportation, preferably by boat, of the huge volumes required of these raw materials is minimized. Major operations are found in such cities as Chicago, Detroit, and Cleveland, utilizing ore from Minnesota and northern Michigan, limestone from quarries surrounding the lakes, and Appalachian coal, all transported on the lakes themselves by long freighters.

Coke may well be produced from coal onsite. First, it is important to realize that any time coal is handled in large amounts, the accumulation of dust in the air generates a risk of explosion. Coal is heated in the absence of air in the coke plant to drive out volatile components. Coke plants have traditionally been dirty operations, adding aromatic organics and other unpleasant or hazardous gases to the air. Modern operations are cleaner as the result of trapping the aromatics, which are salable, and using the coke oven gas, which contains hydrogen, methane, and carbon monoxide, as a fuel in the plant (even in the coke oven itself). Worker exposure to the aromatics is regulated by OSHA using a special index called the BSFTPM (benzene-soluble fraction of total particulate matter) to describe the aromatic emissions. SO_2 is produced as H_2S in the coke oven gas burns and can be a significant hazard if the gas is not scrubbed.

Turning to the blast furnace itself, limestone, coke, and ore are poured in the top. Hot air enters the lower part of the furnace, so that coke burns and raises the temperature. This air may be enriched with a small amount of pure oxygen, which speeds up the iron making process. At these elevated temperatures two reduction reactions occur. In one the coke is the reducing agent:

$$Fe_3O_4 + C \quad --------> \quad Fe + CO_2 + CO \quad \text{(not balanced)}$$
$$\text{ore} \qquad \text{coke}$$

In the other, the carbon monoxide is the reducing agent:

$$Fe_3O_4 + 4\ CO \quad --------> \quad 3\ Fe + 4\ CO_2$$

The high-density molten iron flows to the bottom of the furnace, and is either poured into ingots of "pig iron," or is transported in ladles directly to furnaces to be made into steel. Limestone reacts with the rock, sand, and clay impurities to form a silicate product called slag, which floats on the iron and is removed through a port higher on the furnace than is the iron. Hot gases from the top of the furnaces are used to preheat air entering the bottom of the furnace, and the remaining CO in these gases is burned as a fuel in the plant.

Pig iron is rich in carbon and has some lesser impurities. Most often it is transferred still molten from the blast furnace into a *basic oxygen furnace* where more limestone is added and oxygen is bubbled through to oxidize the carbon.

Scrap steel usually is also part of the charge of the furnace.[4] Silicon, phosphorous, and manganese impurities in the iron are oxidized, and react with the limestone to produce calcium silicates, phosphates, and manganates, which float as slag on the surface of the molten steel. As a result of adding oxygen to the furnace, basic oxygen furnaces produce clouds of iron oxide particulate, which must be collected. This oxide can be fed back into the blast furnace. Steel produced has a range of properties depending on the final carbon content (less carbon yields a more malleable product; more yields a harder product), and the possible addition of other metals can produce an endless array of alloys.

Heat stress and noise are important problems for the workers. Particulate levels and gases such as CO and SO_2 may rise to high levels in plant air, especially in handling the raw materials and if aspects of the operation need maintenance.

Electrolytic Purification

Aluminum and copper are important examples of metals purified by electrolysis, and the details for aluminum are presented. Bauxite (aluminum ore) is surface mined, and must be converted to a relatively pure material for electrolytic processing. Ore is first crushed and ground, then the alumina (Al_2O_3) is "digested" by dissolving it at elevated temperatures in caustic (sodium hydroxide) to produce a solution of sodium aluminate[5]:

$$Al_2O_3 + 2NaOH -----> 2NaAlO_2 + H_2O$$

Suspended solids are removed by sand traps, settling tanks, and filters. The sodium aluminate solutions are cooled, and hydrated aluminum oxide is precipitated in tanks by seeding the solution with crystals of the product and thus reversing the formation of the aluminate salt. The product is dewatered, then dried in high-temperature rotary kilns. Worker heat stress is a potential problem at all stages where elevated temperatures are employed. Throughout this process strongly alkaline solutions and suspensions, sometimes at elevated temperatures, are commonplace. All workers likely to be exposed to these solutions require protective clothing, eye protection, and training. Careful ventilation at sources of dust is important.

The electrolytic method requires first that the source of metal be melted or dissolved. Alumina has a very high melting point, but at lower temperatures it

---

[4]The impurities in such scrap should be known ahead of time, and hazards from oils or other metals in the scrap should be considered during operations.

[5]Heat exchangers used in these processes are cleaned with sulfuric acid by workers in protective clothing.

dissolves in molten cryolite ($AlF_3 \cdot NaF$). The electrolytic cells or *pots* are lined with carbon, which serves as the cathode, and anodes made of carbon bound with coal tar pitch dip into the molten mixture. At the cathode the aluminum ion becomes free aluminum, which collects in a pool at the bottom of the cell to be tapped off, while the oxygen produced at the anode reacts with carbon to become carbon dioxide or carbon monoxide:

$$2Al_2O_3 + 3C \xrightarrow{\text{electrical energy}} 4Al + 3CO_2 \; (+ \; CO)$$

Workers in the "potroom" have the potential of heat stress problems. The possible presence of airborne fluorides from cryolite is a special hazard of this industry. Designing ventilation systems for the pots is complicated by the need for addition of alumina and replacement of spent anodes at frequent intervals. Recent designs enclose the pots, add alumina from the center above, remove spent anodes from the side by opening doors, and run the stream of ventilating air across the top of the pots. General exhaust ventilation of the room is assisted by the high temperatures at which equipment operates. Air entering through floor vents is heated and rises to the room ceiling for exhaust. A creative idea in emission control is the use of alumina destined to be added to pots as the adsorbent in "dry scrubbers." Alumina dust is injected into the exhaust air stream, then is later collected in baghouses. The alumina adsorbs the fluoride without creating a filtration product that has to be disposed of carefully. There may be special filters in the ventilating system with pads of sodium carbonate to react with gaseous fluorides. Where needed, personal respirators are designed with alumina-impregnated filters that absorb the fluorides. Analysis of air samples for fluorides is backed up by testing workers' urine at intervals for fluoride content.

Cells operate at only around 5 volts, but at hundreds of thousands of amps. The electrical demands of this process are very high since each aluminum ion requires three electrons to become an aluminum atom, so these plants are located in regions of inexpensive hydroelectric electricity such as the Tennessee Valley or the Pacific Northwest.

Hydrometallurgy

Metals can be dissolved out of ores in a process called *leaching*. This is done with zinc, silver, gold, magnesium, manganese, and copper. In the case of copper, the metal is dissolved out of the ore using sulfuric acid, creating a solution of $CuSO_4$. The copper sulfate solution is then transferred to a bath where the copper is plated out of the solution onto a copper cathode. Other leaching agents include ammonia and sodium cyanide. When a metal is extracted as a salt, it is sometimes displaced from solution by adding a more "active" metal to the solution. The more active metal goes into solution as ions, and the desired metal precipitates out as free metal.[6] Using copper again as an example, copper is precipitated out of dilute leachate by adding iron to the solution:

$$CuSO_4 + Fe ------> FeSO_4 + Cu$$

Important worker hazards here involve exposure to a variety of chemicals, some toxic.

CASTING

Molten metals are cast (poured into molds) in a *foundry*. Since the early days of the industrial revolution, castings have been commercially made in steel and iron, bronze, and brass. More recently other materials have been handled in this fashion. These include lightweight aluminum and magnesium, high-strength titanium, chromium, nickel, zinc, and a vast array of *alloys*—mixtures of two or more metals. Cast iron is such an important product that foundries are often classified into two groups: ferrous (iron) and nonferrous (all the rest). Some foundries produce small numbers of castings, as in the building of specialized machinery, while others produce a large number of identical products, as would be the case in automobile production. Although casting principles are the same in each case, the process and source of hazard are generally quite different. In either case, however, foundry air may become contaminated with a number of metal products, combustion products, or silica particles when sand molds are used. Metal fume fever is a response to the presence in the lung of oxides of several metals, and has the symptoms of influenza. This is discussed under the headings of individual metals in Chapter 15. Other hazards associated with foundries include heat stress, noise, and vibration.

Especially with some of the more recently employed metals and when the source is scrap metal, there is a problem of fumes—either of the metal itself or of impurities—entering the atmosphere. For example, scrap automobiles often contain traces of lead. This is serious not only because lead is an unusually toxic metal, but because it boils at a lower temperature than iron melts, and so is likely to be a serious air contaminant. Other scrap may be oily and release partially burned products into the air when the scrap is heated.

All foundries melt the metals, requiring the generation of high temperatures. Modern installations often use electric induction furnaces to do the actual

---

[6]Remember the single displacement reactions you learned in general chemistry?

melting. This is a relatively clean operation, but in most areas the cost of electricity makes it an expensive one. For this reason, and to melt the metal more quickly, even these foundries use gas or oil burners to preheat the metal. Furnaces used in a foundry may produce carbon monoxide and sulfur dioxide as combustion products. Furthermore, it is sometimes necessary to dry the surface of the mold with a torch before pouring in the metal. Anytime burning gas strikes a colder surface, gases burn incompletely to produce carbon monoxide. Furnaces burning fossil fuels must be ventilated, and the choices of local or general ventilation are important. General ventilation by way of ceiling-mounted blowers is usually an inefficient method of handling furnace emissions, unless the blower is directly over and close to the furnace. A local system that selectively draws air from the site of combustion is more efficient and cost-effective. However, especially in a system added later to older equipment, local ventilation may need to be disconnected in order to add new metal to the furnace or to transfer the molten metal (*tap* the furnace). In larger-scale operations, metal may be transferred from the furnace to the site of pouring in a ladle carried on an overhead monorail or by a crane. It is difficult to supply local ventilation in such a situation, although moving ducts that follow the ladle have been designed into some operations.

Casting may be done in permanent molds, which are repeatedly reused. This is particularly true for lower-melting-point metals. However, casting is most often done in molds of shaped sand (Figure 14.1), which is high-melting-point silica (quartz). This is very common for casting iron, by far the most frequently cast metal, but is also used for bronze, aluminum, and other metals. Sand is placed in a holder called the *flask,* and an exactly shaped depression is made in the sand using a *pattern.* Patterns are the products of skilled workers, may be made of wood, metal (iron or bronze), or plastic, and are reused many times. If the object to be formed is hollow, a *core* is suspended in the sand hollow formed by the pattern. Often the mold has two flasks that each have a depression in the sand, and are bolted together to provide the complete shape of the final product. In this case the top flask is called the *cope* and the bottom is called the *drag.* A hole to allow the metal to enter the mold, a *sprue,* must either be part of the pattern or must be cut into the sand. *Risers,* channels through which molten metal rises to show that the mold is full, are sometimes added.

Sand is mixed with agents to hold it in the correct shape. Many foundry sands contain clay, either naturally present or added. When the sand particles are coated with clay, and water is added, the particles adhere to produce a strong and accurate form from the pattern. Chemicals such as furan or phenolics are also used, and the shaping is sometimes done with the mixture hot, sometimes cold. Especially when the processing is done hot, or when chemicals contain catalysts and reaction follows, such chemical additives can be the source of gases, so ventilation should be provided.

Sand must be packed tightly into the mold, and several methods are used to accomplish this. In the most labor-intensive approach, sand is shoveled by hand

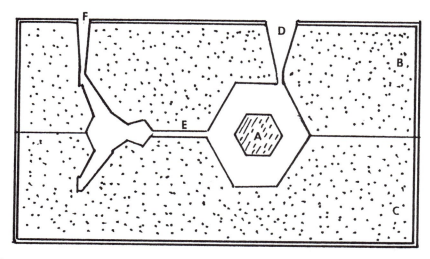

Figure 14.1. A sand mold is shown here ready for use. The shapes of objects to be cast have been pressed into sand using a pattern. A hollow casting is formed by suspending a core (A) into the depression. This mold is constructed of the two parts, the upper half called the cope (B) and the bottom half called the drag (C). Molten metal is pured down the sprue (D), fills the first space, passes through the runner (E) to fill the second space and finally fills the riser (F). When metal appears at the top of the riser, the mold is full. Ater cooling and removal from the mold, metal in the sprue, runner and riser is knocked off.

and rammed by the workers around the pattern. In a *jolt machine,* the pattern on the bottom is covered with sand. The entire flask is then dropped and stopped suddenly, packing the sand down around the pattern. In a *squeeze machine* a ram presses the sand into the flask. In high-production foundries, these operations are combined in a single machine—a *jolt rollover squeeze machine*—that also removes the pattern, leaving the mold ready for the molten metal. Such machines are sources of noise and vibration. Particularly for large castings, sand is taken from a conveyor by a device called a *sand slinger,* and is "firehosed" into the flask using an arm with an impeller that can drive the sand at perhaps 10,000 feet per minute. Sand slingers are an obvious source of silica particulate entering plant air.

Respirable sand particles are always a health threat, creating the possibility of workers developing silicosis. Sand must be handled extensively, being moved from storage to the site where forms are filled, poured, and pressed into the forms by jolt-squeeze machines. After casting, sand must be blown off the casting, and used sand is moved to the next location, perhaps back to be reused. These areas should be ventilated, and using remote-controlled machines to perform the functions keeps workers away from dusty zones. Moving sand about the foundry through pipes by a pneumatic conveying system is a very clean option for this process. Sand spilled on the floor can be a significant threat. Gratings on the floors and spill pits can collect this sand and prevent fans or traffic from moving fine dust into the air.

One problem for designing effective ventilation is that sources of contamination move around in the foundry. For example, the casting process involves pouring molten metal from the ladle into molds. Following this, either the molds or the ladle must move on so that the next mold may be filled. How this is accomplished depends on the size of the molds and the type of production, i.e., whether a few castings are individually produced or a large number in an assembly line fashion. Small molds may move along a conveyor, pausing to be filled, then moving on to a cooling area and eventual removal either of the sand from the casting or of the casting from the permanent mold. When hot metal is poured into sand molds, organic materials used as binders are destroyed, and the resulting smokes and gases, which continue to be released for some time, must be captured, preferably by local ventilation. There is a second release of these air contaminants when the mold is opened, allowing sand that had been distant from the metal, and therefore still contains organic material, to come in contact with the hot metal. Molds on a conveyor should move directly into the hood of a local ventilation system to trap these gases.

In sand casting systems, the sand is removed from the casting in a "shakeout" process. Some sand may remain, to be removed by air hoses. Sand may then be collected for reuse and transported back to the mold forming area. Ventilation is important in all these steps to prevent silica dust from entering the atmosphere.

Castings need to have "appendages" removed. The sprue and riser and any runners or gates that move metal from one part of the mold to another must be knocked off. Where two halves of a mold join together there is a ridge of metal to be removed called the flashing. These steps may be accomplished by use of air-driven chipping hammers and grinders, which produce a lot of dust and noise. Often they are performed in a separate room to facilitate ventilation.

Good housekeeping is an important part of maintaining a safe work environment. Allowing dusts to lie about invites further exposure by inhalation. Clothing can become heavily contaminated. Cleanup should be done with a thorough understanding of the nature of the dusts. Sweeping can add quantities of a fine particulate to the air. Hosing an area can allow reactions between oxides and water to produce harmful gases. Cleaning furnace burners should be done with care to prevent exposure to vanadium oxide residues from the fuel.

FORGING AND STAMPING

Modern industrial forging is an update of the blacksmith tradition. Blacksmiths heated metal and hammered it into horseshoes, swords, or other objects by hand. There are a number of variations of the forging technique, and we will not attempt to cover all of them in this discussion. Instead the focus is placed on a few typical operations. Metal is often, but not always, heated before being forged in order to make deformation into the desired shape easier (Figure 14.2). Deformation typically involves hammering or squeezing to shape a piece of metal by forcing it into a die with one or more blows of a press, by rolling a length of material into a desired cross section between grooved rollers, or by extruding through a die to form wire or rods of a desired cross-sectional shape.

In a typical operation, metal stock is cut to the appropriate-sized piece and is heated to the correct temperature in a furnace. It is then placed in a die, the tool that provides the correct shape for the finished forging. The die has a top half and a bottom half, and includes places for excess metal to flow when the shaping is performed. A single *"hammer"* blow then deforms the metal, or a *press* squeezes the metal into the shape of the die. Afterwards the *flash,* the excess metal, is *trimmed* from the part by placing the part on a trimming die and shearing it off. Internal stresses are produced in metals by the shaping processes. Carefully controlled heating can cause these stresses to be released, followed by controlled cooling, a process called *heat treatment*. Processing metals at high temperatures in air often produces oxide coatings, which can be removed by methods described in the next section.

The most important hazards associated with forging center on heat stress and noise. (See Table 14.1.) Fuel-burning furnaces produce contaminant gases, most seriously carbon monoxide. Operating hammers or presses requires safeguards that the worker does not have a hand in danger during the processing of the metal. Controls that require two hands to activate the machine are useful.

Another common process is stamping. Sheet metal is the usual raw material. Objects are shaped in dies, using presses, as in forging. Heating the metal is omitted in stamping, which avoids many sources of air contamination. Stamping plants generally form large numbers of parts at high speed, feeding rolls of sheet metal into a press. Noise and the possibility of accidental injury are significant hazards.

CLEANING METAL SURFACES

The cleaning of metal surfaces can be an important step in jobs such as applying paint or other coatings, welding, or electroplating. Cleaning is often done by sanding or sand blasting the surface, which can remove oxidation and paint, and can remove surface imperfections to improve the fit of the parts. Hazards include airborne abrasive particles, possibly silica, and airborne particulate of the metal, metal oxides, or a coating such as paint that was on the surface. Noise and vibration are other possible hazards associated with this process. Good ventilation is important, and often hoods placed in the path of flying particulate improve the efficiency of particulate trapping. Good housekeeping is also important, so that residual particulate in the work area doesn't later become airborne.

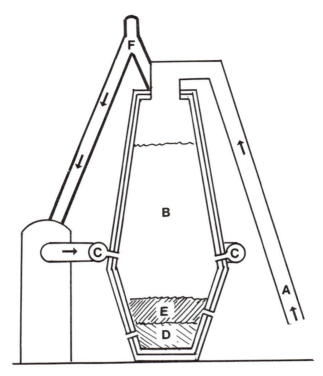

Figure 14.2. Blast furnace: iron ore, coke, and crushed limestone are moved by conveyor (A) into the furnace (B). The furnace has an inner lining of a refractory (fire brick) to withstand the high temperatures. Hot air (over 500°C), possibly enriched with oxygen, enters the furnace (C) and the coke burns, producing carbon monoxide. Carbon and carbon monoxide reduce the iron oxide to iron (D), while the limestone reacts with the impurities to produce slag (E). Hot exhaust gases are used to preheat the air (F). The furnace operates continuously, with iron being tapped every 5 hours and slag every 2 hours at the bottom of the furnace.

Chemical methods such as acid or alkaline cleaning may also be employed. Acid cleaning (pickling, or acid descaling) is usually done with hydrochloric or sulfuric acid, both of which require careful handling. Alkaline descaling usually involves the use of sodium hydroxide (caustic). Protective clothing and eye protection are important, and employees should be trained about the hazards of contact with these chemicals, how to handle them safely, and the emergency procedures to use in case spills or splashes do occur.

Table 14.1. Typical Forging Temperatures (°F).

Material	Temperature
Steel	2200–2400
Stainless Steels	1600–2200
Copper and Alloys (brass and bronze)	1100–1600
Aluminum and Alloys	600–1000
Titanium and Alloys	1500–1900
Magnesium Alloys	500–1000
Molybdenum and Alloys	1900–2700

Degreasing with solvents may be accomplished by dipping in solvent, vapor cleaning, or wiping with solvent-soaked rags. Dipping in a solvent tank is straightforward, but the dirt and grease remain in the solvent and after some time become significant contaminants on the dipped part, as solvent clinging to the surface evaporates from a newly dipped part. As parts are removed from the solvent, there is a possibility of significant amounts of vapor contaminating the air, and handling the newly dipped part can lead to dermal exposure. Ventilation and worker protective clothing are necessary.

Vapor degreasing is done in a degreasing tank. Such a tank has high sides and heating coils on its floor. A layer of solvent is added to the bottom, and heating produces a high concentration of vapors above that solvent. At the top of the tank walls are cooling coils that condense the vapors back to liquid, which then runs down the sides to the liquid layer in the bottom. The part to be cleaned is lowered into the solvent vapors, which condense and run off into the liquid below, carrying the dirt and grease along. Cleaning may be assisted by spraying the part with a stream of hot liquid. An advantage over dipping is that the vapors are free of dirt and grease, which accumulate below. Used properly, a degreasing tank can contribute only a relatively small amount of vapor to the plant atmosphere, and this vapor can be handled by simple ventilation.

Misuse of a degreaser can cause excessive vapor loss to the room. For example, if parts being cleaned have depressions, these can carry liquid solvent out of the tank. Hanging the part with depressions facing down or rotating the part before removal can minimize this problem. Elevating the part too rapidly from the tank can carry excess solvent along. Spraying the part with hot liquid must be done below the level of the cooling coils, and spray must be directed down into the tank, not splashed upwards and out into the room. The cooling system and the thermostat on the heater must both be in good working order to avoid vapors boiling over into the room. Finally, the unit is normally sold set up to be used with a particular solvent. Changing to a more volatile solvent, perhaps a cheaper one, without readjusting the thermostat can lead to the vapor concentration in the tank rising to levels the cooling coils cannot handle.

Degreasing tanks have commonly used such solvents as trichloroethylene, methyl chloroform, perchloroethylene, methylene chloride, and trichlorotrifluoroethane. There is a trend toward using more fluorocarbons which, even though they are more expensive, require less energy to boil the solvent and have a lower escape rate from the apparatus. Substitution of fluorinated hydrocarbons will also be encouraged to reduce chlorinated hydrocarbon use because of ozone layer problems. When changing solvents, the toxic properties of the new solvent should be carefully assessed, since often one of the easiest ways to lower the level of hazard in a plant is to substitute a less toxic material into an operation.

ELECTROPLATING

Electroplating is a method of applying a decorative, corrosion-resistant or wear-resistant finish to metal surfaces, and has been a large industry for many decades. In principle, the object to be plated is hung in a conducting bath as the cathode in a direct current circuit and a sample of the metal to be applied to the surface is hung as the anode. Upon application of the electrical current, metal enters the solution at the anode as ions and is converted back to metal atoms at the surface of the object that is the cathode. Sometimes more than one layer of metal is plated onto the object. For example, in chromium plating, a steel object may first be coated with copper, then nickel, before the final bright chromium surface is added.

Applying a uniform coating is a challenge, especially to irregular objects, since projections coat more heavily and indentations less heavily. It is also important that the surface to be plated be very clean, since impurities, especially grease and oil, block the plating action. For this reason, objects to be plated are generally run through cleaning baths that may contain acid, caustic, or organic solvents before plating. The cleaning agents must be rinsed off before the object enters the plating bath.

Plating, like aluminum purification, is a low-voltage, high-amperage operation. After the object is plated in the bath, it is rinsed in a series of tanks of water. Sometimes a second or even third metal is added, requiring more plating baths and rinse tanks. A typical installation has objects being moved automatically from one to another of a long line of tanks.

Doing good electroplating is a fine art. Correct orientation or movement of the object during plating affects the outcome. Bath temperatures are regulated, and plating baths are prepared by following carefully worked out recipes designed to produce the best plated surface with lower electrical consumption. Assembling the chemicals is an expensive operation, so it is hoped a bath will provide long service. Furthermore, the problem of disposing of the sizable contents of a plating bath is complicated by the fact that many of the bath components are toxic and damaging to the environment: acids, alkalies, cyanide compounds, and many metal salts. Properly treating all these components for disposal is expensive. Even disposing of the water in rinse tanks presents problems.

An ingenious system has been developed to minimize the need for disposal. Since plating baths are often heated, there is significant loss of water due to evaporation. Rather than adding replacement water directly to the plating tank, water is brought back from the first rinse tank, which in turn is replaced from the rinse tank beyond. The water eventually added to replace that lost by evaporation goes into the last rinse tank. Chemicals rinsed from the objects are returned to the plating bath along with the water, rather than becoming hazardous waste for disposal, and the chemical components of the plating tank are therefore not depleted.

Finally, plating baths may develop problems on use. Platers have become skilled in "nursing the bath back to health" by adding chemicals found to be useful for specific problems, thereby extending the life of the bath. Once in operation, a plating bath may be used continuously for long periods.

Hazards to workers focus heavily on the use of toxic or corrosive chemicals and the dangers associated with electrical machinery. Noise is not a significant problem in the plating itself, although other operations such as polishing or other metal finishing that may be associated with the plating may present difficulties. Baths are heated, but temperatures are usually moderate, so heat stress is easier to avoid.

WELDING

Welding may be subdivided into that done with a flame, using such fuels as acetylene, and arc welding, where a high-voltage differential is generated between the work surface and the welding rod. A bond is formed by melting the metal, allowing the two surfaces to flow together. Additional metal may be added from a welding rod that is melted in the flame or is actually one of the electrodes in the arc process. Very strong bonds can be made that are air- and watertight. Welding is most successful with ferrous materials, although it can be done with a number of nonferrous metals. Flame welding torches are also used for cutting, simply by melting completely through the object. In some fabrication and repair jobs both processes are employed in completing a project.

Numerous workers are employed as welders, and there is a wide variation in the degree of hazard they experience on the job. Welding done by such major employers as automobile companies is often performed by automatic machinery, with very low hazard to employees. Welders are also employed by companies too small to be regulated by OSHA, such as auto repair shops, and can potentially involve high levels of exposure to toxic substances. While many jobs in industry are performed at a workstation, and safety features such as guards and adequate ventilation can be installed as part of that station, welding is frequently a repair operation, and welders go to the site of the job to work. Ventilation could well be completely inadequate, and the job could even be in a confined space in which the products of the welding operation accumulate.

Obtaining clean surfaces before welding produces better, stronger welds. Methods described in the cleaning sections may be used. If chlorinated solvents are used, it is important to completely remove the solvent before welding. Such solvents may decompose in the heat of welding to produce more toxic products such as chlorine, hydrogen chloride, or phosgene.

In addition to previously described methods, grease and oxide may be removed by use of a hot flame. Alternatively, flux can be added to a contaminated surface, and successful welding performed. Fluxes contain metal sulfates, chlorides, or fluorides, which dissolve oxides. Much of the flux material is volatilized in the heat of welding, and harmful airborne materials may result. Ventilation is important.

Burning a surface clean or cutting old metal structures can be hazardous because of coatings on the metals. Old paint may contain lead, for example. There is a report of welders cutting apart metal beams in an old rail station in Glasgow, Scotland, and being seriously dosed with lead from the paint. Asbestos should be removed from any structure to be welded or cut. Some plastic coatings, such as Teflon®, produce harmful fumes when vaporized. Welding metal plated with cadmium as a corrosion inhibitor would result in production of potentially very toxic cadmium and cadmium oxide fume. It is also important to be sure tanks and other containers are empty before doing any welding. Toxic, flammable, or explosive contents in a tank being welded could result in a serious incident occurring.

The hazards to the eyes and skin due to sparks and fragments of hot metal are well recognized, and a mental picture of a welder at work always includes a face mask, gloves, and other protective clothing. Dangers from chemical exposure are less well recognized, and vary with the job being performed. First, consider the composition of the metal being welded, and the possibility of fumes arising as the work progresses. The composition of the welding rod is the next consideration. Beyond that, one must imagine all the other sources of airborne contamination. Combustion products of the flame may accumulate. Electric arcs produce the irritating and oxidizing gas ozone.

KEY POINTS

1. Underground mining requires careful ventilation due to particulate from drilling and blasting, and gases from combustion and blasting. Noise is a problem and temperature extremes may occur.

2. Airborne particulates and noise can be hazards in surface mining.

3. Mine safety regulations largely flow from three laws passed in 1966, 1969, and 1977. Standards are found in 30 CFR Chapter 1.

4. Ores are first crushed and ground in a process to increase metal content. Froth flotation is then frequently used to separate metal compounds from dirt and rock. Hazards include airborne particulate, noise, and exposure to dangerous chemicals.

5. Smelting reduces metal oxides and sulfides in furnaces. Iron is smelted in blast furnaces, using coke as the reducing agent and limestone to combine with impurities. Heat stress, noise, and airborne particulates are problems.

6. Iron is converted to steel in basic oxygen furnaces that burn out carbon and use more limestone to remove impurities. Heat stress, noise, and airborne particulates are problems.

7. In electrolytic purification, metals are recovered by plating ions out as free metals on cathodes. Bauxite is crushed and alumina is dissolved in NaOH. Impurities are discarded and alumina is precipitated. This is added to electrolytic cells with carbon electrodes, and aluminum metal is formed. Heat stress, airborne particulates, and particularly airborne fluorides are problems.

8. In hydrometallurgy metals are leached from ores by chemicals. Hazardous chemicals are a threat.

9. Metals are shaped by casting in a foundry. Molds may be permanent or shaped sand. Metal fumes, silica particulate, heat, noise, and vibration are problems.

10. Metal surfaces are cleaned by abrasives, including sand blasting. Alkalies, acids, and organic solvents are also used. Chemical exposures and airborne abrasives, especially silica, are hazards.

11. In electroplating, metal films are electrically deposited on surfaces. Baths include a variety of hazardous chemicals.

12. Welding joins metal surfaces with molten metal. Melting is accomplished by gas torches or high-voltage electric arcs. A variety of agents enter the atmosphere during welding, and hot metallic sparks are produced.

PROBLEMS

1. Open 30 CFR Chapter 1—the first volume—to the table of contents on the second page. Locate safety and health standards for the following: A. Metal and nonmetal surface mines; B. Metal and nonmetal underground mines; C. Surface coal mines; D. Underground coal mines.

2. Turn to Part 56 and scan the contents.

   A. In what section do we find the limits on airborne contaminants, and what are these limits?

   B. Where are respirator standards, and what are they? How could you obtain a copy of those standards?

3. A. In each of the four parts located in Question 1, locate and compare the noise limit standards.

   B. Compare these with the standards discussed in Chapter 10.

   C. What are the standards for sound level meters?

4. Turn to Part 71 and scan the contents.

   A. Find the *respirable dust* standards. Where are they?

   B. What is the maximum permitted level of respirable dust?

   C. Where do we find the modification of this standard when quartz is present in the dust, and what is quartz?

   D. At what level of quartz is the need for lower respirable dust levels triggered?

   E. Using the formula presented, what is the maximum permitted respirable dust level when the dust is 10% quartz?

   F. Calculate the maximum concentration of respirable *quartz* dust permitted in the air when the percent quartz is 4%; is 5%; is 6%.

   G. How often is the mine operator required to test the air at each workstation?

   H. Locate Subpart D. This deals with the procedure following an inspection that has uncovered an unacceptably high respirable dust level.

What is the mine operator required to do and by when? Specifically, who must respond to this submission and how is compliance assured?

5. Some mines, particularly uranium mines, may contain the radioactive gas radon. Such exposure is dealt with in 30 CFR 57.5037.

   A. How must testing be done? Where can descriptions of these methods be obtained?

   B. What is the trigger level for compliance testing?

   C. What is the maximum permitted working level of exposure and the maximum total annual exposure of the worker from this source?

6. Familiarize yourself with 29 CFR 1910.1029.

   A. The chief hazard to workers around coke ovens is exposure to *coke oven emissions*. What is the definition of coke oven emissions?

   B. What is the hazard of "green plush"?

   C. What problems do coke oven emissions cause?

   D. What is the permissible exposure limit for coke oven emissions?

   E. How should sampling of employee exposure be done?

   F. What is a *beehive oven?*

   G. What does OSHA require operators of a beehive oven to do regarding employee safety?

7. Find the Part 1910 Index near the end of the volume. Look up "welding."

   A. Where is this information located?

   B. Locate the requirements for ventilation. What is the general requirement for ventilation air flow per welder?

   C. How does this differ from ventilation regulations where a specifically listed toxic metal is involved. Use indoor welding involving cadmium as an example.

   D. What are the requirements for ventilation for welding in a confined space?

8. Locate regulations governing forging in 29 CFR using the index. What special concerns predominate?

BIBLIOGRAPHY

P. R. Atkins and K. P. Karsten, "Aluminum," in L. V. Cralley and L. J. Cralley, *In Plant Practices for Job Related Health Hazards Control, Volume 1*, John Wiley and Sons, New York, 1989.

M. M. Garcia, "Mining and Milling," in L. V. Cralley and L. J. Cralley, *In Plant Practices for Job Related Health Hazards Control, Volume 1*, John Wiley and Sons, New York, 1989.

J. N. Lockington and D. L. Webster, "Steel," in L. V. Cralley and L. J. Cralley, *In Plant Practices for Job Related Health Hazards Control, Volume 1*, John Wiley and Sons, New York, 1989.

NIOSH, *Occupational Health Control Technology for the Primary Aluminum Industry*, U.S. Department of Health and Human Services, Cincinnati, Ohio, June, 1983.

R. C. Scholz, "Sand Cast Foundry," in L. V. Cralley and L. J. Cralley, *In Plant Practices for Job Related Health Hazards Control, Volume 1*, John Wiley and Sons, New York, 1989.

T. J. Slavin, "Welding Operations," in L. V. Cralley and L. J. Cralley, *In Plant Practices for Job Related Health Hazards Control, Volume 1*, John Wiley and Sons, New York, 1989.

T. J. Walker, "Metal Cleaning," in L. V. Cralley and L. J. Cralley, *In Plant Practices for Job Related Health Hazards Control, Volume 1*, John Wiley and Sons, New York, 1989.

CHAPTER 15

METALS, PART II

DETAILS ABOUT SPECIFIC METALS

Based on the outline of metallurgy presented in Chapter 14, discussion is extended to include uses of and purification from ores of additional metals, and to deal with the specific hazards generated. When reading this chapter, recognize that although intake of a compound of a particular metal may cause a given set of toxic symptoms, it is not safe to generalize that all compounds of that metal give the same symptoms. The intake of lead salts generates symptoms quite different from those produced by tetraethyl lead. There is wide variation in the toxicity of the various forms of mercury. Initially, we must look at each metallic compound separately in dealing with hazards accompanying exposure.

We can be exposed to a metal even though that metal does not have a role in a particular manufacturing process. For example, ores generally contain a variety of metals. In fact, a number of metals are obtained as byproducts of the extraction of some other material. Industrial metals are therefore seldom pure. We may be alert to the toxic consequences of exposure to the metal that is the object of a particular process, only to find that a more serious threat is posed by an accompanying metal.

An additional source of exposure to metals lies in the fact that fossil fuels contain metals. The burning of oil or coal, particularly coal, adds quantities of metal-rich particulate into the atmosphere. Combustion of such wastes as paper and plastics has a similar potential.

ALUMINUM

Once a precious curiosity, today 18,000,000 metric tons of aluminum are produced annually worldwide—4,000,000 metric tons in the U.S. alone. Added to primary production in the U.S. is 2,200,000 metric tons from recycling. Aluminum is heavily used in construction, automobiles, aircraft, and appliances. Where weight reduction is important, as in aircraft construction, the low density of aluminum is an important advantage. Good heat transfer properties make it useful for parts of internal combustion engines, compressors, and radiators.

Low toxicity when ingested and ability to be formed into thin flexible sheets has led to its use in food packaging, and its low porosity makes it particularly useful for wrapping frozen foods. Soft drink cans are a major use, and a major source of aluminum for recycling. Because of the corrosion resistance imparted by the thin oxide coating that forms on the metal surface, aluminum finds use in items exposed to the weather, such as highway signals, lampposts, and signs. Powdered aluminum is used in paints and fireworks. Aluminum oxide is found in abrasives and high-temperature brick.

The winning of aluminum from ore was discussed in Chapter 14. The first stage in using this primary aluminum often involves casting the molten metal. Because of the widespread use of aluminum and the high cost in energy of capturing the metal from its ore, recycling aluminum has high priority. As in steelmaking, the scrap is often added to new aluminum in furnaces as the metal is melted for casting operations. Aluminum *must be dried* by preheating before adding it to the furnace, since even small amounts of water can expand violently as steam in the melt. In furnaces using gas or petroleum, burners present a hazard due to the gaseous combustion products, but also generate a noise problem. Heat stress is a concern around the furnaces.

Oxides and other impurities are removed from the melt by addition of fluxes of sulfates, chlorides, and fluorides, followed by skimming of the dross. Some oxides sink to the bottom, so the taphole is located high enough in the furnace to leave these behind. Hydrogen gas, derived from moisture, oil, or grease in the hot melt, dissolves in aluminum and causes metal embrittlement. Chlorine gas is bubbled through the melt through a "wand," forming chlorides of hydrogen, sodium, calcium, and magnesium. The metal salts become part of the dross and are skimmed off. Chlorine gas is corrosive to systems used to transport it, especially at connecting points, and, given the toxicity of the gas, the chlorine transport system must be maintained with great care. The furnace must be adequately ventilated.

Metals such as copper, zinc, magnesium, chromium, beryllium, nickel, silicon, and titanium are added to the melt to produce desired alloys. The molten aluminum is then cast directly into products or into ingots that are later formed into products by rolling, forging, or extruding. The dross and oxide residue at the bottom of the furnace are rich enough sources of aluminum to justify reprocessing in primary aluminum recovery systems. Dross is often wetted down when it has been removed. It contains an unpredictable mixture of substances, especially if scrap was added to the melt, and must be treated carefully. Water reacts with nitrides to produce ammonia. Many of the metal salts in dross are water-soluble, so runoff must be prevented from soaking into the ground and contaminating groundwater.

The construction and demolition of furnaces requires care. Asbestos is often used as a heat shield between fire brick and an iron vessel forming the outside container. The fire brick may contain silica.

Aluminum is forged in a manner similar to the forging of iron (Chapter 14), although temperatures are much lower. Dies are sprayed with lubricants, often lead or tin metal soaps, leading to the potential of exposure to organic or inorganic lead or tin. Cleaning parts after they are formed can expose workers to the cleaning agents. Problems may be experienced both from the standpoint of inhalation and dermatitis. Heat stress is possible, and forges have serious noise problems.

There have been problems with fibrosis in the lungs of those working with aluminum oxide (Shaver's Disease), particularly from either bauxite dust, or from dust produced in the manufacture of abrasives. Pulmonary fibrosis also results from exposure to metallic aluminum powders (aluminum dust lung). A sensitization has been reported in response to fumes from aluminum soldering, but this has been attributed to components of the flux, rather than to the aluminum itself. Workers in electrolysis plants have suffered from fluorosis, a problem resulting from exposure to the fluorides of the cryolite bath. This problem has been controlled by measuring the urinary fluoride content of the workers, and by monitoring the plant environment (Chapter 14). Problems may arise from the release of coal tar pitch volatiles from Soderberg electrodes. Exposure to these polycyclic aromatic hydrocarbons may increase the incidence of lung cancer.

Aluminum alkyls of many different compositions are finding increasing use in industry as catalysts, for synthesis of other organometallics, and in the production of various organic compounds. These compounds ignite spontaneously, and so represent a fire hazard. The vapors convert rapidly to an aluminum oxide smoke, the inhalation of which can cause metal fume fever. Skin contact with aluminum alkyls causes burns.

ANTIMONY

Antimony is used in alloys as type, bearing surfaces, covering for electrical cable, solder, ammunition, and pewter. Compounds of antimony are found in paints, lacquers, rubbers, glass, and ceramics.

Antimony ores are largely sulfide or oxide deposits. Antimony oxide is volatilized from the ore, and is collected in baghouses or precipitators. The oxide is smelted with coke, potash, or soda ash to obtain the free metal. Metallic antimony may also be obtained from sulfide ores by smelting with iron scrap, forming the iron sulfide.

Antimony spots is a skin rash much like a pox disease. Oxides and halides also irritate the respiratory tract, eyes, and mouth. Antimony sulfide was associated with heart damage to workers in one case, and illness with liver enlargement in others. Pneumoconiosis is reported in workers dealing with the oxide, and studies of smelter workers showed an above normal incidence of lung cancer when exposure levels were higher than standards presently allow.

Problems with exposure to antimony as the ore, or during smelting, are made worse by the presence of other metals, including arsenic and lead.

Stibine, the hydride of antimony—and a very toxic gas—may be formed when antimony alloys are dissolved in acid, or when lead storage cells are over-charged. Welding, cutting, or soldering are other procedures capable of generating stibine. Early symptoms of overexposure include headache, nausea, and blood in the urine.

ARSENIC

Arsenic is certainly recognized as a classic poison, even by people who have never studied toxicology. Its role in the political and marital history of Europe is sizable and secure. Historically, the discovery of an analytical method to detect arsenic poisoning led to a revolution in the legal handling of deliberate poisonings.

Arsenic is used as an alloying agent with lead in type, battery plates, bearings, cable sheathing, and shot for ammunition. Arsenic compounds have been used in pesticides, leading to exposure of workers manufacturing the products and agricultural workers applying them.

It is obtained as a byproduct of the processing of other ores. Arsenic trioxide is driven off during the smelting of gold, lead, or especially copper ores, and can be harvested in cooling chambers. The risk of exposure is serious in the smelting operation. The harvesting and transporting of the oxide must be done cautiously, as must repair work to furnaces, cleaning of flues, or other maintenance operations.

Arsenic causes extensive skin problems. Skin irritation is observed as reddening, possible swelling, and an itching or burning sensation. Reddening and swelling around the eyes is likely. Long-term exposure can lead to a mottling or bronzing of the skin. The nasal passages and upper respiratory tract are irritated by exposure, and the destruction of nasal tissue can lead to puncture or perforation of the nasal septum, the division between the nostrils. Less commonly, exposed workers have suffered gastrointestinal upset. Peripheral nerve damage leading to pain, loss of sensation, and weakness in the limbs has occurred as a result of exposure to arsenate sprays.

Arsine is the hydride gas of arsenic, has a garlic-like odor, and can arise in much the same fashion as does stibine. It is likely to be produced along with stibine, since arsenic and antimony often occur together. There is danger of arsine forming in the processing of metals in which arsenic is an impurity. For example, washing down the dusts resulting from tin refining is reported to have fatally exposed workers in a plant to arsine. A similar wetting of dross in a zinc furnace released dangerous levels of the gas. Arsine causes destruction of the red blood cells, resulting in anemia. The release of hemoglobin from the broken cells can then cause kidney damage.

Arsenic compounds have been implicated as carcinogens. Occupational exposure can lead to skin cancers. In addition, workers in copper smelters, where arsenic exposure can be high, or those exposed to pesticides containing arsenic, have developed lung cancer to a significantly greater degree than normal. This is more serious in cases where exposure is accompanied by the inhalation of cancer promoters, including cigarette smoke or such irritating industrial gases as sulfur dioxide.

BERYLLIUM

Beryllium finds use as an alloying agent, particularly to strengthen copper, but also with aluminum, magnesium, and steel. Beryllium oxide is used in ceramics. However, the uses of beryllium that have increased most rapidly in recent decades are as a moderator in atomic reactors, and as a structural component in space vehicles.

Mining, transport, and other handling of beryl, the ore of beryllium, does not present great hazard. However, becoming exposed to beryllium dust or fume, as when milling or otherwise working with the metal, presents a serious toxic hazard. Insoluble compounds or particulate metal deposited in the body can produce symptoms at a much later date.

Respiratory problems and skin irritations were early recognized as resulting from beryllium exposure. Chronic beryllium disease was first recognized in workers assembling fluorescent lamps, which at that time used beryllium phosphors. Respiratory problems can vary from acute irritation, with fluids collecting in the lungs, to a long-term damaging of the lung tissues. Symptoms of lung damage can appear years after exposure to the metal has ceased. Finally, there is evidence that excess exposure to beryllium increases the likelihood of lung cancer. Skin irritation reflects the occurrence of sensitization. Over an extended period of exposure, damage may occur to a number of internal organs.

Recently, the permitted exposure level has been sharply reduced. Other phosphors have replaced beryllium in fluorescent and X-ray tubes.

CADMIUM

About half the cadmium produced is used to apply anticorrosion coatings to metal products by electroplating. Other uses include alloying with other metals to produce bearings and solders, in nickel-cadmium batteries, and in nuclear reactors. Cadmium oxide (red) and cadmium sulfide (yellow) are used as pigments in paints.

Cadmium is obtained primarily as a byproduct of the smelting of zinc, lead, and copper ores, with zinc ores being the major source. Cadmium is volatilized in the roasting process, and is collected in air filters as cadmium oxide. Such byproduct cadmium oxide is likely to contain other toxic impurities such as lead, arsenic, and selenium. Sulfuric acid is added to the dust to convert the oxide to calcine (cadmium sulfate). Calcine dissolves in sulfuric acid, but many contaminants precipitate and are filtered out of the solution. Hazards here include exposure to mists of sulfuric acid as well as to compounds of cadmium and its impurities. Eye protection, respirators, and protective clothing are needed.

The greatest risks of routine exposure occur in smelting operations, while calcining, and during the distillation of cadmium sponge. Dust or fume formation associated with bearing or battery manufacture presents a hazard. Workers dealing with such sources of airborne cadmium should wear respirators and coveralls that are laundered at the end of each day's work. Plant exhaust ventilation should be equipped with a system to trap particulate. Plants bubbling hydrogen sulfide gas for precipitating cadmium sulfide as a pigment must monitor levels of H_2S carefully.

Some of the worst accidental exposures have occurred during high-temperature operations where the presence of cadmium in the metals being processed was not suspected. These included welding and cutting cadmium-coated pipes, heating plated rivets, and melting scrap.

Workers chronically exposed to cadmium are likely to show symptoms of kidney damage, especially protein in the urine (proteinuria). Cadmium oxide dusts have produced lung damage, including fibrosis and eventual emphysema. In other studies, workers were found to have skeletal problems due to bone demineralization. The extreme example of this is the itai-itai disease, which occurred in Japan among people exposed to cadmium wastes in their water. An increased occurrence of cancer, especially prostate cancer, has been reported in workers exposed to cadmium or cadmium oxide. Cadmium has a very long half-life in the body, so that continuous exposure, even to very low levels, must be avoided.

CHROMIUM

Chromium ore, chromite, is no longer mined in the United States. The ore is reduced with carbon or silicon in an electric furnace to produce an iron alloy called ferrochromium. Several techniques are then employed to obtain either the pure metal, or a variety of chromium compounds.

Chromium is an alloying agent in producing stainless steels, electrical resistance wires and cutting tools. Chromium is plated from a chromic acid bath containing sulfuric acid to form decorative and protective metal surfaces. Chromates and dichromates have a variety of applications, such as tanning, dyeing, and photography, and as zinc chromate in primer paint.

Hexavalent salts are irritating and destructive to tissue. Mists from electrolysis baths and plating baths cause dermatitis and damage to nasal membranes.

Skin irritation and ulceration also trouble lithographers, painters using chromium-containing pigments, workers in magnesium foundries (castings are chromate-treated to improve weathering), workers handling chromium-containing cements or chromium-treated timber, and welders working with stainless steels or using welding rods containing chromium. When dusts, fumes, or mists are inhaled, irritation problems extend to the lungs, and there are numerous reports of cancer in the respiratory tract from hexavalent chromium exposure. Over the years, the worst exposure problem in the United States was from the chromate-chromite mixture intermediate in chromate preparation. Workers suffered bronchial cancer on exposure to this mixture. An acute oral dose causes kidney damage. Because of the toxic nature of chromium plating bath contents, disposal must be done with regard for its serious potential for environmental damage.

COBALT

Cobalt is a byproduct of mining copper and other metals. It is used largely in alloys in such applications as aircraft, turbines, and magnets. There are catalysts containing cobalt, and the oxide is used as a coloring agent in glass. A toxicologically significant application is as a binder for tungsten carbide and titanium carbide abrasives. Any abrasive is likely to become particulate in the air when used.

Cobalt is a byproduct of ores of copper and nickel. Smelted ores produce *matte,* a mixture of the three metals. Crushed and ground matte is leached with hot acid to produce a solution of nickel and cobalt, possibly including copper. Dusts from this operation are hazardous, so workers should have eye and respiratory protection and should wear coveralls. Noise both from crushing and from leaching operations is a problem, and the leaching process may lead to some heat stress. If present in the leached solution, copper is removed, and caustic plus an oxidizing agent precipitates cobaltic hydroxide containing some nickel. Treatment with ammonia and an oxidizer completes the separation of cobalt and nickel. All these chemicals are hazardous, and handling, ventilation, and personal worker protection should be appropriate to the risk. At high pressure and temperature, hydrogen is used to reduce cobalt to the free metal.

Cobalt metal fumes and dust cause nose, throat, eye, and skin irritation. Milling and abrasive tool manufacture expose workers to dust that produces pneumoconiosis. Workers also experience a dermatitis from abrasive manufacture or handling cobalt alloys, clays, or pottery. Cobalt in cement has been reported to cause skin irritation.

COPPER

Copper is used in a vast range of applications, including wiring, plumbing, roofing, and cookware manufacturing. These depend on the high electrical and heat conductivity of copper, and its good weathering properties. Many alloys of copper are produced, far more than just the familiar brass and bronze.

At one time a surprising proportion of copper mined in the world came from underground deposits of metallic copper in northern Michigan. A huge boulder of copper metal from these deposits is on display in the mall in Washington, D.C. Today, however, most copper is surface mined as low-grade sulfide ores. These are crushed and concentrated by flotation. Flotation concentrates not only the copper, but other, more toxic metals found in the ore, such as arsenic and lead. Roasting may be employed before smelting, but more often today the "green charge" is added directly to the smelting furnace. In roasting, fluxes such as limestone are added to the ore, and it is heated at just under 1000°F. In the smelter at above 1800°F, iron and copper combine into a matte, which is drawn off into ladles to be treated in the converter. Here a flux is added and streams of air are blown through the melt using a number of tubes called *tuyeres*. Iron sulfide oxidizes first, and the iron forms a slag with the flux. The copper metal is freed from the sulfide and is collected in a form more than 95% pure. Further purification is accomplished by electrolysis.

Copper may also be extracted by *electrowinning* from appropriate ores. The ore is leached with dilute sulfuric acid as $CuSO_4$. Copper is stripped from this dilute solution into a kerosene solution by a phenolic agent, then reextracted into sulfuric acid as a concentrated solution. Copper is plated out of this solution, and the acid is recycled back through the extraction process. The formation of arsine during this process is a concern, so sampling and analysis should be performed.

Mining and processing carry risks of lung damage from dusts. Noise exposure is high around crushing, grinding, and screening operations. In roasting, smelting, and converting, air may be contaminated with SO_2 and various toxic metals, so good ventilation and monitoring to assess worker exposures is necessary. Burners produce high noise levels, and should be enclosed. At the converter, the tuyeres and any leakage of compressed air connected to them adds to the background noise level, and clearing the tuyeres when cooled copper blocks them is a particular noise problem.

Electrolytic baths can expose workers to acid mists. Cutting, welding, and otherwise raising the metal to high temperatures increases the danger of metal fume fever and allergic dermatitis. Damage to the upper respiratory tract from metal fume and dust takes the form of irritation and atrophy of the nasal mucosa. Complaints of irritation from exposure to dusts of copper oxide and other copper salts have been recorded.

Finally it must be noted that even at very low concentration copper salts are lethal to lower life forms. Copper salts are sometimes used to reduce snail populations in lakes, and mussel populations have been destroyed by levels of copper too low for analysis by atomic absorption. Great care must be taken about release of copper compounds into the environment.

INDIUM

Indium is used as a protective coating on bearings, in solders, and in a variety of high technology roles including special mirrors, transistors, and infrared detectors. It is often found in the ores of other metals, particularly of zinc. The best source of indium is the flue dusts of zinc smelters. Exposures are possible in the course of plating and manufacturing the metal. Many hazards have not been well studied, and the literature on indium problems is sparse.

LEAD

Storage batteries are the largest single use of lead. The production of tetramethyllead and tetraethyllead as antiknock additives in automobile fuels was a major use of lead. However, this industry has been phased out in the U.S. due to environmental restrictions, at first because lead harms the catalytic converters required in automobiles, then later because of concerns about levels of lead in the atmosphere. Alloyed with antimony and tin, lead is used as a sheath to weatherproof electrical cables, as type metal, and in bearings. Lead is found in pigments, in solders, and in ammunition. The use of lead in shot has been vigorously and, more recently, successfully opposed by environmental groups because of long-range negative effects on wildlife.

Considering levels of exposure and levels necessary to produce toxic effects, the public is at greater risk from lead than from any other metal. In the general public, the body burden of this metal is a sizable fraction of those levels at which symptoms of damage are first seen. Because of this high background, any further exposure of individuals as the result of employment should be carefully monitored and reduced to a minimum level. With the phasing out of lead additives in gasoline and other measures to reduce lead exposures, the levels of lead in the U.S. public have been slowly dropping. Blood lead levels and levels of intermediates in heme synthesis are common measures of the lead body burden.

The toxic effects of lead on the body (plumbism) are varied. The nervous system is most sensitive to damage. The symptoms of lead poisoning include convulsions, hallucinations, coma, weakness, and tremors. Lead palsy is seen first as wrist drop, on the right in right-handed persons. Damage to blood cell–forming systems and increased fragility of the red blood cells results in anemia. Kidney damage is also commonplace in lead poisoning. The most common symptom of overexposure to lead is intestinal colic. Constipation is followed by intense abdominal pain. Once exposed to lead, the body carries a residue for a long time. It is incorporated into the bones, and is only released very slowly. Lead causes kidney cancer in test animals, but a connection between lead exposure and cancer in humans is less clear.

Organic lead compounds, particularly tetramethyllead and tetraethyllead, present a different pattern of toxicity. There is no colic, but signs of nervous system damage abound. These include insomnia, restlessness, hallucinations, and delusions.

Lead ore is *galena,* a sulfide ore, which is most often removed from underground mines.[1] Several other metals, including tin, bismuth, arsenic, silver, gold, antimony, and copper, are found in these ores. In fact arsenic, antimony, and bismuth are usually obtained as byproducts of the processing of lead ores.

Lead ore is crushed, ground, and concentrated by flotation. Given the toxicity of lead, control of dust by wet spraying and ventilation during these operations, and during transportation and sampling, is essential. Noise stress is also a problem. Ore is *sintered*—heated with flux to burn off sulfur. The fused sinter is crushed and becomes the raw material for the blast furnace. Sinter, iron, and coke are mixed in the blast furnace and air enriched in oxygen enters the bottom of the furnace. The molten product is poured into containers where slag is skimmed from the metal product. The slag is often used as a source of zinc.

Additionally, the recycling of automobile batteries and solders provides more than a third of the lead used in the U.S. In a process called *secondary smelting,* old batteries are broken up and the lead recovered. Exposures may occur to lead and to sulfuric acid in this process. Further refining is similar to that performed on smelted lead ore.

Lead is melted at a controlled temperature in a *drossing* plant, and higher-melting-point impurities such as copper and arsenic solidify and float to the top. Additions to the kettle separate other impurities, and the dross is removed and remelted. It then separates into three liquid layers. *Matte* is a copper-iron mixture and *speiss* is an arsenide layer. These are shipped to a copper smelter, and the third layer, lead, is returned to the kettle. Further *refining* removes such impurities as silver, antimony, tin, copper, tellurium, zinc, bismuth, and gold by adding various agents to the molten lead and removing the resultant slag.

Chief potential respiratory exposures of concern are to metallic dusts and to carbon monoxide from the furnaces. Arsenic requires special monitoring if present. Noise from burners and heat stress are also concerns. Good housekeeping is important and employees must be trained in the hazards of the materials being processed. Besides inhalation, ingestion of particulate is a threat. Food, beverage, and tobacco products must not be allowed in the work area, and employees should clean and change before going to the lunchroom.

One hazard for exposure occurs when spraying lead-based paints. One of the common lead pigments, white lead, is giving way to zinc oxide and titanium oxide. However, red lead in metal primers continues to find use because of its superior properties with respect to the inhibiting of corrosion. Hazard due to

---

[1]Carbonate and sulfate ores are also found.

lead-based paints continues after the painting process is done. The removal of these paints by abrasion or burning puts harmful particulates into the air. Other sources of hazard from lead include grinding or sanding soldered surfaces and manufacturing batteries. Occasional serious exposure has resulted from cutting or welding alloys containing lead, or cutting up structures painted with lead-based paint using a torch. Any process involving the manufacture or handling of such organic lead compounds such as gasoline additives must be done with extreme care, since these compounds are highly toxic, particularly to the nervous system.

MANGANESE

Manganese oxide (pyrolusite) is the chief ore of manganese. Its primary use is as an alloying metal, particularly in steel. Manganese dioxide is used in the production of dry cell batteries.

Exposure occurs through dusts in mining, transportating, crushing, and sieving ores. There are dust and fumes near reduction furnaces. Chronic exposure can produce a pneumonia-like problem with possible fibrosis in the lungs. Manganism is a disease of the central nervous system resulting from absorption of manganese through the digestive tract from the swallowing of dusts moved up from the lungs over a period of time. Symptoms range from headaches, cramps, apathy, weakness, and insomnia to psychosis with delusions and hallucinations.

MERCURY

Mercury is obtained from the sulfide ore cinnabar. It is used in types of electrical apparatus and in caustic soda plants as part of the electrode. The ability of mercury to form solutions or amalgams with other metals is used to extract gold and silver from ores. As a catalyst in urethane foam production, it remains in the foam product. Mercury has been added to paints to prevent mildew, and in marine applications to inhibit growth on boat hulls. Mercury compounds have been used to discourage mold production in paper pulp and on seed grain.

From the toxicologist's standpoint, mercury is found in three forms, which all have different health effects. First, hazards from mercury include inhalation of metal vapors. These can occur in work areas containing liquid mercury, and in an unsuspected manner by evaporation of spilled mercury that has soaked into crevices in floors. Special risks occur when mercury vapor is driven off an amalgam by heating. This can occur during the application of gold to objects by dipping them in a gold amalgam, then heating them to remove the mercury. Metallic mercury has a high attraction to the central nervous system, and causes a broad range of symptoms, from headache, tremors, weakness, insomnia, or drowsiness to emotional and psychotic disturbance with extreme excitability and irritability. Early symptoms of mercury poisoning include salivation and tenderness of the gums. Second, mercury salts often cause skin problems. This problem is particularly troublesome in handling mercury fulminate, which is used as a detonator in explosives. Third, alkyl mercury compounds present special problems of a somewhat different character than those of the inorganic mercury compounds. Alkyl mercury compounds are used to kill molds, especially in stored seeds. They are very easily absorbed, and cause serious damage to the nervous system. Motor control is affected, and narrowing of the visual field to the point of complete blindness occurs. The damage is long-lasting or permanent.

NICKEL

Steel making uses about half of all nickel produced. Most of the rest is used in other alloys or is plated onto other metals. Lesser uses include as a component of nickel-cadmium batteries, as catalysts, and as a coloring agent in glass.

Canada is the major source of nickel. Mining and purification were partly covered in the discussion of cobalt. Several flotation and magnetic separation techniques are used to purify the ore. Nickel sulfide is reduced using coke. Purification of nickel can be done electrolytically, or by conversion to nickel carbonyl using carbon monoxide, followed by deposition onto nickel pellets.

Hazardous exposure to fumes and dusts has occurred during high-temperature processing of nickel sulfide ores into nickel oxide, resulting in nasal and lung cancer. Not all ore sources seem to present this threat. Nickel carbonyl, which is sometimes an intermediate in the refining of nickel, is a very toxic gas that causes a hemorrhagic pneumonia.

Nickel platers suffer from a dermatitis caused by skin contact with nickel salts. There is also a chronic eczema which is not necessarily produced at the point of contact. High temperatures around the plating baths contribute to the risk. Some individuals are more susceptible to becoming sensitized to nickel than others, and once sensitized they respond even to contact with nickel alloys.

TIN

Major uses for tin include tinplate, solder, and copper alloys. The latter include brass, bronze, babbitt (bearing metal), and pewter. Organotins are used as catalysts, polymer stabilizers, antimicrobials and in marine antifouling paints.

Not much tin is mined in the U.S., and about one quarter of the supply is obtained by recycling. Most ores are sulfides, which are smelted and refined. Long contact with tin oxide dust or fumes from smelting operations causes stannosis, a lung deposition problem that produces little fibrosis and little in the way of serious symptoms.

Organotins are likely to be irritating to skin and eyes. The trimethyl-, triethyl-, and tributyltins are particularly toxic. The EPA has proposed restrictions on the use of tributyltin in antifouling paints, because it is harmful at levels of less than 1 ppb to nontarget fish and shellfish. Presently, 624,000 gallons of such paint are sold annually.

TITANIUM

Pigments containing titanium oxide are used in paints, papers, plastics, rubber, and ceramics. The metal is used in aircraft and missiles, where a high strength-to-weight ratio makes it a useful structural material. Titanium cooling coils are used in power plants.

Titanium is a very abundant metal, the chief ores of which are oxides. Isolation is by conversion of the oxide to a chloride, then reduction of the chloride to the metal. Pure metal is obtained using a consumable titanium electrode in an electric arc furnace.

Titanium chloride causes skin and eye burns, and serious lung damage if inhaled. The oxide seems to be relatively innocuous when inhaled. Reports of problems may relate to other compounds mixed with titanium oxide.

TUNGSTEN

Most of the tungsten produced is used in very hard, wear-resistant alloys. These are used largely in cutting tools. Tungsten carbide is an important abrasive. The chief ores of tungsten are tungstate salts.

Lung problems (hard metal disease) have been experienced by workers with exposure to dusts from tungsten carbide abrasives cemented in place using cobalt. These have already been mentioned under the cobalt heading. Reports of pulmonary problems are numerous. Removal from exposure is usually a sufficient remedy when symptoms such as coughing and wheezing first appear. However, the correlation between length of exposure and severity of lung damage is poor, and in more advanced stages the disease is progressive, and can lead to death. In some workers, early symptoms indicate sensitization. This may be a factor in the progressive form of the disease. Exposure to the dust also produces skin irritation, inferring sensitization of the skin.

URANIUM

Uranium is primarily used to produce fuel for nuclear reactors and to produce weapons. Much of the mining of uranium ores in the U.S. occurs in western states. Early speculators, hopeful of quick riches, established a large number of small uranium mining operations. As the industry matured, these often poorly ventilated small mines gave way to a relatively small number of much safer operations. The uranium is leached from ore using acidic ferric sulfate. In the process, the ferric ion is reduced to ferrous ion. Its reoxidation, allowing its reuse, is accomplished using a microorganism. The uranium is electrolytically reduced, and is precipitated as green cake (uranous fluoride) by the addition of hydrofluoric acid.

Mining hazards include exposure to such radioactive species as uranium itself, radium, or radon gas. Inhalation of these radioactive species increases the risk of lung cancer. Chemical exposures are less important, and include exposures to such other metals found with the uranium as vanadium, lead, thorium, manganese, and arsenic. The silica rock of western mines adds the risk of silicosis. Because the ore is processed by wet techniques, the risks are relatively low. Uranium hexafluoride, a gas produced for isotopic separation in the gas centrifuge, is hazardous if a leak should occur in the apparatus handling this compound.

VANADIUM

Alloys are the chief use of vanadium. The ores are widely distributed, and the metal is usually extracted by leaching. Other metals are usually produced along with vanadium. In one process, uranium is precipitated from the leach solution first, followed by vanadium. In another, vanadium is leached from iron ore or from the slag produced in steel making. Metallic vanadium is produced by carbon reduction of the vanadium oxide, or by a calcium metal reduction process. The oxide is also reduced using silicon or aluminum.

The greatest hazards of mining vanadium are exposures to uranium and radon in the case of carnotite ores, and to the silica in the rocks of the mines. Much of the extraction of vanadium is a wet process, minimizing the respiratory exposure risks. Vanadium oxide can produce a respiratory sensitization with inflammation, irritation, and sometimes pneumonia. Sensitization of the skin produces redness, itching, and eruptions.

A problem not related to the production or use of vanadium occurs in the case of oils with high vanadium content. Workers cleaning burners, or dealing with the oil ash of residual oils, display an inflammatory respiratory problem.

ZINC

Most zinc is used in automobiles. Large amounts are used in building products and appliances, and there are numerous lesser uses. Frequently, the use of zinc is to coat iron or steel in a corrosion-preventing process called galvanization. Zinc compounds are used in paint and in rubber.

Zinc ore is largely mined underground. The ores are enriched, sulfides are roasted to convert them to oxides, and the oxides are smelted with coke or coal, or leached for electrolytic deposition.

Zinc oxide fume causes fume fever. Reports also have appeared of a gastrointestinal disturbance from zinc oxide exposure. Other metals associated with zinc, such as arsenic, cadmium, lead, and manganese, add significantly to risks. For example, dissolving the zinc in acid or alkali in a leaching operation can lead to the release of arsine gas.

KEY POINTS

1. Extraction processes that involve heating in furnaces involve the hazards of heat stress, exposure to combustion products such as CO, and noise from the burners.

2. Ores are often sources for more than one metal. In a given process, exposure to an unwanted metal can be a hazard.

3. Airborne aluminum, aluminum oxides, and fluoride compounds are special problems in aluminum purification and processing. Aluminum alkyls are highly reactive.

4. Antimony causes a rash, and its oxides and chlorides are irritating if inhaled. Stibine is a very toxic gaseous hydride of antimony.

5. Arsenic trioxide, produced in the smelting of other metals, is hazardous. Arsenic compounds cause PNS damage and G.I. tract problems, and are skin and eye irritants. Some compounds are carcinogens. Arsine, the highly toxic gaseous hydride of arsenic, may be produced accidentally in the purification of metals containing arsenic as an impurity.

6. Beryllium causes skin and lung irritation. It may accumulate to cause chronic problems.

7. Cadmium is a common commercial metal that is very damaging to kidneys. Its oxide causes emphysema, and in extreme cases bone demineralization is reported. It may be carcinogenic, and has a very long half-life in the body.

8. Chromium salts are very irritating to skin and lungs, and some are carcinogenic.

9. Cobalt metal fumes cause irritation to nose, eyes, and skin.

10. Copper metal fumes and dusts damage the upper respiratory tract.

11. Indium is not well studied.

12. Lead is broadly dispersed in the environment, and most people already carry a relatively high body burden. It causes nervous system damage (CNS and PNS), kidney damage, and colic. Lead alkyl compounds are potent CNS poisons.

13. Manganese exposure causes CNS and lung damage.

14. Mercury is highly toxic in some forms, causing serious CNS damage.

15. Nickel is a carcinogen. Salts cause skin irritation and can be sensitizers.

16. Tin can cause lung fibrosis, and organotins are very irritating to skin and eyes.

17. Titanium is low in toxicity, but titanium tetrachloride is very irritating.

18. Tungsten particulate causes lung problems and may be a skin sensitizer.

19. Uranium hazards focus strongly on its radioactivity and that of compounds found with it.

20. Vanadium oxide can cause lung sensitization and skin irritation.

21. Zinc oxide causes metal fume fever.

PROBLEMS

1. Outline the conversion of scrap soft drink cans into a casting for an air conditioner compressor, indicating hazards to the workers.

2. Summarize where workers are likely to be exposed to arsenic, including exposure not related to arsenic production.

3.   In what industries is chromium exposure likely in the U.S.?

4.   When cleaning equipment used in the leaching of cobalt, what are the hazards and how may workers be protected?

5.   Locate the standards for working with lead using the index of 29 CFR 1900–1910.

   A.   What is the PEL for lead?

   B.   What must be the accuracy of the monitoring device?

   C.   Who must have blood lead levels measured, and how often?

   D.   What must be the accuracy of the lead blood level measuring device?

   E.   How frequently must employees get a medical exam, and what must be included?

6.   What metals are suspected or proven carcinogens?

BIBLIOGRAPHY

D. J. Burton and J. P. Sieverson, "Primary and Secondary Smelting—Lead, Zinc and Cadmium," in L. V. Cralley and L. J. Cralley, *In Plant Practices for Job Related Health Hazards Control,* Volume 1, John Wiley and Sons, New York, 1989.

C. E. Dungey, "Copper," in L. V. Cralley and L. J. Cralley, *In Plant Practices for Job Related Health Hazards Control,* Volume 1, John Wiley and Sons, New York, 1989.

R. A. Goyer, "Toxic Effects of Metals," in C. D. Klaassen, M. O. Amdur and J. Doull, *Toxicology—The Basic Science of Poisons,* 3rd ed., Macmillan Publishing Co., New York, 1986.

NIOSH, *Occupational Hazard Assessment Criteria for Controlling Occupational Exposure to Cobalt,* DHEW 82-107, October, 1981.

OSHA, *Prudent Practices for Controlling Lead Exposure in the Secondary Lead Smelting Industry—A Guide for Employers and Employees,* U.S. Dept. of Labor, Washington, 1981.

B. R. Roy and S. A. Thielke, "Cobalt-Nickel Refining," in L. V. Cralley and L. J. Cralley, *In Plant Practices for Job Related Health Hazards Control,* Volume 1, John Wiley and Sons, New York, 1989.

U.S. Dept. Health and Human Services, *Occupational Respiratory Diseases,* DHEW 86-102, September, 1986.

T. J. Walker, "Aluminum Metalworking," in L. V. Cralley and L. J. Cralley, *In Plant Practices for Job Related Health Hazards Control,* Volume 1, John Wiley and Sons, New York, 1989.

O. Wong, D. F. Liart and R. W. Morgan, *Critical Evaluation of Epidemiological Studies of Nickel-Exposed Workers,* Environmental Health Associates, 1983.

CHAPTER 16

POLYMERS

Polymers are composed of extremely high-molecular-weight molecules, which are assembled by chemically joining together large numbers of relatively small molecules, *monomers,* in a process called polymerization. Many of the components of living systems are polymers: proteins, polysaccharides, and nucleic acids. A very early human industry involved the utilization of some of these natural polymers, including wool (a protein) and cotton (a polysaccharide), to produce clothing. Only relatively recently has chemical technology reached the stage of generating a variety of synthetic polymers. In 1869 celluloid, a nitrate derivative of the plant polysaccharide cellulose, was the first plastic invented, and in 1909 Bakelite™, a phenol formaldehyde polymer and the first completely synthetic plastic, was patented. Once launched, the growth of production and use of synthetic polymers has been impressive. The production of plastics, elastomers, and fibers tripled in the decade following 1963.[1] Large industries have grown around manmade polymers or commercial utilization of natural polymers. We shall focus attention on some of these industries.

PLASTICS

Plastics production is a giant industry by any measure. In 1986, over 50 billion pounds were produced in the U.S., and the total is expected to reach 75 billion pounds by the year 2000. The amounts of selected products are shown in Table 16.1. According to Chem Systems,[2] packaging uses the largest share of plastics (28%), followed by the construction industry (22%). In fact, plastics have moved into every aspect of daily life, replacing wood, metal, glass, oil-based paint, traditional adhesives, paper, and natural rubber. Substitution of plastics for metals in automobiles has improved styling, eliminated some corrosion problems, improved crash safety, and reduced weight to improve fuel

---

[1]*Chem. Eng. News,* May 6, 1974.

[2]*Chem. Eng. News,* Aug. 24, 1987.

Table 16.1. Leading Plastics Products in 1985.

Plastics	Production, in billions of pounds
Thermoplastics	
Low-density polyethylene	8.9
Polyvinylchloride and copolymers	6.8
High-density polyethylene	6.7
Polypropylene	5.1
Polystyrene	4.1
Thermoset Resins	
Phenolics	2.6
Urea resins	1.2
Polyesters	1.2
Epoxies	0.4
Melamine resins	0.2

Reprinted with permission from *Chemical and Engineering News* (April 21, 1986). © 1986, American Chemical Society.

economy.   Some uses, for example films, represent new areas of materials consumption. The disposal of all this plastic material is of increasing concern, the EPA estimating that by the end of the century, plastics will comprise 9.8% of municipal wastes. Increasing effort is directed toward methods for plastics recycling.

THE CHEMISTRY OF PLASTICS

Plastics are structural polymers, and generally are synthesized using petroleum starting materials. There are, however, some important plastics based on modification of cellulose, a natural polymer, and silicones are based on silicon, rather than carbon, structures. In order to understand the hazards associated with plastics manufacturing operations, it is useful to understand some of the basic chemistry involved.

Linear and Thermoset Polymers

Two major classes of plastics result from the polymerization process. *Thermoplastics* are linear polymers. This means the monomers are joined together linearly, like beads on a string. The final solid plastic could be likened to a bowl of spaghetti, a mass composed of long strands packed together (Figure 16.1a). The name thermoplastic arises from the fact that on heating (thermo) the material softens (becomes plastic). This is because the individual strands of hot polymer are capable of sliding past one another. Such a plastic can be heated, then molded or extruded into desired shapes. On cooling, the mass becomes rigid once again. Such a process can be repeated, so scrap can be remelted and formed again.

The other class of polymers, the *thermoset plastics,* connect the monomers into a three-dimensional grid (Figure 16.1b). It is possible that, once formed, any given part of such a plastic object could be connected to any other part through a continuous succession of chemical bonds, so that in a sense a thermoset plastic object is like a single molecule. Thermoset plastics are very rigid. Any attempt to change the shape of the finished object requires massive breaking of chemical bonds, and thereby destruction of the structural integrity. On heating, thermoset plastics retain their shape until the temperature is high enough to break their chemical bonds. In the presence of air this results in charring as the carbon structures oxidize.

Linear polymers can be altered to assume thermoset properties by *crosslinking* or *curing* the plastic (Figure 16.1c). The linear strands are bridged by a chemical agent to assemble the strands into larger units. Such a modification process can be carried out to whatever degree is desired, resulting in properties that fall anywhere in the range from a meltable and flexible thermoplastic to those of a rigid and hard thermoset. A familiar example is latex, which is found as the weak, extremely elastic uncrosslinked linear polymer of rubber cement, as the still elastic but more rigid product in rubber bands, and as the much more rigid and abrasion-resistant material found in an automobile tire. The difference lies in the degree of curing. This is an example of fine tuning the chemistry of polymers to obtain exactly the compromise one wishes in the final properties.

Addition and Condensation Polymers

The chemical reactions used to join the monomer units together fall into two classes. The common feature of the monomers of *addition polymers* is the presence of a carbon-to-carbon double bond. A number of these monomers are shown in Table 16.2. When the polymer forms, the double bond becomes a single bond. The resulting addition polymers have a continuous row of carbon atoms joined by single bonds, and differ only in the specific groups sticking out from this carbon backbone. Addition polymers are linear polymers.

Condensation polymers are the other major class of polymers. In order to form a condensation polymer, it is necessary to have two chemical groups that react to join to one another. For example, carboxylic acids react with alcohols to form esters and with amines to form amides. Furthermore, it is necessary to

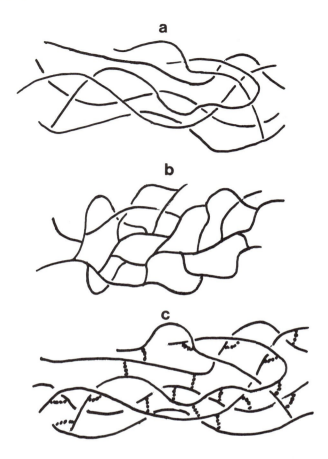

Figure 16.1. Here we see three physical forms of plastic. (a) is a typical thermoplastic in which long strands are mixed, but are independent of one another. On heating, these strands become more mobile, and the shape of the polymer mass changes as they slide past one another. (b) shows a thermoset plastic. Strands are interconnected into a three-dimensional array, and so have no ability to slide past one another. Such structures remain rigid until the temperature is high enough to break chemical bonds. (c) represents the same thermoplastic structure as (a), except that the plastic has been chemically crosslinked (cured). Physically this product more closely resembles (b) than (a).

have two such groups on each monomer. The formation of nylon-6,6 involves the use of two monomers, one with two carboxyl groups and the other with two amine groups. If one carboxyl group reacts with one amine group, two mono-

Table 16.2. Commercial Addition Polymers.

Polymer	Commercial Name	Monomer	Formula
polyethylene	Polythene	ethene, ethylene	$CH_2 = CH_2$
polypropylene		propene, propylene	$CH_2 = CH - CH_3$
polystyrene	Styrofoam	styrene	$CH_2 = CH - C_6H_5$
polyvinylchloride	Vinyl	vinyl chloride	$CH_2 = CH - Cl$
polyacrylonitrile	Orlon	acrylonitrile	$CH_2 = CH - CN$
polymethylmethacrylate	Lucite™ , acrylic plastic	methyl methacrylate	$CH_2 = CH - CH_3$ $C = CH_2$ $C-OCH_3$ $= O$
polytetrafluoroethylene	Teflon®	tetrafluoroethylene	$CF_2 = CF_2$

mers are now joined by an amide linkage. However, at one end of the combination there is an unreacted carboxyl group available to react with another amine, and at the other end there is an unreacted amine available to react with another carboxyl group. If one of the monomers were to have three reactive groups, the polymerization reactions would form the thermoset type of structure. Examples of condensation polymers are found in Table 16.3.

The Relationship of Chemical Structure to Properties

The manner in which such properties as strength, elasticity, and hardness correspond to chemical structure of a polymer is very interesting to chemists. Most significant are the ability of two strands to attract one another, the flexibility of the polymer backbone, and the regularity of the repeating pattern along the polymer strand. Such characteristics differ with the monomer selected.

We can fine tune the properties by mixing more than one monomer together to produce a *copolymer.* The final product should display contributions from each of the monomers. Examples of common commercial copolymers include Saran™ and ABS plastics. Saran™ is a copolymer of vinyl chloride and vinylidene chloride, while ABS has acrylonitrile, styrene, and butadiene in its formula. A longer list of copolymers commonly used in industry is provided in Table 16.4.

Linear polymers show varying degrees of ability for the polymer strands to take on an orderly arrangement. This depends on such variables as the regularity of the polymer structure, the existence and strength of special attractive forces between chemical structures on the strands, and the rigidity of the strands. A plastic that is very disorderly is termed amorphous. Amorphous plastics are more flexible, but have lower resistance to tear. When a plastic has orderly regions, called *crystallites,* it becomes more rigid and stronger (Figure 16.2). The degree of crystallinity can vary for a given plastic. It is influenced by the manner of processing the plastic. When the plastic is cooled slowly after molding, the strands have more time to respond to their mutual attractions, and to arrange themselves into orderly patterns. This could mean a higher proportion of crystalline versus amorphous regions, or it could mean larger crystallites.

Additions to Plastics

A number of compounds are added to plastics to modify their properties. It is also possible to prevent crystallites from forming simply by adding compounds called *plasticizers* to the molten plastic. Plasticizers interfere physically with the alignment of the strands, assuring a more flexible product. Thus, plasticizers facilitate use of vinyl plastics for clothing or upholstery. When vinyls were first used for auto interiors, the plasticizers would "cook out" of the

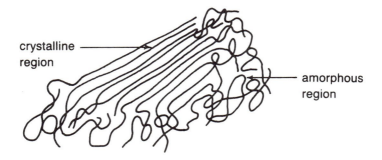

crystalline region

amorphous region

Figure 16.2. Crystallinity. When a heated polymer cools, the polymer strands may arrange themselves into orderly arrays called crystalline regions or crystallites. These impart a strength and rigidity to the plastic that it would not have in the totally amorphous state. To keep plastics flexible, compounds called plasticizers are added to the hot plastic to physically block alignment of the strands.

plastics in hot weather, coating the insides of windows and leaving the vinyl shrunken and brittle. Today, plasticizers are more stable and permanent, and are also generally less toxic.

Converting monomers to a polymer would progress very slowly unassisted. For this reason polymerization reactions are often run at elevated temperatures and *catalysts* are added. Catalysts stimulate reaction rate markedly, even at low concentration. Since they are added to the reaction vessel, they remain in the finished polymer.

Other compounds are added to modify properties of plastics. Most plastics contain *pigments.* If the compound must be durable out of doors, *stabilizers* are included that block UV light and so protect the material from the effects of sunlight. Plastics sometimes contain *fillers,* which are added at the time of polymerization. These may be intended to modify the properties of the product. Fiberglass is an example of a filler used to make the finished thermoplastic product stronger and more rigid. In thermoset plastics, fillers may simply be inert material such as clay that is "diluting" the more expensive plastic components. This can be done successfully because of the great strength and rigidity of thermoset plastics.

Where Do These Operations Take Place?

Plastics (resins) are generally synthesized in a large chemical plant, often located near a refinery, since monomers are often derived from petroleum

Table 16.3. Some Commercial Condensation Polymers.

Polymer	Monomers	Formula
nylon-6	caprolactam	
nylon-6,6	1,6-diaminohexane, adipic acid	
polycarbonate	bisphenol a, phosgene	
polybutylene terephthalate	1,4-butanediol, dimethylterephthalate	
polytetramethylene terephthalate		
polyethylene terephthalate	ethylene glycol, dimethylterephthalate	
wholly aromatic copolyester	p,p′-dihydroxybiphenyl, terephthalic acid	
aromatic polyester	bisphenol a, terephthalic acid	
polyurethane	bisphenol a, toluene diisocyanate, polyglycol	
urea-formaldehyde	urea, formaldehyde	
melamine-formaldehyde	melamine, formaldehyde	
phenol-formaldehyde	various phenols, formaldehyde	

Table 16.4. Commercial Copolymers.

Polymer	Monomers
ABS	acrylonitrile, butadiene, styrene
OAS	olefin, acrylonitrile, styrene
ACS	acrylonitrile, chlorinated polyethylene, styrene
ASA	acrylic, styrene, acrylonitrile
LLDPE (linear low-density polyethylene)	ethylene, hexene, butene
VLDPE (very low-density polyethylene)	ethylene, other olefins
Ionomer	ethylene, methacrylic acid salts
EMA	ethylene, methyl acrylate
EEA	ethylene, ethyl acrylate
EVA	ethylene, vinyl acetate
SAN	styrene, acrylonitrile
SB	styrene, butadiene
VDA copolymers	vinylidene chloride + vinyl chloride, acrylates, or acrylonitrile

sources. Chemical companies produce the polymer for sale to manufacturers. Thermoplastics are usually completely synthesized, mixed with additives, and shipped as bags of pellets. These pellets are dumped into the hoppers of machines in the manufacturing plant, which melt them and form them into a product. This product may then be machined or otherwise reshaped, and may be decorated. In the case of thermoplastics, very little chemistry is performed in the manufacturing plant.

Thermoset plastics cannot be handled in this fashion. Once the full polymer structure is formed, the shape is permanent. The final polymerization is therefore performed in a mold by the manufacturer. The chemical company may prepare and ship low-molecular-weight prepolymers. These are mixed with catalysts (curing agents) and crosslinking structures, which are then fed into the mold. Alternatively, monomeric structures and a catalyst are mixed.

HAZARDS OF POLYMER SYNTHESIS

The synthesis of polymers is a chemical process carried on in plants with towers, tanks, and reactors connected by endless pipes with valves and gauges. As is typical of such operations, chemicals are in closed systems, so they seldom contact workers under normal circumstances. It is important to focus on times when such contact does occur, to be sure all employees are informed of the hazards of the compounds, of necessary personal protection if contact does occur, and of emergency spill procedures.

Chemical exposure is more likely during unloading of chemicals at the plant site. Leaks in pipes, valves, and gauges are always possible. There is special concern any time systems are opened, as during plant maintenance. Routine cleaning of tanks and other vessels involves worker entry into confined spaces and direct contact with chemicals, even if the tank is purged in an appropriate fashion beforehand.

TYPES OF PLASTICS PROCESSING

It is useful to understand what operations occur in a manufacturing facility in order to fully appreciate the generation of hazards. Manufacturing operations range from huge factories, such as those operated by auto companies to produce the plastic components of cars, to very small plants contracting production of a few components or of short runs of a few parts. Often the machinery used is quite standard, and only the changing of a mold or die is required to convert a machine from production of one part to another in a small operation.

Mixing

As has been stated, a finished product is likely to contain a variety of additives to the pure polymer, such as pigments, plasticizers, fillers, and stabilizers. Dry mixing is accomplished with mechanical stirrers, much as a cake batter is made in the kitchen. A better mix is obtained if the plastic is melted, and this may be the total method employed, or it may be preceded by dry mixing. The plastic may be formed immediately into product once melted and mixed, or it may be formed into strips or sheets, which are then cut into pellets to feed into the hopper of a molding or other forming machine later. Particulate in the air is the most serious hazard of mixing processes.

Casting and Injection Molding

When molding is done without pressure, it is termed *casting*. Casting could involve pouring the plastic into a mold. There are also continuous casting processes where a sheet or film is produced by pouring the plastic onto a belt moving through an oven. Casting can be done with polymers, or with polymer-monomer mixes that complete polymerization in a hot mold.

The most common way to form plastic into products is *injection molding*. Pellets are melted, then the melt is forced into a mold. The mold is cooled and the product removed, often by use of a high-pressure air jet. The shapes of the sprues, runners, and gates are trimmed off and recycled as scrap, but otherwise the surface is very well finished and usually needs no further work. Injection molding of thermoplastics can form large objects and can be run at a high production rate.

Thermoset plastics can be injection molded, but the process is more difficult. There is a need to adjust the heating process to soften the reactants before molding, but the product must not harden in the heating chamber.

Molding requires that the dies be opened to remove the product, then closed against the machine again for the next cycle, which is a noisy process. This is laid over the general noise level of the hydraulic pumps, valves, and lines. When an air jet is used to remove the parts from the mold, the noise produced is quite loud, and the cutting off of scrap can be noisy. Noise levels are a major problem in an injection molding operation. Heat can volatilize materials from the melt, but such vapors are usually easily controlled.

Blow Molding

Blow molding is used to produce hollow objects such as bottles and containers. In *extrusion blow molding,* a hot plastic tube is formed and the end is pinched shut, usually by the mold. The molds are water-cooled, and the plastic tube is inflated to fill the mold cavity, then solidified against the cool surface of the mold. Scrap is cut off and recycled. In *injection blow molding* a quantity of plastic is injected on a rod and forced into a mold, where the container is blown. It differs from extrusion blow molding in that the cavity is the shape of the final product, so there is no scrap.

As with injection molding, noise levels are high. The use of compressed air is primarily responsible for the noise.

Vacuum and Pressure Forming

In *vacuum forming,* a thin sheet of plastic is heated and laid over a mold, either male or female. The space between the sheet and the mold is then evacuated, forcing the sheet to conform to the shape of the mold. The mold is cooled to set the plastic in the desired shape, and the part is removed. *Pressure forming* is a very similar process, the only difference being that pressure is applied outside the sheet rather than a vacuum being formed between the sheet and the mold. These processes adapt well to high production rates. Vacuum molding is also used to package objects, especially fragile objects, by placing a plastic sheet over the object and evacuating the space between the object and the sheet. Once again, pumping air generates noise in all these processes.

Extrusion and Calendaring

A number of products, including pipe, tubing, sheets, films, and fibers, are prepared by *extrusion.* The molten plastic is forced continuously through a die to generate the desired cross section. Some films are prepared by blowing a bubble, then drawing the bubble away from the die continuously. When the desired final surface is obtained by squeezing a sheet between heated rollers, producing whatever cross section and surface are desired, the process is called *calendaring.* Coating textiles or other substrates with plastic can be done on a calendaring machine. Several components of extruders generate noise.

Molding Thermoset Plastics

Usually the starting material for thermoset molding is a mixture of low-molecular-weight polymers, filler, and crosslinking agents. This molding is more challenging, because the material first melts and flows throughout the mold cavity, then hardens into its permanent shape under the influence of heat. Melting must be done to allow the material to flow into the mold, but hardening must not occur until the material is in place. Often it is also necessary to open the mold before hardening to allow gases to escape. Clearly, timing and temperature must be carefully adjusted. The plastic may be added to the mold, then the mold is closed and the material melts and flows (compression molding), or it may be melted first, then forced into the mold (transfer molding). Thermoset plastic scrap normally cannot be recycled. Machinery noise is a problem for the workers.

Foam Processing

Foams—plastics with gas bubbles throughout the finished product—can be produced using either thermoplastic or thermoset resins. Foams are strong for their mass and have excellent thermal, electrical, and acoustical insulation properties. Bubbles are produced either because the polymerization reaction releases a gas or because a foaming agent (chemical *blowing* agent) is added to the plastic that generates a gas as the plastic is heated. For example, nitrogen may be added to the polymer melt under pressure. When the pressure is released, the gas leaves solution as a multitude of tiny bubbles.

Liquids such as Freons™ (chlorinated-fluorinated hydrocarbons) have been very popular blowing agents because of their low toxicity, but these will need to be replaced as the ban on such compounds is imposed to reduce their damaging effect on the atmospheric ozone layer. Several organic compounds that decompose to gases when heated are used. These include azodicarbonamide, 1,1′-azo-bisformamide, p-toluene-sulfonyl semicarbazide and p-toluene-sulfonyl hydrazide. These compounds are often strong irritants, and present a hazard to workers if they become airborne during mixing. Azo compounds are incompatible with acids or the ketone peroxides, which may be involved in curing thermoset plastics. Contact with traces in the plastic initiates the desired nitrogen formation, but direct mixing produces a highly exothermic reaction and combustible gases.

Foams can be injection molded, extruded, or coated onto plastic sheets. The latter can later have embossed patterns pressed into them with hot patterned rollers or dies. An example is the padded decorative door panel material used in cars.

HAZARDS IN PLASTICS MANUFACTURING

In the manufacturing plant, thermoplastic polymers come to the plant already synthesized. The least hazardous material encountered by workers is the polymer itself, since it cannot enter the body. Huge polymer molecules are not volatile, so entry as vapors into the lung is ruled out. Contact with the skin, lining of the gastrointestinal tract, or walls of the respiratory passages similarly does not lead to entry into the body, since such large molecules cannot cross the membranes blocking such access. However, thermoplastics include substances that may well be irritating or otherwise produce unwanted effects.

It is possible to have unreacted monomers in a resin sample escape as the polymer is heated to be processed, since monomers are volatile compounds. Shaping operations, such as cutting or sanding, may also release unreacted monomer. Most monomers are volatile compounds. In the case of highly hazardous monomers such as vinyl chloride, plastics are "stripped" after synthesis to remove most unreacted monomer. Additives, such as dyes, blockers, and plasticizers, may also be released during manufacturing operations, and their hazard must be considered. However, the beads of thermoplastic used in the manufacturing plant usually present little hazard.

This generalization about the safety of thermoplastic materials does not extend to thermoset plastics. Much more handling of potentially hazardous chemicals is involved in thermoset plastic manufacture. Generally, in forming such a plastic, one of the starting materials is a polymer and the curing operation is one of crosslinking the material into a three-dimensional rigid array. In that case, the curing agent is a small and reactive molecule. There are also many examples where the reactants in forming thermoset plastics are all low-molecular-weight, potentially hazardous substances. Finally, additives, including dyes, screening agents, and fillers, are often added at the time of manufacture.

After forming the product, whether thermoplastic or thermoset plastic, any operation involving sawing, grinding, milling, or sanding the product, creates dusts. Excess plastic that is trimmed off a molded product is often ground up, possibly mixed with new pellets, and recycled back into processing. Once again, there is opportunity for dust to enter the air. Such operations may also release volatile chemicals, such as unreacted monomer trapped in the body of the plastic.

Particulates of any sort entering the lung are of concern. What seem to be completely inert substances can stimulate scarring of the lungs and loss of lung capacity. Unanticipated problems can result from the intrusion of any substance into the lung, especially since some portion of the material so inhaled may well remain for some time in the lung passages.

Any step in processing the plastic product or disposing of scrap that involves heat can lead to decomposition of the material. Such breakdown of plastics produces a variety of particularly offensive compounds, including such gases as HCN, HCl, and phosgene. Even the relatively noncombustible Teflon® polymers release a fume that causes an influenza-like condition, with high fever, headache, and a cough. Once again, the problem arises of the worker who smokes cigarettes. Particulate in the air from polymer grinding can coat the tobacco, and this particulate will decompose when the cigarette is burned.

COMMENTS ON SPECIFIC COMPOUNDS

With more than one hundred polymers and many more copolymers in commercial use, it is not within the scope of this book to provide a comprehensive list of the hazardous substances that could possibly be encountered. However, some general comments are certainly appropriate, and discussion of some of the more commonly encountered serious hazards should be useful. These are organized by the function of the compound in producing the final polymer.

Monomers

A few monomers for which special problems exist are discussed below. (See also Table 16.5.)

A. **Vinyl Chloride.** Polymerization of vinyl chloride is accomplished in a variety of ways using peroxide catalysts. Vinyl chloride does not have good warning properties, such as odor or irritant action. Historically, it was considered to be a safe compound with a TLV of 500 ppm. Workers are reported to have leaned over open vats and inhaled deeply to experience its narcotic effects. Its use as an anesthetic had been studied. In 1961 it was suggested by Dow Chemical that the TLV should be lowered based on preliminary studies. A series of problems surfaced in the period that followed. Workers who cleaned vinyl chloride polymerization vats suffered degeneration of bones (acroosteolysis) in their hands. In 1972 evidence of liver damage was observed at levels greater than 300 ppm. Then in 1974, studies showed that workers displayed a higher than expected level of a very rare cancer, angiosarcoma of the liver. The OSHA standard for exposure to vinyl chloride became 1 ppm with a 5 ppm ceiling as of January, 1981. Stripping techniques are now used to remove unreacted monomer from the polymer product.

B. **Acrylonitrile.** This is the monomer in the production of Orlon,™ and is one of the components of the copolymers ABS (acrylonitrile-butadiene-styrene) and SAN (styrene-acrylonitrile). Acrylonitrile can be absorbed through the lungs or skin. It penetrates rubber gloves, making these inadequate as a protective measure. In the body it serves as a source of cyanide ion, producing asphyxiation. It has been shown to be a cause of cancer in both the lung and the bowel. The OSHA standards (January 1981) were 2 ppm with a 10 ppm ceiling.

C. **Styrene.** This compound is employed not only to prepare polystyrene and styrofoam, but as one of the copolymers in ABS, styrene-butadiene, and SAN plastics. The vapors are irritating, in fact irritating enough that dangerous exposure is not likely to be tolerated, and there is the usual skin irritation due to removal of skin oils. Styrene affects the central nervous system, producing depression. A problem called "styrene sickness" involves nausea, vomiting, dizziness, and fatigue. OSHA standards limit exposure to styrene to 100 ppm,

Table 16.5. Toxicity of Selected Monomers.

Monomer	Toxicity (Airborne)		Exposure Limits	
	LC50		PEL	
	ppm	mg/m³	ppm	mg/m³
Acrylamide	—	—	0.3	—
Acrylonitrile	425	—	2	—
Butadiene	—	259,000	1000	2200
Chloroprene	—	2,300-11,800	25	90
Diglycidyl ether	30	—	0.5	2.8
Epichlorohydrin	250	—	5	19
Ethylene	950,000	—	—	—
Formaldehyde	—	92-400	1 (10 ppm peak)	—
Isopropyl glycidyl ether	1,100-1,500	—	50	240
Maleic anhydride	—	—	0.25	1
Methyl acrylate	1,350	12,800	10	35
Methylene bisphenyl isocyanate	—	178	0.02	0.2
Phenol	—	74-177	5	19
Phosgene	—	1,000-10,000	0.1	0.4
Phthalic anhydride	—	—	2	12
Styrene	—	9,500–24,000	100, 200 (ceiling)	—
Toluene diisocyanate	10	—	0.02	0.14
Vinyl chloride	180,000	—	cancer suspect agent	
Vinylidene chloride	6,350	—	1	4

with an acceptable ceiling of 200 ppm and an acceptable maximum peak of 600 ppm for no more than 5 minutes in any 3-hour work period. Styrene is shipped mixed with polymerization inhibitors such as butylcatechol and hydroquinone to prevent spontaneous premature polymerization. Both of these are sensitizers.

D. **Epoxy Resin Monomers.** Toxicologically, this is a very troublesome group of compounds. Epichlorohydrin is an irritant to the skin, causing burning, itching, and redness, with pain and blistering appearing after contact. It can be absorbed through the skin, and is irritating to the eyes, causing damage at higher concentrations. Epichlorohydrin is severely irritating to the lungs, producing pneumonitis (fluid in the lungs) hours after exposure. In the body, liver and kidney damage result, and sterility is caused. Finally, it is a sensitizer, leading to later allergic response upon contact with even small quantities of epichlorohydrin. The PEL is 5 ppm. Compounds used with epichlorohydrin to produce epoxy resins include bisphenol A, glycidyl ethers, and aliphatic poly-amines such as p-phenylenediamine, diethylenetriamine and triethylenetetra-mine. All these are irritants and sensitizers. The polyamines are particularly hazardous, causing severe irritation, chemical burns, reddened and itching skin, blistering, facial swelling, and asthma. They can cause bronchospasms and coughing for days after exposure. Sawing or machining finished plastic prod-ucts can release amines and unreacted epichlorohydrin.

E. **Phenolic and Amino Resin Monomers.** The phenolic thermoset plastics are produced by the reaction of a variety of phenols with formaldehyde or furfural. The phenols include phenol itself, cresol, xylenol, p-t-butylphenol, and resorcinol. All these compounds, the phenols and the aldehydes, are potent irritants. Above 3 ppm, formaldehyde is irritating to the eyes and upper respira-tory tract. Furthermore, formaldehyde is a sensitizer. Other findings about formaldehyde, including the possibility that it is a human carcinogen, are under review, and standards for its safe use were revised down to a level of 1 ppm in 1987. Amino resins are similar to phenolics, substituting urea or melanine for the phenols. Here too, formaldehyde is a serious problem. Hexamethyltetra-mine, used in forming those polymers, decomposes to give formaldehyde and ammonia, both of which are irritating.

F. **Diisocyanates.** Polyurethane and polyisocyanurates are produced by reacting diisocyanates with polyesters or polyethers having terminal alcohol groups on the chain. The diisocyanates include TDI (toluene diisocyanate) and MDI (methylene diphenyl diisocyanate), and these compounds present a serious threat. The isocyanate chemical group was responsible for the massive loss of life in the chemical escape at Bhopal, India. The difference between the methyl isocyanate that escaped at Bhopal and these compounds is basically one of volatility. Isocyanates react rapidly with water, and so are irritants on moist surfaces such as the eyes and respiratory tract. In addition they are potent sensitizers. The PEL for MDI is 0.02 ppm and for TDI is 0.05 ppm.

G. **Polyester and Alkyd Monomers.** Polyesters and alkyds may utilize acid anhydrides, such as phthalic anhydride or maleic anhydride, along with polyal-cohol structures, as monomers. The acid anhydrides react vigorously with small amounts of water, producing heat and concentrated acid. They cause burns and irritation on contact with eyes, nasal passages, throat, and moist skin. Unlike many of the compounds in this chapter, they are not volatile liquids. However, they can enter the workplace air as dusts produced during transfer and handling. Polyethylene terephthalate is made by reacting dimethyl terephthalate with ethylene glycol to polymerize at higher temperatures by ester exchange. Antimony oxide and zinc acetate are commonly used in this process, and dusts result as the bags are opened and dumped.

H. **Acrylic Monomers.** Acrylic acid and a wide variety of its derivatives, especially esters, are used as monomers.These are skin irritants, and the vapors produce nervous system effects such as headaches, irritability, numbness in the extremities, slurred speech, and fatigue. The nervous system symptoms for acrylamide are particularly severe, and the PEL for this solid is 0.3 mg/m^3 in the air.

I. **Polycarbonate Monomers.** The preparation of polycarbonate plastics involves the use of bisphenol A, dioxane, and phosgene. Phosgene gas has been described in Chapter 6 as a lower respiratory tract irritant. As such, it must be handled with extreme care. The odor and immediate level of irritation experienced on contact with this compound are insufficient warning to workers to assure they will not voluntarily allow dangerous doses into the lungs. Dioxane has been classed as a carcinogen.

J. **Nylon Monomers.** Nylons may be prepared using diacids or diacid chlorides, which are reacted with diamines. The diacids are not serious problems, but diacid chlorides are similar to acid anhydrides in properties, and can cause severe skin or eye damage. The diamines are irritants and sensitizers, and so also must be carefully handled. The amino acid caprolactam, used to make Nylon 6, causes nose and throat irritation, irritability, and nervousness.

K. **Vinylidene Chloride.** Most commonly, this monomer is copolymerized with vinyl chloride. It is irritating to the skin and eyes, and may include a phenolic inhibitor, which increases the irritant potential. (See Styrene.) When inhaled, it produces a state of drunkenness.

L. **Vinyl Acetate.** This monomer is used to prepare polyvinyl alcohol. The ester (acetate) groups are removed after polymerization. The vapors of the monomer are upper respiratory tract irritants, and are narcotic.

M. **Ethylene.** Ethylene is obtained from a refinery cracker, and is polymerized using organic peroxide catalysts. Ethylene has low toxicity, but a high fire and explosion hazard.

N. **Viscose Rayon Components.** One starts here with the polymer (cellulose) and modifies it. The most hazardous compound in the process is carbon disulfide, which is extremely volatile and has a very low flash point. It affects the central nervous system, producing anything from dizziness and fatigue to psychological disturbances. It is also a respiratory irritant.

O. **Cellulose Acetate Components.** Cellulose (wood pulp or cotton) is mixed with glacial acetic acid, sulfuric acid, and acetic anhydride to produce this polymer. Some ester linkages are then removed by acid-catalyzed hydrolysis to adjust polymer properties. Milling the cellulose source is a noisy process, and dust from the cellulose source and acid vapors are obvious problems. Washing with water generates high humidity conditions, adding to the potential for heat stress in hot weather. Drying and handling the final product is another time of potential dust problems.

Catalysts (Curing Agents)

A sampling of commonly used catalysts is shown in Table 16.6. Reactions to form addition polymers very often proceed by a free radical mechanism. Such a mechanism requires a source of free radicals—structures with unpaired electrons—to initiate the chemical reaction. Most often organic peroxides are used. Around fifty different such compounds are available, primarily peresters and percarbonates. Benzoyl peroxide and MEK peroxide are representative and frequently encountered examples. Peroxides share similar properties, all being strong oxidizing agents capable of reacting violently with appropriate substances. Skin irritation and burns are likely on contact with peroxides, and severe eye damage is possible. Benzoyl peroxide is a sensitizer, with chronic exposure leading to an allergic rash. It has a PEL of 5 mg/m^3, and an IDLH level of 1000 mg/m^3. Both from the standpoint of hazard due to direct contact, and because of a potentially violent reaction, waste or spilled peroxides should be diluted with water, and any rags used in the cleanup of spills similarly should be put in water. The rate of chemical decomposition of peroxides increases with temperature, and contact with combustibles such as paper or wood can lead to fires.

The aluminum alkyl catalysts used with polyolefins are physically very hazardous. They react violently with water, and burn spontaneously in air. In solution, these compounds can cause serious burns, and fumes they produce are damaging to the lungs.

Organic metal salts or tertiary amines are used as catalysts in polyurethane or polyisocyanate synthesis. Both penetrate the skin and may be sensitizers.

Accelerators

Although not catalysts themselves, accelerators make catalysts more effective. For example, cobalt naphthanate stimulates the decomposition of peroxides into the very reactive free radicals.

In general, these compounds are irritants, airborne samples causing irritation of eyes, nose, and throat, and direct contact with the skin causing a rash. They are frequently sensitizers. Thiram and disulfiram share with Antabuse™ the ability to block the metabolism of ethanol and aldehydes. Nausea and vomiting follow intake of ethanol after contact with these compounds. The PEL for the

Table 16.6. Commonly Used Catalyst Systems.

Product	Catalysts
ethylene	
high-density polyethylene	aluminum alkyls plus titanium tetrachloride
low-density polyethylene	chromium oxides on alumina or silica *tert*-Butyl peroctoate; other peresters
linear low-density polyethylene	aluminum and magnesium alkyls plus titanium halides
propylene	aluminum alkyls plus titanium trichloride
polystyrene	benzoyl peroxide; peresters
polyvinyl chloride	percarbonates
elastomers	aluminum alkyls plus vanadium trichloride or oxychloride

Reprinted with permission from *Chemical and Engineering News* (Feb. 17, 1986). Copyright 1986, American Chemical Society.

solid thiram is 5 mg/m^3. Dimethylaniline is used as an accelerator in polyester synthesis. It absorbs readily through the skin, and once in the body is a central nervous system depressant. Dimethylaniline also causes kidney and liver damage. It has a PEL of 5 ppm and an IDLH level of 100 ppm.

Stabilizers

Stabilizers or screening agents are added to plastics to slow degradation of the plastic caused by heat or light. Metal soaps are commonly used, and many have serious toxicity problems. Lead, barium, cadmium, calcium, magnesium, and tin compounds are used. Lead and cadmium are serious hazards to health. Once in the body they incorporate into bone structure and remain there for long periods, releasing slowly into the blood. Lead causes damage to internal organs such as liver and kidney, and causes lasting damage to the central nervous system. Cadmium remains in the body even longer than lead. It damages the kidney, and at high levels has caused skeletal problems. Organotin compounds are strong irritants, and can cause chemical burns. Liver and urinary tract damage results from contact. After contact, there is evidence of central nervous system damage, including headaches, loss of visual acuity, and motor control problems. The PEL for organic tin (as tin) is 0.1 mg/m^3 and the IDLH level is 200 mg/m^3. By contrast, calcium, magnesium, and barium soaps have low toxicity.

Plasticizers

Plasticizers are added to linear polymers to prevent crystalline regions from forming by physically interfering with the lining up of the polymer strands into crystallites. Early plasticizers were sometimes quite toxic. Tricresyl phosphate attacks the peripheral nervous system, and some of the chlorinated hydrocarbons used cause liver damage. Today most plasticizers are phthalate esters. These are generally of low toxicity and seldom cause problems. For example, dibutyl phthalate is relatively nonirritating and nonvolatile. In animal testing, high doses are necessary to produce toxic effects. There is some discussion about the possible carcinogenicity of di-2-ethylhexyl phthlate to humans. Esters of adipic acid and of citric acid are also used, and are of low toxicity.

Fillers

Used most frequently with thermoset plastics, fillers are generally stable and inert solids. Clays, silicates, and glass fibers are typical of the materials used. Hazard is largely present when particulates from these enter the air, either while being added to a reaction mixture, or when a finished product is milled or sanded. Silicates can cause fibrosis of the lungs on chronic contact. Irritation from glass fibers is a special case of problems with fillers. Restrictions on mineral dusts depend on the composition (crystalline silica content) but are typically 20 mppcf (millions of particles per cubic foot of air).

Other Additives

Many other compounds can be added to plastics, depending on properties desired in the product. These include pigments, fire retardants (in construction plastics), antioxidants, and agents to help lubricate during extrusion or ease release from a mold. Some flame retardants, for example tris-(2,3-dibromopropyl) phosphate, are cancer suspect agents. The monobenzylether of hydroquinone is used as an antioxidant, and is a sensitizer. Chloronaphthalene is used as a mold release compound, and can cause chloracne. Most pigments are inert and present little hazard. However, they can generate a dust problem during transfer to the plastic preparation.

Summary

These discussions serve only to point up classes of compounds and several specific problem compounds. Toxicity information supplied with chemicals used in a plant or available from such a source as an MSDS should be consulted for all chemicals used in an industrial facility. The possibilities for trouble should be anticipated so as to prevent a serious accident or injury.

ELASTOMERS

Elastomers are polymers that possess a special property. When distorted by stretching and the distorting force is then removed, they return to their original shape. Natural rubber (latex) is a polymer of the diene isoprene. The importance of elastomers to mechanized warfare was an important early stimulus to research in Europe to develop synthetic rubbers to eliminate dependence on the shipping of natural latex from distant, tropical sources. Efforts focused on variations of 1,4-addition polymers of butadiene or a butadiene derivative, since the surviving double bond in such a polymerization was early recognized as central to elastic properties in the product. The new elastomers developed had advantages in properties and price that made the continuation of this effort there and elsewhere commercially viable.

THE USE OF ELASTOMERS

Worldwide, the use of elastomers is estimated to be 14.9 million metric tons in 1992, and about 65% of the elastomer used is synthetic rubber.[3] North America consumes about 28% of the synthetic rubber and 19% of the natural rubber used in the world. Tires alone account for three quarters of the natural rubber and more than half the synthetic rubber use. Of the several synthetic rubbers produced (Table 16.7), styrene-butadiene rubber (SBR) has the greatest share of the market at about 35%. Some properties of SBR are not as good as those of natural rubber or some other synthetics, but it has a relatively low price, and is therefore very competitive. Second to SBR is polybutadiene (BR), with 17% of the market. Other important elastomers include ethylene propylene copolymer (EP) and terpolymer (EPDM), polyisoprene (IR), isoprene isobutylene copolymer (butyl rubber, IIR), polychloroprene (neoprene, CR) and acrylonitrile-butadiene copolymer (nitrile, NBR). There are some low-volume, usually expensive elastomers that are produced for special purposes. These include silicone rubbers, chlorosulfonated polyethylenes, acrylics, copolyesters, epichlorohydrins, polysulfides, and urethanes.

RUBBER PROCESSING

The processing of rubber is similar to that of plastics. Hazards at various steps of the process are also similar.

Tires

Given the importance of tires in the overall elastomer picture, a separate word about tire construction is worthwhile. Tires are built around a body of layers of rubber-coated fabric reinforced with steel, glass, or polymer strands (which nowadays are generally run at right angles to the tread) termed the *plies* of the tire. The *sidewall* is an extruded rubber layer outside the plies, and the tread is the extruded strip that is the part of the tire contacting the road. At the inner edge of the plies is the *bead,* rubber-covered steel wires that are at the point where the tire seals against the rim. Finally, a calendared sheet called the *liner* forms the inner layer of the tire and serves to retain air. The uncured assembled tire[4] is sprayed with mold release and placed in a molding machine. A bladder inserted into the tire is expanded, forcing it against the mold that includes the tread and sidewall pattern. The mold is heated and the tire is cured.

---

[3]*Chem. Eng. News,* May 10, 1993.

[4]An uncured product is termed "green."

Table 16.7. Production and Uses of Selected Rubbers.

Elastomer	% Total Production	% Used for Tire Production
Natural rubber	40	68
SBR	30	70
Polybutadiene	12	75
EPDM	5	7
Neoprene	4	–
Butyl Rubber	4	75
Nitrile rubber	3	–
Polyisoprene	2	50

Compounding and Mixing

The first step in processing elastomers is compounding a mix of elastomer (polymer), plasticizer, curing agent, catalyst, antioxidant, and other components. Sometimes agents, especially toxic agents, are shipped in preweighed plastic containers that are added, plastic and all, without opening. Materials delivered in drums may have to be hand-weighed and hand-added, an opportunity for worker exposure to the chemicals. Large operations such as a tire plant may automate parts of the mixing, commonly done when adding carbon black.

Mixing is usually done in two stages. The first stage omits curing agents and catalysts, and involves thoroughly breaking up the elastomer stock and mixing it with the other agents. This may be stored until needed. Then, shortly before processing, mixing is done again, this time adding agents required for curing. The mixer may be cooled to prevent premature curing.

After mixing, the stock is *milled,* which further mixes the components and delivers the mix as a sheet that is coated with a chemical to make the surface less sticky (antitackifying). Compounded rubber is stored as folded sheet awaiting the second mixing, or is transported to be processed after the second mixing.

Extrusion

Extrusion is a common rubber forming process, and is similar to plastics extrusion. The product is a strip of material with a particular cross-sectional shape. Temperature control is important to adjust the fluidity of the rubber mix without triggering premature curing. In tire production, different rubber stock for sidewalls and tread are fed together into the extruder to form a layered strip in a single operation.

Calendaring

In calendaring, fabrics are coated with rubber. Rubber sheets are formed, the fabric is coated, and the rubber is forced into the fabric strands by *frictioning* with a hot roller moving faster than the coated fabric sheet. Tackifying oils may be added to soften the rubber surface facing the fabric. Both sides of the fabric are coated in tire production.

Forming and Curing

The uncured rubber intermediate prepared by extruding and/or calendaring is now molded or assembled into the shape of the final product. Compression molding, the method used to make tires, places the stock between heated dies to which pressure is applied. Excess stock leaves through channels. In transfer molding, the stock is rammed into the mold, and in injection molding it is heated to become fluid and injected into the mold.

Curing (vulcanization) generates a product with the desired properties. The mix contains a crosslinking material and an accelerator such that under heat and pressure, applied for the correct time interval, the polymers crosslink to the correct degree. When the product is molded, curing occurs in the heated mold. Articles otherwise assembled are usually cured using high-pressure steam (autoclaving), or dry-heat in an oven.

HAZARDS

Hazards include exposure to noise from the mechanical devices used during compounding, extruding, calendaring, and molding. Engineering controls and careful maintenance of the machinery are important. The use of compressed air and high-pressure steam add to noise levels. Airborne particulate may be generated at several points in the transfer of mix components, weighing, addition to the mix, and mixing. Carbon black is especially troublesome because of its very small particle size. Well designed equipment, good maintenance, ventilation, and careful housekeeping minimize this problem. As the stock is heated to be extruded or calendared, volatile components of the mix may enter the plant air. This potential is even greater during molding and curing the final product. Cements and tackifiers (agents to make rubber stickier) may include volatile solvents, which are released after application. Since some operations are run at elevated temperatures, there is a potential for heat stress.

Chemical Hazards with Specific Elastomers

Comments about the relative safety of polymeric material made regarding plastics are applicable here. The operations involving chemical processing in a manufacturing plant preparing elastomers are generally the same as in a plant producing thermoplastics. Many of the chemicals are the same, but some are unique to elastomer production (Table 16.8).

A. **Natural Rubber.** Latexes are used in a number of processes. Since latex is a natural polymer, workers are not exposed to the monomer form or the polymerization process, and the polymer itself presents little hazard. When dipping gloves, footwear, and the like, dilute acids such as acetic or formic acid are present. The chief problem lies in diluting these acids from the concentrated form, which can involve the possibility of acid burns, heat production, and inhalation of excessive amounts of acid vapors. Hydrazines may be used in foam manufacture, and are toxic materials. Ammonia fumes may be present, but usually are at low levels. The manufacture of latex adhesives involves the use of quantities of organic solvents, often chlorinated hydrocarbons. Natural rubber is included in tires, and organic solvent fumes must be removed when compounding tire stock.

B. **Polyisoprene.** Methods have been devised to stereospecifically[5] polymerize the latex monomer isoprene to generate a product with properties like those of natural rubber. Isoprene itself is not well studied, and a TLV value has not been established. It is an anesthetic, as are most unsaturated hydrocarbons. Synthetic polyisoprene does not exactly duplicate the properties of natural rubber. In fact, tire manufacturers continue to use natural latex because of the properties imparted by some of the natural rubber impurities.

C. **Styrene-Butadiene Rubber and Polybutadiene.** A low-molecular-weight hydrocarbon is usually used as a solvent for solution polymerization of cis-polybutadiene, and the polymerization catalyst includes diethylaluminum chloride and cobalt octanoate. SBR is polymerized as a soap emulsion of a mix that is about three-quarters butadiene. Peroxide catalysts (potassium persulfate, alicyclic hydroperoxide) are used.

Butadiene is weakly irritating and narcotic. For the most part, unsaturated aliphatic compounds are not serious problems. However, butadiene may be weakly carcinogenic. Current exposure levels may be causing up to 23 excess cancer deaths per year among the approximately 5900 workers exposed. ACGIH has recommended a 10 ppm standard, contrasting sharply with the present PEL of 1000 ppm. Styrene was discussed earlier in this chapter.

---

[5]It is necessary to form an all cis polymer in order to have elastic properties.

Table 16.8. Elastomer Chemicals—TLV Values.

Chemical	Hazard When Exposed by:			TLV (ppm)	Comments
	Lung	Skin	GI Tract		
acrylonitrile	X	X	X	20	irritant, reacts with air and water
alkyl aluminum	X	X	X	(2 mg/m³)	suggested new standard 10 ppm
butadiene	X			1000	
chlorine	X			1	irritant
chloroprene	X	X	X	25	irritant
ethylene	X			1000	
ethylene dichloride	X	X	X	10	PEL 50 ppm, 100 ppm ceiling
hexane (solvent)	X			50	neurotoxic
propylene	X				
styrene	X	X	X	100	irritant
sulfur dioxide	X		X	5	irritant
toluene diisocyanate	X	X	X	0.005	sensitizer

D. **Nitrile Rubber.** Acrylonitrile and butadiene are copolymers in nitrile rubber. Polymerization is essentially the same as for SBR.

Acrylonitrile is clearly the more dangerous of the two compounds. It absorbs through the skin, lungs, or G.I. tract, and produces cyanide poisoning. An overdose is treated exactly as one would treat cyanide poisoning. Care must be taken handling acrylonitrile, especially since rubber protective wear does not exclude it. It may also be carcinogenic, as indicated by epidemiological studies of workers.

E. **Neoprene.** Chloroprene, the monomer of neoprene, is irritating to the respiratory tract and to the skin, and is a nervous system depressant. It causes damage to the kidney and to the liver of test animals, and exposed workers have experienced temporary loss of hair.

F. **Butyl Rubber.** Polyisobutylene is prepared largely from isobutene, but another monomer, such as isoprene, butadiene or chloroprene, is added to modify the properties. The polymerization is solution polymerization in methyl alcohol, with aluminum chloride as the catalyst. Isobutene, like many other unsaturated aliphatic hydrocarbons, has anesthetic properties and low toxicity.

G. **Silicones.** Chlorosilanes are very irritating and react at moist surfaces such as mucus. The alkoxysilanes are less reactive and damaging.

H. **Chlorosulfonated Polyethylenes.** In the production of these elastomers, polyethylene is reacted with sulfur dioxide and chlorine gases. These gases are both irritating, producing acid in the lungs. Of the two, chlorine is less irritating. As a result, we willingly inhale more chlorine, and allow it to penetrate more deeply into the lungs. There, chlorine reacts slowly with moisture to produce acid.

I. **Polysulfide Rubbers.** Polysulfide elastomers are formed from polysulfides and ethylene dichloride (ethylene chloride). The polysulfides are low-risk materials. Ethylene dichloride is very irritating to eyes and lungs, causing coughing, watering eyes, and salivation. It has an anesthetic effect, and it damages the kidney and the liver.

J. **Polyurethane.** Diisocyanates are the main concern in the production of polyurethanes. (See polyurethanes in the Plastics section.) Exposure is regulated at extremely low levels for these compounds because they are potent sensitizers.

Other Toxicity Problems

As with plastics, a number of compounds are involved in processing in addition to the monomers themselves.

A. **Catalysts.** A variety of peroxides are involved in catalysis of polymerization. Metallic catalysts are used with olefins, including diethylaluminum chloride or vanadium oxychloride (EP and EPDM), and aluminum chloride (butyl rubber). These compounds are highly reactive, and are able to cause serious chemical burns. Handling bags of aluminum chloride creates the risk of exposure to dust if a bag leaks. Moisture reacts with $AlCl_3$ to release HCl, creating a serious risk to eyes and lungs.

B. **Crosslinking Agents.** Vulcanization (crosslinking) of elastomers involves sulfur in the cases of natural rubber and a number of the synthetics. The sulfur itself presents little risk. Hydrogen sulfide or sulfur oxides that may result from such processing present some hazard, and should be controlled by ventilation. Neoprene is cured using zinc or magnesium oxides, which present few problems. EP rubbers utilize peroxides for crosslinking. All these crosslinking agents are solids, and transfer operations should be designed to avoid dusts entering the plant air.

C. **Accelerators.** Accelerators are also employed in vulcanization. These compounds, listed in Table 16.9, are very effective sensitizers, so worker contact should be minimized. There are special problems with thiram and disulfiram, which block the metabolism of alcohol or aldehydes. They produce severe reactions in workers using beverage alcohol or being exposed to paraldehyde. Occasionally, metallic accelerators such as lead oxide are employed. Contact with any lead compounds should be minimized, not only because of the immediate toxic effects of the metal, but also because it is retained for periods of time in the body so that regular exposure can result in an elevated body burden of lead. The use of amines, nitrosamines, and other nitroso compounds as accelerators can lead to potentially carcinogenic N-nitroso compounds in plant air.

D. **Antioxidants.** Rubbers, particularly tires, often contain antioxidants to prevent ozone damage resulting from static charge buildup. These antioxidants are generally aromatic amines or hydroquinone derivatives, and can be irritants or sensitizers. In addition, hydroquinone monobenzylether causes loss of skin pigments.

Table 16.9. Some accelerators.

hexamethylene tetramine	diphenylguanidine phthalate
thiocarbanilide	mercaptobenzothiazole
tetremthylthiuram disulfide (thiram)	benzothiazolyl
tetraethylthiuram bisulfide (disulfiram)	

E. **Fillers.** A number of fillers are added to rubbers. Carbon black is a powder composed of very small particles (25–100 micrometers in diameter). Such small particles can readily penetrate deep into the lung, causing damage that leads to a loss of lung capacity. At levels two to three times the present TLV of 3.5 mg/m^3, workers displayed problems, both with a loss of lung capacity and, in a few cases, with the appearance of fibrosis. At levels below the TLV, the problems were sharply reduced, although they were not completely eliminated. Carbon black is known to contain carcinogenic polycyclic aromatic hydrocarbons, but it has been concluded that these are so strongly adsorbed to the surface of the carbon that they pose no threat to health. Studies show no increase in cancer, either in test animals or in workers with a history of high exposure to carbon black. With respect to this unusual ability of carbon black to adsorb substances, it has been reported to release adsorbed materials such as carbon monoxide when stored in a poorly ventilated location.

Silica and asbestos are also used as fillers. If they enter the atmosphere as dusts, they cause lung fibrosis. Asbestos also is a carcinogen. Most other fillers and pigments are low in hazard. Such compounds as aluminum or calcium silicate, calcium carbonate clays, and magnesium, titanium, or zinc oxides present a nuisance as dusts, but do not pose major health threats.

F. **Lubricants.** Talc is often used as a lubricant to prevent rubber from sticking to itself, or to facilitate its extrusion. Chemically, talc is hydrated magnesium silicate. Long-term inhalation of talc, even at the recommended TLV level of 1.0 mg/m^3, produces increases in bronchitis, cough, and other respiratory problems due to obstruction or irritation of the respiratory tract. At higher levels, fibrosis can occur. This is especially serious if the talc contains traces of asbestos type fibers, as some sources do.

G. **Blowing Agents.** A number of blowing agents have irritating or toxic properties. Azocarbonamide produces respiratory sensitization when inhaled as a dust. Benzenesulfonyl hydrazide and p,p′-oxy-bis(benzenesulfonyl hydrazide) are skin or eye irritants. Azo bis-isobutyronitrile breaks down above 26°C to give tetramethylsuccinonitrile, which has been responsible for headache, nausea, and convulsions in workers making PVC foam. The TLV is 0.5 ppm. In general, care should be taken to avoid inhaling any of these products as dusts.

OTHER POLYMER INDUSTRIES

TEXTILES

Textiles are prepared from a number of natural or synthetic polymeric materials. The industry produces cloth for clothing, bedding, curtains and drapery, towels, wall coverings, and upholstery. A large and closely related industry is the production of carpeting. In spite of the rise in use of synthetic fibers in the period following World War II, there is still a large market for natural fibers such as cotton and wool. In this section we focus on natural fibers and the processing for textile use of the polymers described earlier for use as plastics and elastomers. Textile manufacturers are a major end user of such polymers as polyesters, nylons, and polyolefins (polyethylene and polypropylene).

Cotton

Because harvested cotton contains seeds, leaf and stem parts, and other impurities, and because some sorting of fiber by length is needed, an elaborate process is involved in taking cotton from the bale and spinning it into yarn that includes blowing, brushing, and twisting. After cotton is twisted into thread, it is woven or knitted into cloth. The fabric is then cleaned (scoured) using caustic and detergents. If the cloth is to be dyed, it is treated with chlorine or peroxide bleach. Mercerization involves soaking the cloth in about 20% caustic, after which the appearance, strength, and ability to take dye are improved. Dye is applied in vats such that the length of exposure controls the depth of color obtained, then excess dye is washed off. Patterns may be printed onto the cloth using either rollers or screens and water-insoluble pigments.

Finally, a series of finishing steps may be used. A crosslinking agent may be added to make the cotton "wash and wear." Flame retardants may be added to the cloth. The surface of the cloth may be brushed or sueded (to form a soft nap of loose fibers), or glazed (rubbed with a fast-moving roller).

Hazards of Cotton Processing

A serious lung problem called byssinosis, or brown lung disease, has historically been associated with cotton processing. Early stages of the treatment of cotton generate high levels of cotton dust, particularly the step called carding. Some toxic agent in the smaller dust particles (<15 µm) is damaging to the upper respiratory tract. Damage accumulates with time and dosage of exposure, and is irreversible. The PEL for respirable cotton dust is 200 µg/m^3. Good air filtration and ventilation are important.

Another respiratory problem is *gin mill fever,* which happens when workers first are exposed to cotton processing, or are returning to the job after a long absence. It has the symptoms of influenza, and seems to be very much like metal fume fever.

The "projectile" in weaving machines is oiled. This generates a mist of oil in the atmosphere that can reach measurable levels.

Noise is a serious problem in cotton processing, the result of high-speed machinery. New machinery designs and abatement methods are gradually reducing noise levels in the early stages, but noise associated with weaving is still a serious problem.

Chemical exposures are possible to dyes and printing inks. Of special concern are benzidine dyes, which cause bladder tumors, and ß-naphthylamine. Formaldehyde resins used in crosslinking may have carcinogenic potential.

Wool

Wool is the fleece of sheep. It is washed in soap (scoured), rinsed, and possibly bleached with 35% H_2O_2. Carding—combing with fine wire teeth—removes bits of vegetation and arranges the fibers in parallel array. Wool is spun into yarn and may at this point be dyed. Cloth is woven or knitted, and may be dyed later.

Problems with worker health are not common in wool processing. Handling raw wool carries some risk of infection, and there may be exposure to insect-killing sheep dips. Dyes formerly included some with serious toxic potential, but these have fallen out of use.

Synthetic Fibers

Plastic chips are heated and forced through an extruder (spinnerette) to form a fine continuous filament, which is immediately cooled and lubricated. The strands are heated and stretched, which pulls a portion of the polymers into parallel array in the direction of the strand and so increases the strength of the filament. Some of the strands are wound on bobbins, others may be "crimped," given a wave pattern, heated to set the wave, and cut into lengths of a few inches. This product becomes yarn. The product may be blended with another fiber such as cotton or wool. Dyeing, weaving, and knitting operations are as described above.

As in much of the textile industry, noise is a serious concern, given the use of high-speed machinery. Heat exposure at the spinnerettes, especially during service, and where the strands are heated to be drawn or crimped, can be a problem. In order to reduce static electricity generated by the rapid movement of dry filament through the machinery, plants are sometimes maintained at high levels of humidity, adding to heat stress. Chemical exposures may occur to release agents used to smooth flow through the spinnerette and to monomers and other plastic components at the melting process. Lubricants applied to the fibers can cause contact dermatitis.

ADHESIVES

Industry is increasingly using adhesives in assembly operations. To plant engineers with a long history of joining objects mechanically, the use of adhesives may come as a novel idea, and they are surprised to learn it can serve their purposes. Increased use is also a response to the availability of more effective agents to bond objects together. Finally, it may be a response to the need to automate in order to save on production costs, and such bonding methods lend themselves well to this purpose.

Principles of Adhesive Bonding

It is useful to lay a groundwork of understanding about the nature of attractive forces in order to understand why some substances are easily bonded together and others are not. Basically, all attractions in chemistry are electrostatic; that is to say, they are the result of the attraction between opposite charges. However, within that framework we can recognize subclasses of forces. We shall list four here in order of decreasing strength:

1. *Covalent bonds.* Full chemical bonds between molecules are very strong attractions. These can form only when the adhesive and the surface have chemical groups that are capable of reacting with one another. For example, if the surface included alcohol or amine groups, an epoxy resin could attach itself covalently to that surface. Such a circumstance is unusual.

2. *Hydrogen bonds.* This type of bonding is weaker than full covalent bonding, but if a large number of these bonds can form, the total force can be great. In general they occur when an oxygen, nitrogen, or fluorine atom can come close to an oxygen or nitrogen that has a hydrogen attached to it. Both atoms attract the hydrogen toward them so that in effect the

hydrogen is partially bonded to each. Polyesters, polyamides, polyacrylates, polyacrylamides, and epoxy resins are examples of plastics capable of forming hydrogen bonds with an appropriate other surface.

3. *Polar bonds.* Many times atoms that are chemically bonded to one another do not share equally the electrons that hold them together. In such a case, the electrons are found closer to the atom with the greatest attraction for them. The bond has a negative charge at that end, and a positive charge at the other end. Such bonds can line up so that the positive end of one is next to the negative end of another, producing attraction. Structures in which carbon is attached to oxygen, chlorine, or fluorine display polar bonds. The list includes those listed above as being capable of hydrogen bonding, as well as structures with aldehyde or ketone groups.

4. *Van der Waals forces.* Because of dislocations or vibrations of their electron clouds, all atoms display temporary polarity and attract one another weakly if they are very close together. Although each attraction is very weak, a large number add up to a measurable force.

All but the van der Waals forces depend on the presence of specific functional groups. Sometimes in plastics, small amounts of an appropriate monomer are added to a polymer formulation just for this purpose. Acrylic acid, with its carboxylic acid structure, or acrylonitrile, with its cyano group, are examples of monomers used in this fashion. Some plastics by their chemical nature do not adhere well. Polyolefins and Teflon® are inherently very poor candidates to be used as or with adhesives.

Solvents used in adhesives must be able to "wet" the surfaces to be joined. Solvents like water, alcohols, or ketones (acetone) wet surfaces containing polar bonds. We can see whether or not this is happening by placing a drop of solvent on the surface. If it does interact with the surface, the drop spreads out; if not, it remains as a bead. The surface may be wettable based on its chemistry, but have a coating of a greasy (nonpolar) impurity that blocks intimate contact between the solvent and the surface. Therefore, testing a solvent for its ability to wet should only be done on a very clean surface.

A number of factors are important in the joining of two surfaces. The chemistry of the surfaces to be joined is the first consideration. Some sort of interaction between the surface and the adhesive is necessary to permit attachment. This is often an electrostatic attraction. The physical nature of the surfaces is also considered. A smooth surface is an advantage when the adhesive does not flow well into irregularities. On the other hand, an etched or otherwise roughened surface has a greater surface area, and, when the adhesive binds to the irregularities, produces a greater bonding force. Finally, the cleanliness of the surface is important, since impurities may greatly weaken the bonding by preventing intimate interaction of the adhesive with the surface. All this infers the need for processes in addition to the actual use of the adhesive, including etching, roughening, or solvent cleaning, that may introduce their own specific hazards.

General Hazards

From the standpoint of industrial toxicology, introducing the use of adhesives can bring unanticipated hazards into the workplace. This is especially true if a plant has been running in a safe fashion using mechanical joining of the parts, and then switches to adhesives. The adhesive may not be looked upon with alarm because there is a relatively small amount in use initially, or simply because we are less alert to hazards in an ongoing, trouble-free operation. Any time there is a change in process, the hazards must be reevaluated.

Evaluation of the hazards introduced by adhesive bonding may be more difficult than for some other new processes that are introduced. This is because the adhesive may have a complex formulation, so that several new chemicals are entering the plant at once. Furthermore, it may not be clear what these chemicals are in such a premixed medium. The need for local exhaust ventilation may be introduced in a plant where, up to that time, general circulation of air was adequate to protect workers. The manner of use of the adhesive may affect the risk. If the bonding process is being run at room temperature, and it is found to work more effectively at an elevated temperature, we must be alert to the probability that the levels of volatile components will increase sharply in the plant air.

Polymerization Adhesives

Adhesives may be classified by their mechanism of bonding. In many cases the components necessary to form a polymer are mixed at the site where joining is to take place. The surfaces are pressed together onto this mixture and the material is polymerized or crosslinked in place. Obvious advantages to this process include the opportunity for intimate contact and close fit between the polymer and the surface, and often the lack of a need to remove solvent from the joint.

Perhaps the best known such adhesives are the epoxy resins. The polymerization reaction is initiated by mixing two preparations—the epoxy monomer, and a mixture of polyamine or polyamide—with the catalyst. Unreacted epoxy or amine groups may react with groups on the surfaces being joined to provide an exceptionally strong attachment. That failing, the groups at least provide sites for electrostatic attraction between the adhesive and the surface. Cure times can be adjusted by the choice of catalyst or by the temperature selected. Epoxy phenolics are used where strength at higher temperatures is important. Such strength is also characteristic of polyimines and of silicones.

Polyurethanes are useful for difficult bonding applications. These might occur where the surfaces to be joined have poor qualities of adhesion, but do have some polar structures. Joining such plastics as polyesters, polyamides, polycarbonates, and polyurethanes themselves are examples of good polyurethane applications. Polyesters are used as adhesives where toughness is less important, and a less expensive material is desired.

Acrylate type polymers are often used as adhesives. In order to obtain a viscous, sticky material to apply to the surfaces, a monomer-polymer mixture is sometimes employed. Either a peroxide catalyst or a peroxide and an activator may be used to initiate the polymerization reaction. In the latter case, the mixture with the catalyst may be applied to one surface while the mixture with accelerator is applied to the other. The reaction begins when the surfaces are pressed together. Cyanoacrylates can be used in much the same way, since they do not polymerize until air is absent, and use moisture from the surface as a catalyst.

Hazards of Polymerization Adhesives

The dangers here relate to the fact that instead of the polymers themselves being employed, the considerably more toxic monomers, catalysts, and accelerators are being handled as the adhesive is applied. Special note should be taken of the properties of epoxy monomers, amines, formaldehyde, and other volatile or toxic monomers. Catalysts, such as peroxides, are dangerous both because they cause irritation and burns on contact, and because they react violently on contact with certain other chemicals. Accelerators are notorious irritants. If unreacted adhesive contacts the skin and the reaction begins, there is a possibility of both chemical damage and burns from the heat produced by the polymerization reaction.

Polymers as Adhesives

Solutions of polymers are used as adhesives. They are generally applied either in a volatile solvent or as a water suspension, and in each case the liquid must evaporate to complete the bonding. Preformed polymers are less likely to form a strong joint, since the closeness of fit between the polymer and the surface is not likely to be as good, and attractive interactions between the polymer and chemical groups at the surface are far less likely as a result of this poorer fit.

Substances applied in volatile solvents include nitrocellulose (the familiar model airplane glue), elastomers such as natural rubber (rubber cement), and acrylics. Elastomers and other polymers may be suspended as droplets in water, for example in adhesives used in construction. Natural rubber, casein, and polyvinyl acetate are used in this fashion. Sometimes the properties of such an adhesive are improved if the finished joint is heated to fuse the polymer together, and to drive out the last of the solvent.

Hazards of Solution Adhesives

Overall, this approach is much safer than the polymerization techniques. The polymers themselves are generally not dangerous. Care must be taken about the properties of the solvent used, and any necessary ventilation must be provided.

Welding

Thermoplastics may be used as adhesives by melting them, then joining the two surfaces while the polymer is still soft. Polyamides and polyesters have been used in this fashion. Thermoplastics also can be joined to one another without an adhesive by heating the surfaces to be joined until they melt, then pressing them together.

Hazards of Molten Plastic Adhesives

Any time plastics are heated to the melting point there is the danger of volatile chemicals entering the atmosphere. Primarily these would be unreacted monomer, but other volatile chemicals may be trapped in the solid plastic. If above-minimum temperatures are employed to do the melting, there are increased possibilities for degradation of the plastic, plasticizers, or other chemicals in the plastic, accompanied by a release of volatile liquids or gases. The manner in which the welding is done influences the possibilities for contamination of the workplace atmosphere. For example, if a sheet of plastic is placed between two objects to be joined, and those objects are heated, the hot plastic has minimum exposure to the atmosphere. On the other hand, if the plastic is melted on an open surface, the possibility of the air becoming contaminated is much greater. Adequate ventilation is the most important safeguard at the workstation.

Solvent Bonding

When two plastic surfaces are to be joined together, it is sometimes possible just to soften the plastic with a solvent, press the surfaces together, and wait for the solvent to evaporate. Obviously this is only possible if the polymer dissolves in a volatile solvent. The joint so produced has properties that depend on the intimacy of mixing of the polymers at the surfaces, the elimination of gaps, and the success of the solvent removal. At best, the polymer strands retain a random orientation at the interface, so that the joint remains the weakest part of the combined structures.

Having a solvent of the correct volatility is important. If it is too volatile, it may evaporate before the joining is accomplished. However, if it is too low in volatility, an excessive length of time is required to remove it from the assembly. Commonly employed solvents include toluene, xylene, methyl ethyl ketone, methyl alcohol, and methylene chloride.

Hazards of Solvent Bonding

Clearly, the hazards of solvent bonding arise from the release of solvent vapors into the workplace. The solvents used are commonplace chemicals with well known toxicity characteristics. Ventilation is essential.

Coupling Agents

A number of compounds are produced called coupling agents. These are linear organic chains with a silane group on one end and some chemical functional group such as a double bond, an epoxy group, or an amine group at the other end. The silane group adheres well to glass or metals. By coating fillers such as glass fibers with these substances, a functional group handle with which a polymeric adhesive can react is added to the filler. These agents are chemically quite reactive, and so are generally toxic and irritating. The manufacturer's recommendations should be consulted regarding handling and personal protection when coupling agents are used.

Surface Preparation and Its Hazards

A clean surface is essential to tight bonding, but the cleaning procedures often involve the use of quite hazardous chemicals. Removal of oily residues from the surface may be done either with a detergent or with a solvent. Solvents are most frequently used, and introduce the problems associated with solvent vapors in the air. Skin irritation due to the removal of skin oils and the passage of solvent through the skin into the body are added risks.

Surfaces are often roughened or cleaned by abrasion before bonding. Measures must be taken to prevent the inhalation of particles broadcast into the air by such operations. Metallic surfaces may also be treated to remove oxide coatings that would reduce adherence. Although abrasive cleaning can remove oxides, etching the surface with chromic acid does a more satisfactory job. Such cleaning results in bonding to adhesives that is half again as strong as solvent cleaning and sanding, and as much as 8–10 times as strong as solvent cleaning alone. The etching solution contains sulfuric acid, a very strong acid capable of producing serious chemical burns, and a dichromate salt, which is a very strong oxidizing agent. The combination is deadly and must be treated with extreme care. Not only is there the potential for chemical burns, but particulate dichromate or droplets from the bath entering the atmosphere are very irritating to the nasal cavities, and on long exposure can cause bronchitis. Chromium has been listed as a carcinogen, with respiratory tract tumors occurring in workers who inhale chromium compounds. The use of chromic acid baths should be accompanied by extensive safeguards.

KEY POINTS

1. Plastics are solid polymers that divide into two groups; linear polymers or thermoplastics, and cross-linked (three dimensional) or thermoset plastics. Thermoplastics can take on thermoset properties by crosslinking (curing) the plastic.

2. Additional polymers are formed by joining unsaturated monomers by addition reactions, eliminating the double bonds. Condensation polymers are formed by reaction of functional groups to join monomers together. Copolymers have more than one type of monomer.

3. Polymer properties are affected by the degree of crystallite formation in the solid. Plasticizers are inert additives that prevent crystallite formation.

4. Other plastics additives include pigments, stabilizers to block UV light, and fillers to extend the plastic and/or modify properties.

5. Monomers are mixed and catalysts added to stimulate polymerization. This is done to thermoplastics before shipping, and to thermoset plastics at the time of molding.

6. Plastics may be shaped by injection molding, casting, blow molding, vacuum and pressure forming, extrusion, and calendaring. Plastics may also be formed and shaped.

7. Hazards in plastics manufacturing arise from the chemicals and from particulate produced by grinding, sawing, or otherwise shaping.

8. Elastomers are linear polymers which return to their original shape when distorted. They are generally addition polymers and may be copolymers. A major use is the manufacturing of tires.

9. Manufacture of elastomers is similar to that of plastics, and many of the same hazards are involved.

10. Textile polymers include natural polymers (cotton, wool) and synthetic polymers (rayons, nylons, orlons). Cotton processing involves a special hazard from airborne cotton fibers, which produce brown lung disease. Synthetic fibers have hazards similar to those of the corresponding plastics.

11. Adhesives join surfaces by generating attractive forces at the interface of the pieces. Polymerization adhesives form polymers from monomers at the interface. Solutions of polymers lose solvent at the interface and join surfaces by their mutual attraction to the polymer. Polymers may be melted in the interface to join the surfaces. Solvents may dissolve the surfaces, usually plastic surfaces, at the interface to join the surfaces. Finally, compounds that react with functional groups at the interface surface can join the faces covalently. Hazards include the chemicals used and the formation of particles as the surfaces are prepared for joining.

PROBLEMS

1. Recycling plastics is a relatively new industry. Imagine inspecting a plant that is recycling polypropylene milk cartons, converting the scrap into fishing tackle boxes. The cartons are compressed into blocks and chipped into small pieces. These are washed with strong detergent in hot water, rinsed, and dried in a stream of hot air. The chips are then transferred to a grinder where they are mixed with pigments and transferred to the hopper of an injection molding machine where they are converted to the product. What hazards would you look for in this operation?

2. Thermoset plastics are rigid because of a massive progression of covalent bonds with fixed angles. Consider what generates the degree of rigidity found in a thermoplastic.

    A. Why would nylon be more rigid than polyethylene, if the only differences were the structure of the polymer.

    B. Why would polypropylene that had been cooled slowly be more rigid than a sample that had been cooled quickly?

    C. How does crosslinking affect rigidity?

    D. How does a plasticizer affect rigidity?

3. You are the safety officer in a plant that assembles display shelving for retail stores. The company is considering switching from fasteners (screws and bolts) to adhesives to lower production costs. The plant engineer has decided that a polymerization adhesive would be stronger, producing a better product. You are asked to prepare a comparison of the advantages and disadvantages of using polymers in solution versus polymerization adhesives from the standpoint of the cost of modifying the plant to accommodate the new process safely.

4. You have been hired as an industrial hygienist at a textile mill that converts raw cotton to woven cloth.

   A. Use the index in the 29 CFR 1910 to locate the standards for cotton dust.

   B. What is the definition of respirable cotton dust?

   C. What is the standard method for its measurement?

   D. What is the PEL for cotton dust?

   E. How often must monitoring be done?

   F. Read the details of the sampling procedure in Appendix A.

5. Locate vinyl chloride in the index of 29 CFR 1910.

   A. What is the PEL, ceiling, and action level of vinyl chloride?

   B. Contrast controls on exposure to the polymer forms of acrylonitrile (29 CFR 1910.1045) and polyvinylchloride.

6. Select and read an article from this list and indicate (1) what workplace problems are discussed and (2) what analytical methods are employed: *Am. Ind. Hyg. Assoc. J.* (1983), **44**, 521; *ibid* (1977), **38**, 205; *ibid* (1980), **41**, 204; *ibid* (1980), **41**, 212; *ibid* (1977), **38**, 394.

BIBLIOGRAPHY

AIHA, *Toxicological, Industrial Hygiene and Medical Control of Polyure-thanes, Polyisocyanurates, and related materials*, 2nd ed., Akron, OH, 1983.

L. F. Dieringer, "Rubber," in L. V. Cralley and L. J. Cralley, *In Plant Practices for Job Related Health Hazards Control, Volume 1*, John Wiley and Sons, New York, 1989.

International Agency for Research on Cancer, *Monographs on the Evaluation of Carcinogenic Risk of Chemicals to Humans: The Rubber Industry*, Vol. 26, World Health Organization, 1982.

International Labour Office, *Encyclopedia of Occupational Health and Safety*, Geneva, Switzerland.

E. B. Katzenmeyer, Jr., "Butadiene Rubber," in L. V. Cralley and L. J. Cralley, *In Plant Practices for Job Related Health Hazards Control, Volume 1*, John Wiley and Sons, New York, 1989.

A. J. Kinloch, *Adhesion and Adhesives*, Blackie and Son (Chapman and Hall), N.Y., 1987.

J. R. Lynch, "Butyl Rubber," in L. V. Cralley and L. J. Cralley, *In Plant Practices for Job Related Health Hazards Control, Volume 1*, John Wiley and Sons, New York, 1989.

M. Morton, Ed., *Rubber Technology*, 2nd ed. Van Nostrand Reinhold, New York, 1985.

J. E. Mutchler, "Plastics," in L. V. Cralley and L. J. Cralley, *In Plant Practices for Job Related Health Hazards Control, Volume 1*, John Wiley and Sons, New York, 1989.

J. D. Neefus, Ed., *Cotton Dust Exposures, Vols. I and II*, American Industrial Hygiene Association, Akron, OH, 1984 and 1987.

NIOSH, *Control Technology in the Plastics and Resins Industry*, Publication No. 81-107, DHHS, Washington, D.C., 1981.

A. R. Nutt, *Toxic Hazards of Rubber Chemicals*, Elsevier Applied Science Publishers, N.Y., 1984.

Rubber and Plastics Research Association, *Clearing the Air—A Guide to Controlling Dust and Fume Hazards in the Rubber Industry*, Shrewsbury, England, 1982.

I. Skeist, *Handbook of Adhesives*, Van Nostrand Reinhold, N.Y., 1977.

P. F. Woolrich, "Polyurethanes and Polyisocyanurates," in L. V. Cralley and L. J. Cralley, *In Plant Practices for Job Related Health Hazards Control, Volume 1*, John Wiley and Sons, New York, 1989.

U. S. Dept. HEW, *Health and Safety Guide for Plastic Fabricators*, NIOSH, Cincinnati, OH, 1975.

ANSWERS

Chapter 1

2. A. Having a three-symbol code and 4 different symbols means we can have 4^3, or 64, different code words, shown in the table. These represent 20 different amino acids.

    B. Leu-Gln-Thr-Pro-Arg-Arg-Ile-Glu-STOP. The next Gly is beyond the stop signal.

    C. Pro is replaced by Leu in the sequence.

    D. The code word still indicates Arg. This is a silent mutation because there is no change in the protein sequence even though the DNA sequence has been altered.

    E. The sequence now reads: Leu-STOP. The protein has been shortened by the elimination of the rest of the amino acids.

    F. The sequence now reads as follows: Leu-Gln-Thr-Pro-Arg-Arg-Ile-Glu-Trp-Gly-.... The stop signal is gone, and if there is DNA to the right of the problem sequence, the protein will be extended according to the continuing DNA sequence until one is encountered.

    G. The sequence now reads: Leu-Lys-His-Pro-Asp-Gly-Ser-Asn-Glu-.... In other words, the sequence is completely changed.

    H. The "frame," the group of three bases read to determine each amino acid, is no longer the same, so that what was the second base in the code word before is the first base now. All code words are changed. It is very unlikely that a protein resulting from a frame shift mutation will be functional *unless the shift occurs very near the end of the sequence.* A few such mutated hemoglobins have been found that still function.

I.
```
┬┬┬┬┬┬┬┬┬┬┬┬┬┬┬┬┬┬┬┬┬┬┬┬┬┬┬┬┬┬┬┬┬┬┬
AATGTTTGTGGGTCTGCCTAGCTTACTCCC
TTACAAACACCCAGACGGATCGAATGAGGG
┴┴┴┴┴┴┴┴┴┴┴┴┴┴┴┴┴┴┴┴┴┴┴┴┴┴┴┴┴┴┴┴┴┴┴
```

J.   The 5th base is a T. The complementary strand has an A at that point, and the repair enzyme would replace the damaged base with the complement of A, which is T.

3.   Teratogens would be the most important chemicals uniquely dangerous to a woman, or more correctly to the fetus of a pregnant woman. Men might be banned from jobs handling chemicals that damage sperm production.

Chapter 2

1.   The child should not be given the same dose. Dose should be measured as mg/kg, and since the weight of the child is about one-third of adult weight, the dose should be about 100 mg/day. This should not be given as one 100-mg pill, because this would be a very high dose at one time. Depending on how the drug is packaged, it might be 2×50 mg or 4×25 mg.

2.   It would still be 75 ppm. Toxicants taken by inhalation do not depend on body mass, since the lungs are proportionately smaller.

3.   The likely routes of intake are inhalation of the vapors and intake through the skin from handling wet parts. D is more toxic by the dermal route, and A is more toxic by inhalation. Furthermore, A has a lower boiling point, indicating that vapors of A will be present in the air over the solvent at higher concentration. In choosing between B and C, we see that C is more toxic by the oral route. However, oral intake is unlikely. More importantly, the 90-day oral study reveals that B accumulates in the body, and B has a lower boiling point than C. C is the best choice.

4.   A.   TEPP to males; Phorate to females.

B.   Malathion to males; Ronnel to females.

C.   Chlorthion, Dicapthon (oral only), Fenthion, Guthion, Malathion, Methyl parathion (dermal only), Methyl trithion, NPD, Phosdrin, (dermal only), Phosphomidon (oral only), Trichlorfon.

D.   No. Phorate is twice as toxic to females, while for Trichlorfon the difference is slightly more than 10%.

E.   Ronnel and Schradan are more than twice as toxic to males, while Carbophenothion, Co-ral, Demeton, Di-syston, EPN, Ethion, Parathion, and Phorate are more than twice as toxic to females. Females.

F.   Oral.

G.   Diazinon is more toxic to females, and it is more toxic orally. Schradan is more toxic orally to males. Phosdrin is more toxic dermally to females. If a compound is more toxic dermally, it crosses the skin barrier more readily than the mucosal barrier of the G.I. tract.

H.   Yes.

5.   A.

$$1 \text{ m}^3 \times \left(\frac{100 \text{ cm}}{\text{m}}\right)^3 \times \frac{1 \text{ L}}{1000 \text{ cm}^3} = 1000 \text{ L}$$

B.

$$\frac{V_1}{T_1} = \frac{V_2}{T_2}; \quad \frac{22.4 \text{ L}}{273 \text{ K}} = \frac{V_2}{298 \text{ K}}; \quad V_2 = 24.5 \text{ L}$$

C.

$$1 \text{ ppm} = \frac{1 \text{ L}}{10^6 \text{ L}} \times \frac{10^3 \text{ L}}{\text{m}^3} \times \frac{1 \text{ mol}}{24.5 \text{ L}} \times \frac{100 \text{ g}}{\text{mol}} \times \frac{1000 \text{ mg}}{\text{g}}$$

$$= 4.09 \text{ mg/m}^3$$

D.

$$2 \text{ ppm} = \frac{1 \text{ L}}{10^6 \text{ L}} \times \frac{10^3 \text{ L}}{\text{m}^3} \times \frac{1 \text{ mol}}{24.5 \text{ L}} \times \frac{320 \text{ g}}{\text{mol}} \times 1000 \text{ mg}$$

$$= 26.1 \text{ mg/m}^3$$

$$4 \text{ ppm} = \frac{1 \text{ L}}{10^6 \text{L}} \times \frac{10^3 \text{L}}{\text{m}^3} \times \frac{1 \text{ mol}}{24.5 \text{ L}} \times \frac{150 \text{ g}}{\text{mol}} \times \frac{1000 \text{ mg}}{\text{g}}$$

$$= 24.5 \text{ mg/m}^3$$

6. A. NOEL/100

   B. Make the test level 100 times the concentration anticipated in work-place air.

   C. Testing only one animal carries the risk that this is a highly resistant individual.

   D. We do not know what the true safe level is, and we do not know if that individual animal is highly sensitive to the chemical.

7. A. Dosing in the drinking water is easy to do, but we do not know what volume of water the animal actually drank and what was spilled.

   B. 2.5 mg/kg

   C. 1.25 mg/kg

   D. LD_{50} is approximately 4.0–4.4 mg/kg

   E. Extrapolating the line of the probit plot to interception with the x-axis is the easy way to determine a threshold value, here close to 1.25 mg/kg.

$$\sigma = \frac{(\text{deviation from average})^2}{\text{number of samples} - 1} = \frac{(X_i - X_{ave})^2}{n - 1}$$

Chapter 3

1. A 70-kg worker would need to transfer 13.7 g of the compound onto the apple to ingest the lethal dose—a highly unlikely event. However, if the compound accumulated in the body effectively, the worker could add small amounts to the body burden each day and perhaps reach a level that was hazardous, in time. A 90-day LD_{50} would give an indication of the likeli-hood of this happening. The LD_{50} only measures systemic toxicity, and other toxic characteristics are also of concern.

2.   Anything that decreases the ability of skin to block the chemical raises the hazard of handling this solvent. This includes cuts and abrasion of the skin, and previous prolonged skin exposure to hot, soapy water or to organic solvents (possibly including this solvent) that loosen the skin keratin layer. One must always consider that the compound may be entering the body by another route at the same time, perhaps by inhaling its vapors, adding to the body burden.

3.   A. $E > C > A > B > D$.

B.   Yes

C.   A, 0.052 mL/min.; C, 0.066 mL/min.

D.   No.

E.   Compounds in the series may be metabolized to more polar structures at different rates. Some may bind to proteins, and others do not.

Chapter 4

1.   Some states already had a high level of worker protection required in their state laws, and needed to change to only a small degree, while others needed to change a great deal. Since providing protection for workers is expensive to manufacturers, a state could use lax rules to attract new industry by advertising themselves as "low cost" or "friendly to industry."

2.   OSHA may publish an early warning to invite reaction to a proposed new standard, and must publish the new standard as a proposed rule, then wait at least 30 days before instituting the standard. Public hearings provide opportunities for input.

Chapter 5

1.   Through the skin of the back of the hand, where it is thinnest.

2.   Cuts and abrasions, loosening of the keratin layer by solvents or water.

3.   A.   They will respond strongly to even slight contact, and should be shifted to a job away from the chemical.

    B.   Becoming sensitized depends on having the chemical bind to a protein in the skin, which then stimulates an immune response. That complete series of events only occurred in those who were sensitized.

    C.   The initial reaction with a skin protein must be most likely to happen with that particular chemical.

4.   A.   Workers 1, 2, 4, 5, 6, 7 probably require hand protection.

    B.   3, 4

    C.   The hazard to worker 4 is contact with the solvent. Reference must be made to a table listing the glove material suitable for that solvent in order to choose the correct gloves.

5.   The chemicals being handled should be cleaned off of the gloves after each day's use. Gloves should be inspected for leaks by inflating them. This could be done in a simple fashion by folding the glove shut at the cuff and rolling back toward the fingers. If the surface of the gloves becomes tacky or shiny, or if the gloves become stiff, this is a sign if impending failure.

Chapter 6

1.   A.   Starting with the nose and extending all the way to the entrance to the alveolus, the passages are lined with wet, sticky mucus. If a particle impacts the mucus it is caught like a fly on flypaper. The pathway to the alveolus includes numerous turns, at any one of which the particle can "spin out" onto the sticky surface. Further, the particle remains suspended by the movement of the air, and the farther into the lung it travels, the slower the air moves. Finally, in the alveolus if the particle touches down on the wet surface, it may be engulfed by phagocytic cells.

    B.   Particles that weigh more are more likely to be deposited at turns in airways or to settle out of slower moving air. Particles weigh more if they are larger in diameter for a given composition or have a higher density for a given diameter.

    C.   The key to being stopped by the mucus lining of the air passages is the ability of the chemical to interact with water, since mucus is highly hydrated. Methane cannot hydrogen bond with water, and is not polar so it is not soluble in water. If it does strike the lining, it simply migrates out again and continues moving with the air.

   D. HCl does dissolve in water, and it dissolves into the mucus on first contact.

2. This is an acidic irritant that reacts with water to form H_3O^+ and $CF_3\text{-}CH_2\text{-}O^-$. The H_3O^+ lowers pH in the passage lining, which stimulates (1) coughing and sneezing, which expel air from the lungs, and (2) production of additional mucus, which is carried out with the coughing.

3. "...there are now fewer vessels carrying the blood leaving the heart, so each vessel carries more blood, producing back pressure on the heart."

4. A. (2) The forklift trucks are a source of CO.
         (3) Heated metals may be a source of metal oxide fume, and gas flames are a source of CO.
         (5) Metal particulate could get into the air from drilling (low risk).
         (6) Metal particulate and abrasive particulate could be getting into the air from grinding (high risk).
         (7) Metal particulate could be stirred up by the compressed air.
         (8) Chlorinated solvent may be entering plant air.
         (11) Metal particulate and abrasive could be entering plant air.
         (12) Paint aerosol.
         (12, 13) Paint solvent may enter plant air.

    B. The chemicals carried to the rinse water would have to be replaced in the plating tanks; now they are replaced from the rinse tanks. If the cyanide solutions in the alkaline copper plating solution were carried into the acidic nickel plating tank, HCN gas would be produced.

5. A. A known volume of air is pumped through a filter, and the number of fibers of particular dimensions are counted in sample fields under a microscope.

    B. This is the upper size for "respirable" particles.

    C. Crocidolite is the carcinogenic type of asbestos.

6. Most agree: ACGIH is more conservative.

Chapter 7

1. A. 0.29%

    B. Now, 94% of the particulate is small particles. The small particles are the respirable particles, and almost none of them have been removed by this ventilation improvement. If hazard depends on particles reaching the alveolus, the health problem is unchanged.

C. Filter weight changes would not have changed significantly, but to the degree that small particles were captured, the hazard would have been reduced.

2. A. Two things must be measured before someone enters the tank: (1) The tetrachloroethylene, which would be a vapor, must be at a safe level, and (2) there must be enough oxygen for breathing. Tetrachloroethylene would be tested using a grab sample, and gas chromatography would probably be used for the analysis. Meters are available to measure oxygen directly.

B. Cadmium would be present as a particulate. An integrated sample would be obtained using a personal air sampling pump equipped with a filter. A grab sample would be obtained at the time of welding to assure that the ceiling value is not exceeded at that time. Analysis would be by atomic absorption spectrometry.

C. As long as the standards for the solvent specified TWA values for an 8-hour day, you would do integrated sampling, probably with a personal air sampling pump and an adsorbent tube. If there was a ceiling value specified, a grab sample should be taken at the touch-up site during spray painting. Analysis would be by gas chromatography.

3. A. (1) Determinate error, producing values that are too low.
    (2) Indeterminate error.
    (3) Determinate error, producing values that are too low.

B. (1) Determinate error, producing values that are too low.
    (2) Values at low concentrations are reliable, but at the high end there is determinate error, producing values that are too low.
    (3) Indeterminate error.
    (4) No effect. The lowered flow rate affects the passage of both standards and samples through the column.
    (5) Determinate error, producing values that are too high.

4. A. 0.00984

B. 0.00222

C. 0.00276

5.   147 ppm

6.   Asbestos: Polycarbonate filters
     Fe_2O_3 fume: Cellulose acetate filters
     Benzene vapor: Charcoal adsorption tubes
     Total mass of particulates: Polyvinylchloride filters

Chapter 8

1.   A.   (1) Silica particulate (silicosis).
          (2) Metal and metal oxide particulate in the air from handling hot metal. Unless alloys used are more exotic than iron for cast iron, metal fume fever should not be a problem. Furnaces produce CO.
          (3) Metal and metal oxide particulate.
          (4) None.
          (5) Anhydrous HCl gas and aqueous HCl aerosols may enter the air from the bath.
          (6) Silica particulate, a greater problem quantitatively than in (1).
          (7) Metal and abrasive particulate.
          (8) None.
          (9) Solvent vapors and paint aerosols.
          (10) CO from internal combustion engines.
          (11) Solvent vapors from degreasing operation, higher during use and highest when the unit is serviced.

     B.   1. 35004, 35005 and 35007 would contain filter elements. An element to adsorb vapors, such as activated charcoal, should be part of 35001, 35006 and 35007. 35002 would have an alkaline-reactive medium and 35003 would have an acid-reactive medium.

          2. Stations (1, 3, 6, 7, 10) 35004 or 35005 depending on size of particulate; stations (4, 8) none; stations (5) 35002; station (9), 35007; station 11, 35001.

          3. You need to [1] color code all masks before they go to the dispensing room, [2] instruct workers about the importance of using the right mask, [3] place signs at the dispensing room and in each workstation indicating the correct color to use in that station.

     C.   Shift exhaust fans on the east wall and the air inlets to the west wall so that HCl fumes and solvent vapor are not drawn past the rest of the workers before leaving the room.

D.   Areas (5), (6), (7), (9) and (11) could benefit from local ventilation. Air quality studies should be done to determine the sites of greatest need.

E.   In the degreasing room. A spill there would quickly saturate the air with vapors.

F.   Replace the forklifts with electric units.

2.   A.   Since air is moved into the building, the building will have a "positive" pressure, that is, higher than atmospheric.

B.   1920 ft^3/min.

C.   108 mg/min.

D.   Yes.

E.   0.92 mg/m^3.

F.   13.9 min.

G.   Locate the adhesive operation and an exhaust vent as close together as possible.

3.   A.   2.3 ft.

B.   320 ft/min.

4.   Baffles add to the resistance of the system, requiring more energy to move the air.

5.   A.   There is the possibility of a dangerous atmosphere, so a permit is needed, and welding is to be done, so a hot permit is needed.

B.   The air should be tested for oxygen level, levels of flammable vapors, and levels of toxic vapors or gases. The literature should be consulted to learn the toxicity of the fuel vapors and the LFL and LEL of the fuel stored in the tank.

C.   An entry supervisor must sign the permit. An attendant must be outside the tank, and a rescue team must be alerted and available.

D. The tank could be flooded with an inert gas to prevent ignition of the remaining fuel. The welder will then require a breathing apparatus to supply oxygen.

Chapter 9

1. A. Look at the description of boiling point in the text: the temperature at which the vapor pressure equals the atmospheric pressure. Vapor pressure is a constant at a given temperature, but if the atmospheric pressure varies, the vapor pressure necessary varies with it, and thus the temperature (boiling point) needed to obtain the correct vapor pressure changes.

   B. The definitions say the same thing. It is possible to measure such a value in a variety of ways, not all of which may produce the same experimental value for a given substance. Defining the assay method assures the recording of a single value, avoiding confusion in the literature. ASTM stands for American Society for Testing and Materials.

   C. A combustible liquid has a flashpoint at or above 100°F (37.8°C), while a flammable liquid has a flashpoint below 100°F. The lower the flashpoint, the greater the volatility of the liquid, and consequently the greater the fire hazard.

   A Class 1B liquid is a *flammable* liquid that has a flashpoint below 73°F (22.8°C) and a boiling point at or above 100°F (CFR 1910.106 (a)(19)(ii)). This allows regulations to be adjusted to the level of hazard. As an example see CFR 1910.106 (b)(2)(iv)(*f*) and (*g*)—normal venting for ABOVEGROUND tanks.

   D. CFR 1910.106 (b)(1)(i) "Tanks shall be built of steel, except as provided in (*b*) through (*e*) of this subdivision." CFR 1910.106 (b)(2)(ii)(*a*) "The distance between two flammable or combustible liquid storage tanks shall not be less than 3 feet"; (*b*) "Except as provided in subdivision (*c*) of this subdivision the distance between any two adjacent tanks shall not be less than one-sixth the sum of their diameters...." 2 × 21 ft/6 = 7 ft.

2. A. bottom

    B. top

    C. vapor is not flammable

    D. gas is not flammable

    E. bottom

    F. gas is not flammable

    G. gas is about the same molecular weight as O_2 and N_2, so would not layer.

3. A. The small fibers expose a higher proportion of the combustible molecules to the air, and so would burn more rapidly.

    B. The larger particles would burn more slowly to complete consumption and would produce more heat, assuming the oxygen supply were sufficient.

    C. Slow oxidation of the plant material would consume oxygen and reduce the risk of combustion. Circulation of air replaces the oxygen. However, if the sealed bin supported anaerobic breakdown of the plant source with methane production, the methane could rise to the top to reach flammable or explosive concentrations.

    D. Flattened particles expose a higher percentage of their molecules to the air, and would thus present a greater fuel-effective concentration at a given total concentration. They would have a lower limit.

4. The compound or mixture must supply fuel, an oxidant, and a reaction that produces large volumes of gases rapidly.

Chapter 10

1. A. The eustachian tube equalizes the pressure on the eardrum by allowing air to enter or leave the middle ear.

    B. The cochlea is a fluid-filled structure in which vibration transmitted from the eardrum is converted into nerve signals.

C. The stirrup is the last of the three bones connecting the eardrum to the cochlea.

D. The semicircular canals have nothing to do with hearing, but rather are associated with balance.

2. A. The peaks are compressions of air in the atmosphere.

B. The height of the wave.

C. The pitch is related to the wavelength or peak-to-peak distance.

D. Loudness is related to amplitude.

E. Frequency = (speed of sound)/wavelength.

F. These are independent variables.

G. Frequency units are Hz, (s^{-1}).

H. 2000Hz; 250 Hz.

I. 6.5 in.; 0.172 m.

3. A. 20–20,000 Hz

B. 500–3000 Hz.

C. 4000 Hz.

D. No, since the frequency is above the range of human speech.

E. No. Greatest acuity is at 4000 Hz.

4. A. Sound power is the expenditure of energy generating the sound, with watts as the units. Sound pressure is the pressure of the compressions in the air (amplitude of the sound wave) and has pressure units (N/m^2).

B. Sound pressure level is 20 times the log of the ratio of two pressures, so the units cancel. It is expressed in decibels (dB).

C. 2×10^{-5} N/m² is chosen as the quietest sound detected by a person with perfect hearing.

D. $L_p = 20 \log (40 \text{ N/m}^2/2.0 \times 10^{-5} \text{ N/m}^2)$ dB = 126 dB

E. You would set the range at 120–130 dB. The scale would read 6.

5. A. The x-axis is frequency by octaves, usually from 250 Hz to 8000 Hz. The y-axis is the lowest sound pressure level in dB detected by the person being tested.

B. The graph would have a straight line drawn through 0 dB.

C. Hearing is tested with an audiometer, and the person tested indicates the lowest sound level detected at each frequency.

D. A loss of 25 dB at any frequency is considered impairment, and impairment is most serious in the range of human speech.

E. F.B.T. may be experiencing normal high-frequency hearing loss as age increases, but J.J.H. may have occupationally induced hearing loss.

6. $2/8 + 1/2 + 0.5/1 = 1.25$. The site is not in compliance.

7. (1) Replace the press with an automatic hydraulic press accepting steel from a roll and feeding parts into the bin, eliminating the full-time worker at the site.

(2) Place acoustic tile around the machine and on the ceiling.

(3) Place the press on a resilient pad.
(4) It sounds like enclosing the press would be difficult, but a heavy barrier directly between the press and the workers might be possible.

Chapter 11

1. $V = 23p$; $Ag = 47p$; $K = 19p$.

$^{15}N = 7p, 8n$; $^{23}Na = 11p, 12n$; $^{235}U = 92p, 143n$.

2. $^{26}Na \text{ ------> } ^{26}Mg + {}^0\beta$

   $^{212}Rn \text{ -----> } ^{208}Po + {}^4\alpha$

   $^{111}Sb \text{ -----> } ^{111}Sn + {}^0\beta$

   $^{16}O + {}^1n \text{ -----> } ^{16}N + {}^1p$

3.   A.   One rem equals: 1 roentgen of X- or γ radiation; 1 rad due to X-, γ, or β radiation; 0.1 rad due to neutrons or high-energy protons; 0.05 rad of particles heavier than protons and with sufficient energy to reach the lens of the eye.

   B.   The quality factor of X-, γ or β radiation is 1, of neutrons or high-energy protons is 10, and of particles heavier than protons and with sufficient energy to reach the lens of the eye is 20.

4.   The ratio is: 100/58 rem/calendar year, or 1.7 times as much.

5.   A.   5 rem/calendar year ÷ 0.1 rem/calendar year = 50 times greater.

   B.   Hands, forearms, feet, and ankles are not highly sensitive to radiation, but the named tissues are radiation-sensitive. Dosing active blood-forming organs (bone marrow of long bones) can cause loss of blood cells. Dosing the lens of the eye can cause cataracts. Dosing the gonads can reduce fertility.

   C.   29 CFR 1910.96 (b)(2)(i-ii): Whole body exposure can be as high as 3 rem/calendar quarter if the accumulated occupational exposure is less than 5(N–18) rems, where N is the current age.

   D.   No, she need not be: 5(45–18) = 135 rem.

   E.   29 CFR 1910.96 (b)(3): 10% of values on Table G.18.

6.   A. $0.105 \text{ m}^{-1.}$

   B.   2325 cpm.

Chapter 12

1. A. Portable wooden ladders.

    B. 1910.25 (c)(2–10); 9 kinds.

    C. See 1910.25 (c)(18): Wane is bark or the lack of wood for any cause on the corner of a piece of wood.

    D. See 1910.25 (c)(16): A shake is a separation along the grain, most of which occurs between the rings of annual growth.

    E. No, it is all definitions.

    F. (c) describes wooden and (d) describes metal ladders. ...(1) is repeated so that each section is complete and independent of the other.

    G. This section is regulations regarding wooden ladders.

    I. The ladder conforms to OSHA standards in 1910.25.

2. A. An injury to a worker as a result of an accident, such as a fall.

    B. White finger syndrome: A vibration injury to the hands involving nerve and blood vessel damage.

    C. An inflammation of the tendons.

    D. A tenosynovitis and nerve damage caused by abrasion in the carpal tunnel (passage through the wrist) due to working with the wrist bent.

    E. A tenosynovitis due to continuous use of the forefinger to run a tool.

    F. A tenosynovitis due to continuous twisting to the forearm, as when a screw is tightened.

    G. Damage due to a loss of blood circulation in a tissue, which could be caused by continuous pressure of the tissue.

3. A. Examples: suspension system failing, as from a broken rope or failed mount on the roof; worker falling from the platform; wind blowing platform about; power system for moving platform failing.

B. The July 1, 1992 edition has 34 pages.

C. [1910.66 (f)(7)(i)] Each specific installation shall use suspension wire ropes or combination cable and connections meeting the specification recommended by the manufacturer of the hoisting machine used. [1910.66 (f)(7)(iii)] Suspension wire rope grade shall be at least improved plow steel or equivalent. [1910.66 (f)(7)(ii)] Each suspension rope shall have a "Design Factor" (ratio of rated strength of suspension wire to rated working load) of at least 10.

D. [1910.66 (f)(7)(ii)] $F = S(N)/W$
$$= (5000 \text{ lb})(2)/(2000 \text{ lb})$$
$$= 5 < 10 \quad \text{System is not in compliance.}$$

E. [1910.66 (f)(5)(iii)] A thorough inspection of suspension wire ropes in service shall be made once a month. [1910.66 (f)(7)(vi)] A corrosion resistant tag shall be attached to one of the wire rope fastenings... [1910.66 (f)(7)(vii)] A new tag shall be installed at each rope renewal. [1910.66 (f)(7)(viii)] The original tag shall be stamped with the date of resocketing...

F. These control lateral movement of the platform in the wind.

G. [1910.66 (f)(1)(iii–iv)] Equipment that is exposed to wind when not in service shall be designed to withstand forces generated by winds of at least 100 mph... Equipment that is exposed to wind when in service shall be designed to withstand forces generated by winds of at least 50 mph... [1910.66 (i)(2)(v–vi)] The platform shall not be operated in winds in excess of 25 mph... On exterior installations, an anemometer shall be mounted on the platform...

4.

CAUTION		CAUTION	DANGER
THIS EQUIPMENT STARTS AND STOPS AUTOMATI- CALLY	EYE WASH FOUNTAIN	WEAR HEARING PROTECTION IN THIS AREA	OXYGEN IN USE–NO SMOKING OR OPEN FLAME
yellow & black	green & white	yellow & black	red, black, & white

5. A sabre saw produces vibration. Gloves probably should be used to prevent slivers in the hands from handling the wood. The amount of force

required to push the saw along the template should be checked. The curved templates require constant turning of the saw, and with the saw at chest height the wrist will be bent and will need to twist, and the forearm will need to rotate. This could cause carpal tunnel syndrome and epicondylitis.

The work should be lowered, and two handles perpendicular to the wood surface should be attached to the saw handle to allow the saw to be operated with two hands whose wrists are straight and to turn the saw by extending one arm or the other. The saw should be mounted on a suspension device so that its weight is borne at the end of a cut. The automatic shutoff when the switch is released is a good safety feature; however, constant pressure with the forefinger could produce ischemia in the finger pad. The switch should be replaced by a pressure switch in one of the handles so that when the worker releases the tool it stops.

Chapter 13

1. A. Moving the air increases the rate of evaporation from the skin, providing cooling.

    B. Air in the building becomes hotter than outside air because of the sun heating the blackened roof. An exhaust fan removes this heated air, which must then be replaced by the entry somewhere of outside air. Air inlets would facilitate the entry of outside air, speeding the process of removing the heated air. By locating these inlets on the opposite wall of the room, the entire room is swept by entering outside air, minimizing pockets of hot air remaining in the room.

    C. R and C. The roof, heated by the sun, is radiating heat to objects in the room (R). It is also heating the air, and in South Carolina on a hot summer day, air is very likely to reach temperatures above body temperature (C). Painting the roof with a reflective paint, perhaps aluminum paint, would reduce effectiveness of the sun at raising the temperature of the roof.

2. Worker 1 is continuously in the high-temperature zone, so this person should be watched for heat stress. Worker 2 only spends part of the time in the hot zone. However, this worker is climbing steps carrying heavy bags, and is wearing a coverall garment that prevents cooling of the body by circulating air. Add to this the time spent in the high-temperature zone. This person should be watched closely for signs of heat stress. Worker 3 has only brief encounters with the high-temperature zone, and is probably not threatened by heat stress.

3. Worker 1 is losing body heat due to the constant evaporation of sweat. However, by being unprotected from the sun, this worker is taking in radiative heat energy. Worker 2 is a similar case to worker 1, however, because some of the sweat is not evaporating from this worker's body, water is being lost without corresponding cooling being achieved. Worker 3 has blocked much of the radiation energy from the sun, and is still getting a large degree of benefit from the evaporation of sweat. Worker 4 has blocked out the sun, but has also blocked air movement across the body. Worker 4 may be at greatest risk.

4. A. Air at a higher temperature than the body transfers heat to the body (C). However, heat loss due to evaporation of sweat depends not on temperature differences, but on the relative humidity (water content) of the surrounding air. Dry air accepts water evaporating from the skin more rapidly, producing a higher rate of heat loss. The higher the skin temperature, the more rapid the evaporation, and again the more rapid the heat loss. Therefore, exposure of skin is in this case an advantage.

   B. T_G, when compared to T_A, measures the contribution of radiative heating to body heat burden. Since the contribution of heat radiation is not changed by the spray cooling of the air, when T_A is lowered, T_G is also lowered. T_{NWB} starts at T_A, and is then lowered by the cooling effect of evaporation. T_A is lowered by the water spray, but the humidity is raised, reducing the cooling effect of evaporation. Because of these counterbalancing effects, it is hard to predict the effect on T_{NWB}.

   C. $WBGT = 0.7T_{NWB} + 0.2T_G + 0.1T_A$

   Before installation:

   $WBGT = 0.7(24.2) + 0.2(46.1) + 0.1(42.8) = 30.4$

   After installation:

   $WBGT = 0.7(24.4) + 0.2(35.0) + 0.1(29.4) = 27.0$

5. T_G should show the greatest change.

   $WBGT = 0.7T_{NWB} + 0.2T_G + 0.1T_A$

   Before installation:

$$WBGT = 0.7(36.3) + 0.2(71.7) + 0.1(47.8) = 44.5$$

After installation:

$$WBGT = 0.7(29.8) + 0.2(43.3) + 0.1(43.3) = 33.9$$

Chapter 14

1. A. Part 56; B. Part 57; C. Part 71; D. Part 70.

2. A. 30 CFR 56.5001. The limits are the ACGIH TLV values of 1973.

    B. 30 CFR 56.5005. The standards are those set by the American National Standards Institute: ANSI 288.2—1969. The standard can be obtained by contacting ANSI at 1430 Broadway, New York, NY, 10018.

3. A. 30 CFR 56.5050; 30 CFR 57.5050; 30 CFR 70.500; 30 CFR 71. They are all the same.

    B. They are the same.

    C. The standards are presented by the American National Standards Institute: ANSI S1.4—1971.

4. A. 30 CFR 71.100.

    B. 2.0 mg/m^3.

    C. 30 CFR 71.101, silica.

    D. 5%.

    E. $10 \text{ mg/m}^3/10 = 1 \text{ mg/m}^3$.

    F. At 4% and 5%, 2 mg/m^3 is permitted. At 6%: $10 \text{ mg/m}^3/6 = 1.7 \text{ mg/m}^3$ is permitted.

    $$0.04 \times 2 \text{ mg/m}^3 = 0.08 \text{ mg/m}^3$$
    $$0.05 \times 2 \text{ mg/m}^3 = 0.10 \text{ mg/m}^3$$
    $$0.06 \times 1.7 \text{ mg/m}^3 = 0.10 \text{ mg/m}^3$$

    G. 30 CFR 208. Bimonthly.

H. The operator must identify the mine and workstation in violation, must indicate the measures to be taken to achieve compliance, and indicate the time, place, and manner in which these measures will be employed. In 30 CFR 71.301 we see that the District Manager, the manager of the Mine Safety and Health District in which the mine is located, must approve the plan, and may take air samples to assure that the measures bring the mine into compliance.

A. Test methods and equipment are outlined in ANSI N13.8—1973, which may be obtained either at the Metal and Nonmetal Mine Safety and Health Subdistrict office or from the American National Standards Institute, Inc., 1430 Broadway, New York, NY, 10018.

B. "If concentrations of radon daughters in excess of 0.1 WL are found in an exhaust air sample...."

C. The maximum permissible concentration (30 CFR 57.5039) is 1.0 WL and the maximum annual exposure (30 CFR 57.5038) is 4 WLM.

A. 29 CFR 1910.1029 (b) Definitions: "Coke oven emissions" means the benzene-soluble fraction of total particulate matter present during the destructive distillation or carbonization of coal for the production of coke.

B. 29 CFR 1910.1029 (b): "Green plush" means coke that when removed from the oven results in emissions due to the presence of unvolatilized coal. The hazard would therefore be a release of coke oven emissions.

C. 29 CFR 1910.1029 Appendix A (II): Exposure to coke oven emissions is a cause of lung cancer and kidney cancer in humans. Although there have not been an excessive number of skin cancer cases in humans, repeated skin contact with coke oven emissions should be avoided.

D. 29 CFR 1910.1029 (c) or 29 CFR 1910.1029 Appendix A (I)(C): 150 micrograms per cubic meter of air determined as an average over an 8-hour period.

E. 29 CFR 1910.1029 Appendix B (I)(A): Samples collected should be full shift (at least 7-hour) samples. Sampling should be done using a personal sampling pump with pulsation damper at a flow rate of 2 liters per minute. Samples should be collected on 0.8 micrometer pore size silver membrane filters (37 mm diameter) preceded by Gelman® glass fiber type A-E filters encased in three piece plastic (polystyrene) field monitor cassettes. The cassette face cap should be on and the plug removed. The rotameter should be checked every hour to ensure that proper flow rates are maintained.

A minimum of three full-shift samples should be collected for each job classification of each battery, at least one from each shift. If disparate results are obtained for particular job classification, sampling should be repeated. It is advisable to sample each shift on more than one day to account for environmental variables (wind, precipitation, etc.) which may affect sampling. Differences in exposure among different work shifts may indicate a need to improve work practices on a particular shift. Sampling results for each job classification should not be averaged. Multiple samples from the same shift on each battery may be used to calculate an average exposure for a particular job classification.

F. 29 CFR 1910.1029 (b): "Beehive oven" means a coke oven in which the products of carbonization other than coke are not recovered, but are released into ambient air.

G. 29 CFR 1910.1029 (f)(1)(iii)(*a*): The employer shall institute engineering and work practice controls on all beehive ovens at the earliest possible time to reduce and maintain employee exposures at or below the permissible exposure limit,...

29 CFR 1910.1029 (f)(1)(iii)(*b*) If, after implementing all engineering and work practice controls required by paragraph (f)(1)(iii)(*b*) of this section, employee exposures still exceed the permissible exposure limit, the employer shall research, develop, and implement any other engineering and work practice controls necessary...

Wherever the engineering and work practice controls which can be instituted are not sufficient to reduce employee exposure to or below the permissible exposure limit, the employer shall nonetheless use them to reduce exposures to the lowest level achievable by these controls, and shall supplement them by the use of respiratory protection which complies with paragraph (g) of this section.

7. A. 29 CFR 1910.251–.257

    B. 29 CFR.252 (c)(2)(i): Mechanical ventilation shall be provided when welding or cutting is done on metals not covered in paragraphs (c)(5) Through (c)(12) of this section... (A) in a space of less than 10,000 cubic feet (284 m^3) per welder, (B) in a room having a ceiling of less than 16 feet (5 m), (C) in confined spaces or where the welding space contains partitions, balconies or other structural barriers to the extent that they significantly obstruct cross ventilation. (ii) Such ventilation shall be at the minimum rate of 2000 cubic feet (57 m³) per minute per welder...

C. 29 CFR.252 (c)(9)(i) *General.* Welding or cutting indoors or in confined spaces involving cadmium-bearing or cadmium-coated base metals shall be done using local exhaust ventilation or airline respirators unless atmospheric tests under the most adverse conditions have established that the workers' exposure is within the acceptable concentrations defined by §1910.1000 of this part.

D. 29 CFR.252 (c)(4) *Ventilation in confined spaces-* (i) *Air replacement:* All welding and cutting operations carried on in confined spaces shall be adequately ventilated to prevent the accumulation of toxic materials or possible oxygen deficiency.

8. 29 CFR 1910.218: Guards to prevent physical injury during operation or changing of dies.

Chapter 15

1. Cans are added to molten primary aluminum. Moisture in cans may cause explosive steam production. Heat stress, noise exposure, and metal and metal oxide fumes may be produced. If chlorine is bubbled through the melt, exposure to the gas through leaks in the system or as it exits the melt are possible. If fluxes or alloying metals are added to the melt, dusts may reach the workers. Dross from the furnace contains an unpredictable mixture of compounds, some of which could be volatile.

2. During its purification. During the smelting of gold, lead, copper, cadmium, and zinc. When alloying with lead. In preparation of arsenic pesticides.

3. Chromium plating, stainless steel production, welding and cutting, painting with chromate pigments, printing with chromate dyes, tanning leather, applying wood preservatives, cement production and handling, magnesium casting, and chromate/dichromate production.

4. Three metals may be present in quantity: nickel, copper, and cobalt. Any dry residues in the vessel could produce airborne dust rich in metal salts. The leaching is done with hot acid, so there are acid residues. Workers should use respirators and should wear protective clothing.

5. A. 29 CFR 1910.1025 (c). PEL is 50 µg/m³ TWA.

B.   29 CFR 1910.1025 (d)(9) ...which has an accuracy (to a confidence level of 95%) of not less than 20% for airborne concentrations of lead equal to or greater than 30 µg/m^3.

C.   29 CFR 1910.1025 (j)(1)(i) ...all workers exposed above action level for more than 30 days per year. (j)(2)(i)(A) At least every 6 months for each employee covered under paragraph (j)(1)(i). (j)(2)(i)(B) At least every 2 months for each employee whose last blood sampling and analysis indicated a blood level at or above 40 µg/100 mL. (j)(2)(i)(C) At least monthly during the removal period of each employee removed from exposure to lead due to an elevated blood level.

D.   29 CFR 1910.1025 (j)(2)(iii) ...have an accuracy (to a confidence level of 95%) within plus or minus 15% or 6 µg/m^3, whichever is larger.

E.   29 CFR 1910.1025 (j)(3)(i) ...to each employee covered under (j)(1)(i) of this section (A) At least annually...

6.   As, Cr, Ni, Be, Cd, Pb.

Chapter 16

1.   Some biological hazard may exist due to residues of milk in the cartons having been subject to bacterial degradation. Transferring dyes into the mix may involve contact with potentially toxic chemicals. Proper guards must be placed around the chopping, grinding, and molding machinery. Heat stress is possible from the washing and drying operations and from the injection molding machines. Noise would be a problem in compression and chipping, blowing the chips dry, grinding, and from the injection molding machinery. Skin exposure to hot detergent solutions should be avoided. Final trimming of the fishing boxes may generate particulates.

2.   A.   Nylon (polyamide) strands can hydrogen bond to one another, whereas only very weak attractions can form between polyethylene strands.

B.   Slow cooling allows a greater proportion of the plastic to be in the form of crystallites.

C.   Crosslinking sharply increases rigidity.

D.   Plasticizers reduce crystallite formation, lowering rigidity.

3.   Polymerization adhesive. Advantages: No solvent to evaporate. Disadvantages: Possible worker contact with reactive and toxic prepolymerization chemicals. Polymers in solution. Advantages: Far fewer (possibly none) of the reactive or toxic compounds used in the polymerization technique. Disadvantages: Evaporated solvent may require removal by ventilation.

4.   A.   29 CFR 1910.1043

     B.   29 CFR 1910.1043 (b). "Lint free respirable cotton dust" means particles of cotton dust of approximately 15 micrometers or less aerodynamic equivalent diameter.

     C.   29 CFR 1910.1043 (c)(1) ...measured by a vertical elutriator or a method of equivalent accuracy and precision. 29 CFR 1910.1043 (b) "Vertical elutriator cotton dust sampler" means a dust sampler that has a particle size cutoff at approximately 15 micrometers when operating at the flow rate of 7.4±0.2 L/min.

     D.   29 CFR 1910.1043 (c)(1)   The PEL for yarn manufacturing is 200 $\mu g/m^3$ of lint-free respirable dust mean concentration averaged over an 8-hour day; for slashing and weaving to 750 $\mu g/m^3$ of lint free-respirable dust mean concentration averaged over an 8-hour day; in other operations to 500 $\mu g/m^3$ of lint-free respirable dust mean concentration averaged over an 8-hour day.

     E.   29 CFR 1910.1043 (d)(3)(i) The employer shall repeat the measurements required by paragraph (d)(2) at least every six months.

5.   A.   29 CFR 1910.1017(c)(1) PEL ...1 ppm averaged over any 8-hour period; 29 CFR 1910.1017(c)(2) ...5 ppm averaged over any period not exceeding 15 minutes; 29 CFR 1910.1017(b)(1) Action level ...0.5 ppm averaged over an 8-hour workday.

     B.   29 CFR 1910.1017(a)(2) Applies to the polyvinyl chloride not yet processed into product. 29 CFR 1910.1045(a)(2)(i) Excludes polymers including acrylonitrile as a monomer from coverage. Acrylonitrile 1910.1045.

Glossary

accelerator — a compound that improves the effectiveness of a catalyst

acceptable ceiling concentration — a ceiling level of exposure that is permitted only for a specified time interval

acclimatization (to heat) — adjustment of the body to working at high temperatures

ACGIH — The American Conference of Governmental Industrial Hygienists

acid descaling (pickling) — cleaning a metal surface with acid

acne — a swelling due to the blockage of flow of oil out of a skin pore

acroosteolysis — the degeneration of bones in the hands as the result of exposure to vinyl chloride

action level — a level of an airborne toxicant equal to half the PEL. When the workplace is below this level, regulation is simplified.

acute lethal dose — the LD_{50} of a compound when tested for a very short period — **usually as a single dose**

acute testing or dosing — testing of the toxic effects of a compound for a short **period of time** — usually a single dose

addition polymer — a polymer formed by an addition reaction from monomers with a carbon to carbon double bond

adsorbent tubes — tubes for collection of gas or vapor for analysis that adsorb gas or vapor on the surface of a solid

aerosols — airborne liquid droplets

AIHA — American Industrial Hygiene Association

air flow — the volume of air per unit time moving through a duct

alkaline descaling — cleaning a metal surface with caustic

allergic contact dermatitis — skin inflammation due to allergic response to a chemical upon skin contact

allergic response — the activation of the immune system by contact with a chemical to produce swelling, irritation, asthma, fluid retention, etc.

alloy — a deliberate mixture of metals prepared to obtain desired properties

alopecia — loss of hair due to toxic response to a chemical

alpha radiation (Â radiation) — radioactive particle emitted from the atomic nucleus composed of two protons and two neutrons

aluminum dust lung — lung fibrosis from inhaling aluminum powder

alveoli — the air sacs at the ends of the finest branches of the lungs in which gases exchange between air and the blood

amplitude (sound) — when sound is visualized as a wave, amplitude is the peak height and is proportional to loudness

anemia — inadequate capacity to transport oxygen in the blood

anemic hypoxia — reduced oxygen supply due to impaired oxygen transport by the blood

anoxic hypoxia — reduced oxygen supply due to interference with respiration

ANSI — American National Standards Institute

antibodies — proteins called immunoglobins that bind to antigens and may stimulate allergic responses

antigen — a chemical to which an antibody binds which may stimulate an allergic response

antimony spots — a skin rash caused by exposure to antimony

asbestosis — a lung disease caused by inhaling asbestos fibers

asphyxiant — a gas that causes a reduction of the oxygen supply to tissues

atomic absorption spectrometry — an analytical method for metals in which light passing through a hot gas stream is selectively absorbed by specific metal atoms or ions

audiometer — a device that generates pure tones at known sound pressure levels to test hearing

auto ignition temperature (ignition temperature) — the lowest temperature that causes combustion of a gas or vapor without a spark

bag house — a chamber with filters at the exhaust point of a plant ventilation system in which particulate is removed from the air

barrier cream — a skin cream containing oils or silicones that helps prevent contact of the skin with chemicals

basal cell carcinoma — a type of skin cancer

basic oxygen furnace — a furnace for converting iron to steel where oxygen is passed through molten iron and resulting oxidized impurities react with limestone

becquerel (Bq) — a unit for measuring radioactivity equal to 1 dps.

behavioral toxin — a toxic agent that damages or otherwise affects the nervous system

benefication — removal of clay, dirt or rock from ore after its removal from the mine

beta radiation (ß radiation) — radioactive particle emitted from the atomic nucleus composed of an electron

bias (determinate error) — error in an analysis such that the measured value is always higher or lower than the true value

black lung disease — (coal miners pneumoconiosis) a disease of the lungs caused by inhaling coal dust

blast furnace — an iron smelter using coal and limestone to convert ore to iron

blood brain barrier — refers to the relative difficulty of transporting chemicals from the blood into the central nervous system

blow molding — formation of a hollow object such as a bottle by inflating plastic inside a mold

blowing agent — a reagent added to plastic to generate gas, creating a foam

Bq (becquerel) — a unit for measuring radioactivity equal to 1 dps.

breakthrough (sample collection) — the amount of sample collected at the point when the collecting device is saturated and allows sample to pass through

bronchial tree — the branched structure of air passages in the lung

bronchioles — smaller branches of the respiratory tract

brown lung disease (byssinosis) — a lung disease caused by inhaling cotton fiber

BSFTPM — the OSHA index for worker exposure to the benzene soluble fraction of total particulate in a coke oven

bubbler (impinger) — a gas or vapor collecting device which bubbles the air sample through a container of solvent

capillary (anatomy) — the smallest diameter blood vessels through the walls of which nutrients, oxygen and wastes are transported to and from the tissues

byssinosis (brown lung disease) — a lung disease caused by inhaling cotton fiber

calendaring — coating a fabric with an elastomer

capture velocity — the air velocity needed to move molecules or particles in the atmosphere into a duct

carcinogen — a compound which on exposure increases the likelihood of cancer

carpal tunnel — a passage through the bones of the wrist for the tendons and median nerve

carpal tunnel syndrome — inflammation of the median nerve in the carpal tunnel resulting in a loss of sensation and grip

carrier gas — the moving phase in gas chromatography

CAS number — a number assigned to a chemical for identification by the Chemical Abstracts Service

catalyst — a chemical that accelerates the rate of a chemical reaction

cataract — the clouding of the lens of the eye

caustic — NaOH

ceiling value — the maximum level of exposure to a chemical permitted in the workplace, with the duration of this exposure being unspecified

centrifugal (cyclone) collector — air contaminated with particles is swirled in a cone shaped chamber, spinning particles out against the chamber walls

CFR — Code of Federal Regulations. A government publication listing the federal regulations

chemical asphyxiants — compounds that block the transportation or utilization of oxygen

chloracne — the formation of acne-like skin eruptions as the result of exposure to certain chlorinated hydrocarbons

chromatography — a separation method in which the compounds under study are partitioned between a stationary and a moving phase

chromatography column — the holder of the stationary phase in chromatography

chromosome — one of the several DNA double helices found in the nucleus of a cell

chronic testing or dosing — testing of the toxic effects of a chemical for an extended period of time, usually the lifetime of the test animal

Ci (curie) — a measure of radioactivity, one Ci equaling 3.7×10^{10} dps

cilia — hairlike structures under the mucous layer of the respiratory tract that move mucous toward the nose

cirrhosis — scarring of tissue due to a toxic agent or agents

classifier — a screen used to sort particles of crushed ore by size

clearance — the removal of substance from the blood by the kidneys

CNS — central nervous system, including the brain and spinal cord

coal miners pneumoconiosis — (black lung disease) a disease of the lungs caused by inhaling coal dust

cochlea — the nerve ending lined fluid filled spiral where nerve impulses of hearing originate

coke — coal roasted to remove volatile components

colic — constipation followed by intense abdominal pain (as in lead poisoning)

collection efficiency — the percentage of gas or vapor that a sampling device collects

combustion lean limit mixture (lower flammability limit) — the lowest concentration of a gas or vapor able to sustain continuous combustion

combustion rich limit mixture (upper flammability limit) — the highest concentration of a gas or vapor able to sustain continuous combustion

comminution — breaking ore into a powder

complementary DNA — the two strands of DNA in the double helix are complementary in that each base on one strand always pairs with a specific but different base on the other strand

condensation polymer — a polymer formed from monomers each of which has two functional groups, and these groups are capable of reacting to join the monomers together

conjugation — the attachment of chemical structures to foreign substances in the body to facilitate their removal by the kidney

conjunctivitis — irritation of the eye by exposure to UV light

contact dermatitis — dermal response to a chemical upon direct application of that chemical to the skin

continuous sampling (integrated sampling) — sample collectors that gather the sample continuously for a period of time

control group — a set of test animals in a study which are not dosed, but are otherwise treated the same as dosed animals

convective heat transfer — transfer of heat to or from the body by contact with hotter or colder matter

cope — the top flask in a two piece mold

copolymer — a polymer composed of more than one kind of monomer

core temperature — the internal body temperature, taken at the mouth, ear or rectum

cornea — the transparent front surface of the eye

corrosive — a chemical that destroys tissue on contact

covalent bond — a system for strongly joining atoms together based on a sharing of electrons in which the distances between and angles formed by the atoms are relatively fixed

crosslinking (curing) — joining strands of polymers together to create a three dimensional polymer

crystallite — an orderly region within the mass of a polymer

cumulative trauma — physical injury due to repeated insults to the body over a period of time

curie (Ci) — a measure of radioactivity, one Ci equaling 3.7×10^{10} dps

curing — (plastics) crosslinking or catalyzing the formation of a thermoset plastic; (elastomers) crosslinking an elastomer, vulcanization

cyanosis — the condition of an organism experiencing a shortage of oxygen

cyclone (centrifugal) collector — air contaminated with particles is swirled in a cone shaped chamber, spinning particles out against the chamber walls

decibels — units of sound pressure level

decay constant (radioactivity) — the fraction of atoms that decay per unit time

deflecting vane velometer — a device used in the analysis of a ventilating system which measures the air velocity

dermis — the second layer of the skin which includes blood vessels, nerves and collagen fibers

desorption efficiency — the percentage of a collected sample recovered from the collector

determinate error (bias) — error in an analysis such that the measured value is always higher or lower than the true value

DF (radioactivity) — distribution factor: the relative dosage absorbed by a specific organ or tissue

dilution ventilation (GEV, general exhaust ventilation) — movement by a ventilation system of the entire mass of air through a plant, diluting and eventually removing impurities

diploid — cells which contain two sets of DNA, one from each parent, are diploid

distribution factor (DF, radioactivity) — the relative dosage absorbed by a specific organ or tissue

DNA — deoxyribonucleic acid — the very large molecules in the nucleus of the cell carrying the genetic information

DNA double helix — the spiral assembly of a pair of related DNA molecules, which is the form in which cell DNA occurs

dosimeter (radioactivity) — a portable device worn by workers handling or near sources of radioactive materials that measures accumulated exposure

dpm (radioactivity) — disintegrations per minute

dps (radioactivity) — disintegrations per second

drag — bottom flask in a two piece mold

dust — airborne particulate generated by grinding or agricultural processes

dynamic precipitator — a motor driven cyclone collector for removing airborne particulate

dyspnea — shortness of breath

ear drum — the membrane separating the outer and middle ear that vibrates in response to sound waves

edema — the retention of fluid in an organ or in the body as a whole

electrolytic purification (metallurgy) — pure metal is obtained at the cathode from ore or from an impure metal anode

electron capture detector — a gas chromatography detector in which a constant stream of electrons from a ß source is interrupted by electronegative atoms in the compounds being analyzed

electrostatic precipitators — a system to remove particulate from air utilizing charged plates

electrowinning (metallurgy) — obtaining a metal from ore by leaching followed by plating the metal onto a cathode

emergency standard — the exposure standard established by OSHA without following standard procedures upon discovery of a serious threat by a chemical in common use

emphysema — the loss of lung capacity due to tissue damage

enclosure (in ventilation) — a construction at the site of contamination generation which is swept by air entering a hood

enzyme — a catalyst for chemical reactions produced by the cells

EP — ethylene propylene copolymer

EPA — U. S. Environmental Protection Agency

EPDM — ethylene propylene terpolymer

epicondylitis (tennis elbow) — inflammation of the elbow due to excessive rotation of the radius bone in the forearm

epidemiology — the study of the relative characteristics of large exposed and non exposed populations for the purpose of detecting toxic effects

epidermis — the outer layer of the skin, the outer portion of which is interlocked dead cells

ergonomics (human factors engineering) — the study of designing objects such as tools used by workers to prevent cumulative trauma

extrusion — forming bars or tubes by forcing plastic or metal through a die

Federal Register — the published record of the activities of congress

fibrosis — the formation of scar tissue, especially in the lungs

FID (flame ionization detector) — a detector in gas chromatography in which ions produced by burning the compound under study then pass between plates with a high voltage between them

fillers — compounds added to plastics to influence the properties or allow use of less plastic

flame arrestor — a screen placed across a duct or opening to a container to cool approaching burning gases below ignition temperature

flame ionization detector — a detector in gas chromatography in which ions produced by burning the compound under study then pass between plates with a high voltage between them

flammability limits — the concentration of a gas or liquid at the low or high extremes of ability to sustain continuous combustion

flashpoint — the minimum temperature to raise the vapor pressure of a flammable liquid to the lower flammability limit

flask — a container in which a sand mold is shaped

fluorosis — a disease resulting from exposure to airborne fluoride experienced primarily by aluminum workers

follicle — the passage for a shaft of hair passing through and exiting the skin

foundry — a plant where metals are cast into products

frequency (sound) — when sound is visualized as a wave, frequency is the number of waves passing a point per unit time and determines the tone

frictioning — forcing elastomer into a fabric with a hot roller moving faster than the sheet of fabric

froth flotation — use of foams to float out minerals from powdered ore

FTIR — fourier transform infrared spectrophotometry: a rapid method of infrared spectrophotometry

fume — airborne particulate produced by high temperature metal processing

gamma radiation (radiation) — very high frequency electromagnetic radiation emitted by the atomic nucleus

gas chromatography — a chromatographic separation method in which the moving phase is a gas

gastrointestinal tract — the digestive system including everything from the mouth through the stomach and intestines

GC — gas chromatography

gene — the message for assembling a single protein coded as a base sequence on DNA

Geiger-Muller counter — a device for measuring radioactivity based on the creation of ion cascades in a gas by ionizing radiation

general exhaust ventilation (GEV, dilution ventilation) — movement by a ventilation system of the entire mass of air through a plant, diluting and eventually removing impurities

GEV — general exhaust ventilation, dilution ventilation

G.I. tract — the gastrointestinal tract

gin mill fever — an illness like metal fume fever experienced by new or returning workers in a cotton processing facility

glomerulus — the structure in the kidney at which the fluid portion of the blood is filtered into the tubules

grab sample — sample collecting in which a sample is collected in one pass

gray (Gy) — an SI unit of radioactive exposure equal to 100 rads

half life (toxicology) — the time required for half of a dose of a chemical to leave the body

half life (radioactivity) — the time required for half the radioactive atoms of a given isotope to disintegrate in a sample

half thickness (radioactivity) — the thickness of a shielding material required to block half of a specific kind of emission

hazard (of a chemical) — the likelihood under conditions of use of harm being caused by a chemical

Hazard Communication Standard — an OSHA standard established in 1983 requiring all employers to inform employees of the hazard of chemicals in the workplace and of the steps necessary to avoid harm

heat cramps — muscle cramps due to loss of fluid and salt while working in a hot environment

heat exhaustion (heat prostration) — fatigue, weakness and/or fainting due to reduced blood supply to the brain as a result of working in a hot environment

heat stress — increase in core body temperature resulting from working in a high temperature environment

heat stroke — a very high core body temperature leading to tissue damage resulting from working in a high temperature environment

hepatotoxin — a systemic toxin whose target organ is the liver

histotoxic hypoxia — a shortage of oxygen due to inability of tissues to use delivered oxygen

hertz — the units of frequency (s^{-1})

high efficiency cyclone collector — a cyclone collector with a narrow cone, swirling the air faster and improving the efficiency of separation

HMIG — Hazardous Materials Identification Guide

hood (in ventilation) — the site of air uptake in a local exhaust ventilation system

HPLC — high performance liquid chromatography: a chromatography method in which the liquid moving phase is pumped through a column containing stationary phase

human factors engineering (ergonomics) — the study of designing objects such as tools used by workers to prevent cumulative trauma

humerus (anatomy) — the bone of the upper arm between the shoulder and the elbow

hydrogen bond — a moderately strong attractive force between a pair of atoms (oxygen, nitrogen, or fluorine), both of which are attracted to a hydrogen between them

hyperthermia — body core temperature above normal

hypokinetic hypoxia — a shortage of oxygen due to interference with blood flow

hypothermia — body temperature below normal (below 35°C)

hypoxia — a shortage of oxygen in an organism

IDLH — Immediately Dangerous to Life and Health: the maximum level of exposure from which a person could exit in 30 minutes without escape-impairing symptoms or irreversible health effects

ignition temperature (auto ignition temperature) — the lowest temperature that causes combustion of a gas or vapor without a spark

impinger (bubbler) — a gas or vapor collecting device that bubbles the air sample through a container of solvent

indeterminate error (random error) — error in analysis where the measured values scatter around the true value randomly

inflammation — reddening and swelling due to chemical contact

infrared spectrophotometer (IR spectrophotometer) — an analytical instrument that measures the absorbance of specific frequencies of infrared light by a compound under study

inner ear — the part of the ear including the cochlea that converts vibration to nerve impulses

injection molding — forming plastic by forcing molten plastic into a mold

integrated sampling (continuous sampling) — sample collectors that gather the sample continuously for a period of time

ionizing radiation — radioactive emissions that drive electrons off atoms as they pass

IR spectrophotometer (infrared spectrophotometer) — an analytical instrument that measures the absorbance of specific frequencies of infrared light by a compound under study

irritant — a compound that causes inflammation and edema

irritant contact dermatitis — irritation due to skin contact with a chemical

ischemia — loss of circulation and numbness due to blockage (as by compression) of blood vessels

isotopes — atoms of the same element with differing numbers of neutrons

itai-itai disease — bone embrittlement from exposure to cadmium

jolt machine — a machine for packing sand in molds by dropping and suddenly stopping the flask

jolt rollover squeeze machine — a machine that forms sand molds in a single operation at high speed

keratin layer — the outer layer of dead cells in the epidermis

layering (gases) — separation of gases in a contained environment by density

LD_{50} — the dose lethal to 50 percent of the animals in a test group

LD_{LO} — the lowest dose of a compound proving to be lethal

leaching (metallurgy) — metal ions are dissolved from ore using acid, alkali or a complexing agent

liquid trap — (impinger, bubbler) — a gas or vapor collecting device that bubbles the air sample through a container of solvent

local exhaust ventilation — air is drawn into a hood at the site of generation of contamination

lower respiratory tract irritant (whole lung irritant) — a gas whose irritating properties develop slowly, so that we are willing to allow it to enter the entire lung

lower back pain — pain due to excessive lifting, carrying, pushing or pulling — **particularly lifting**

lower flammability limit (combustion lean limit mixture) — the lowest concentration of a gas or vapor able to sustain continuous combustion

L_p **(sound pressure level)** — relative sound pressure, compared to a standard

MAC — Maximum Allowable Concentration — a recommendation for the highest safe level of exposure to a chemical

malignant cell — a cancer cell

malignant mesothelioma — a bronchial carcinoma (cancer) caused by inhaling asbestos

Material Safety Data Sheets — Pamphlets required by the Hazard Communication Standard that list the identity of chemicals, hazard information, and information about the manufacturer

mCi; µCi — respectively, millicurie and microcurie

melanin — the pigment in skin

melanoma — skin cancer of the pigment producing cells

membrane — a sheetlike lipid structure serving as a non-polar barrier around cells and cell substructures

metabolism — the system of chemical reactions by which chemicals are altered in the body

metal fume fever — a disease like influenza caused by inhaling certain chemicals, usually metals

mg/kg — milligrams of compound per kilogram of animal weight, the usual units for expressing the levels of dosing of test animals

mg/m³ — milligrams per cubic meter of air: a unit used to express concentrations of particulates in the air

middle ear — the region of the ear containing tiny bones that transfer vibration from the ear drum to the inner ear

mold — a form used for casting or molding metals or plastics

MSDS — Material Safety Data Sheets: informational sheets about chemicals required to be in the workplace by the "right to know" law

mucus — a moist sticky layer of protein/carbohydrate composition lining internal organs, including the respiratory tract

mutagen — a compound causing genetic (DNA) damage

mutation — an alteration of the genetic message on DNA

nasal cavity — the portion of the respiratory tract between the nostrils and the pharynx

NCI — National Cancer Institute

necrosis — death, here particularly cell death

NED — Normal Equivalent Deviations (standard deviations): the change in dose that produces one standard deviation change in the likelihood of an effect on an organism of a toxin

negative pressure (in ventilation) — in a ventilation system with an exhaust blower, the static pressure in the vent is lower than atmospheric pressure, thus has negative pressure

NEL — No Effect Level: same as NOEL

Neonates — newborn animals

nephron — the functional tubular unit of the kidney

nephrotoxin — a systemic toxin whose target organ is the kidney

neurotoxin — a systemic toxin whose target organ is the nervous system

neutron — a component of the atomic nucleus sometimes emitted as radioactivity

NFPA — National Fire Protection Association

NIOSH — National Institute of Occupational Safety and Health: a research arm of the Federal Bureau of Health and Human Services studying the health hazards found in the workplace

NOAEL — No Observed Adverse Effect Level: same as NOEL

NOEL — No Observed Effect Level: a measure of the highest level of dosing that produces no symptoms

non-ionizing radiation — electromagnetic radiation no more energetic that UV light

NTP — National Toxicology Program

occupational asthma — an allergic (asthma) response to an allergin in the workplace

occupational dermatosis — dermatosis caused by exposure on the job

occupational physician — a physician who specializes in occupational illness

octave bands — a dividing of the range of audible sound into octaves to discuss hearing problems

octave interval — a change in tone such that the frequency exactly doubles

oncogene DNA — segments of DNA involved in the conversion of a normal cell to a cancerous cell

open pit mine (surface mine, strip mine) — a mine in which minerals are removed from a hole in the surface of the ground

OSHA — Occupational Safety and Health Administration: a branch of the Federal Department of Labor established to enforce the safety standards set by the government for the workplace

outer ear — the ear canal connecting the outside with the ear drum

PAH — polycyclic aromatic hydrocarbons

PAT — the Proficiency Analytical Testing program for assessing the accuracy of analytical methods run by NIOSH

pattern — a shape used to prepare the depression in a sand mold

PEL — Permissible Exposure Limit: a recommendation by OSHA for the highest safe level of exposure to a chemical

permanent threshold shift — a permanent drop in hearing acuity

phagocytic cells — cells in the body that engulf small objects, particularly foreign material

photodermatosis — skin problems caused by a chemical that is activated by light

pickling (acid descaling) — cleaning a metal surface with acid

Pitot tube — a water manometer modified to measure both static and velocity pressure in a vent

plasticizer — compound added to a thermoplastic polymer to prevent the formation of crystallites

plumbism — toxic effects of lead

pneumoconiosis — damage to the lungs resulting from inhaling solid particles

pneumonia — a disease in which fluid collects in the lungs

pneumonitis — fluid in the lungs

PNS — peripheral nervous system: the nerves outside the CNS that control muscles and glands and receive sensory signals

polar bond — when two atoms which are joined covalently do not share the electrons between them equally, the atom with the greater attraction has a small negative charge and the other has a small positive charge

pore — an opening in the skin arising from the sweat gland and sebaceous gland

positive pressure (in ventilation) — in a ventilation system with a blower forcing air into the duct, the static pressure in the duct is higher than atmospheric pressure, thus has positive pressure

ppm — parts per million, as grams per one million grams or mG per kG: a common concentration unit for dilute samples of dissolved substances or airborne substances

precancerous cell — a cell that has been transformed such that it is predisposed to be further converted to a malignant cell

prepolymers — short polymer chains that are crosslinked into a thermoset plastic

pressure (in ventilation) — the effect of air molecules colliding with a surface

pressure forming — shaping plastic by forcing a sheet against a mold with air pressure

probit — the change in dose that produces one standard deviation change in the likelihood of an effect on an organism by a toxin

procarcinogen (secondary carcinogen) — a compound that is converted to a carcinogen by cell metabolism

proposed rule — a proposal to establish or change an exposure standard published by OSHA in the Federal Register

pure tone — a tone of a single frequency

push-pull ventilation system — air is blown across a contaminated site toward a hood with negative pressure that picks up the now contaminated air

QF (quality factor — radioactivity) — the relative hazard of a type of radioactive emission

rad — radioactive emission energy absorbed by a target _100 erg/G

radiative heat transfer — transfer of heat to or from the body through space

radioactive isotopes — isotopes of an element that spontaneous disintegrate, releasing a particle or energy from the nucleus

radius (anatomy) — the bone of the forearm that rotates in the elbow to allow arm and hand rotation

RAL — recommended WFGT alert limits for healthy non-acclimatized workers

random error (indeterminate error) — error in analysis where the measured values scatter around the true value randomly

Raynard's disease (white finger disease) — damage to nerves and blood vessels in hands due to the vibration of hand tools

RCRA — Resource Conservation and Recovery Act (1976): federal plan for regulation of handling of hazardous waste.

receiving hood — a hood placed in the path of propelled impurities

refining (metallurgy) — converting ore to metal

REL — recommended WFGT alert limits for healthy acclimatized workers

rem (radioactivity) — dose equivalent for energy of radioactive emission absorbed for a specific organ or tissue [= rad(QF)(DF)]

renotoxin — a systemic toxin whose target organ is the kidney

replication — the synthesis of new DNA using existing DNA as a sequence template

resistance (in ventilation) — the relative difficulty of moving air through ducts

resonance frequency — for each object there is a resonance frequency such that exposure to that frequency results in the object vibrating

respirable particles or dust — particles generally under 1 μ in diameter capable of reaching the alveoli when inhaled

respirator — a device worn over the nose and mouth designed to provide air free of contaminants

riser — a channel through which metal rises to show the mold is full

roentgen (R) — a unit to measure energy absorbed by air from a radioactive source equaling .83 erg/G

rotating vane anemometer — a device used in the analysis of a ventilating system which measures the air velocity

runners — channels to move metal within a mold

sand slinger — a high pressure hose spraying sand into a mold

scintillation counter — radioactivity counter in which light is produced in a (scintillation) fluid by the emission, then light is measured to estimate the level of radioactive emission

sebaceous glands — glands in the skin at the base of the pore that produce skin oil

secondary carcinogen — (procarcinogen) a compound that is converted to a carcinogen by cell metabolism

secondary emissions — radioactive emissions resulting from changes brought about by the original or primary radiation

sensitizer — a compound which generates an immune response, triggers allergies

settling chamber — air in a duct enters a large chamber, causing the air velocity to drop below that sufficient to keep larger particles suspended

Shaver's disease — fibrosis in the lungs from exposure to aluminum oxide

shielding (radioactivity) — interposed matter which blocks radioactive emissions

SI units — Système Internationale — an internationally established set of units of measurement used in most of the world

SIC — Standard Industrial Classification — a list coding industrial work into a series of numbered categories

siderosis — accumulation of iron compounds in the lung

sievert (Sv) — 100 rem

silicosis — lung disease caused by inhaled silica particles accumulating in the lung

silicotic nodules — characteristic pockets of silica that appear in X-rays of the lungs of individuals with silicosis

simple asphyxiant — a gas that dilutes the oxygen in the air

sizing — separation of crushed ore by size

slag — waste in metal purification, especially the silicate waste formed in a blast furnace

smelting — conversion of ore to free metal by heat and reduction

smoke — airborne particulate arising from combustion of organic material

sound level meter — a device to measure sound pressure levels

sound pressure — the fluctuation in pressure caused by sound waves

sound pressure level (L_p) — relative sound pressure, compared to a standard

sprue — a hole to allow molten metal to enter the mold

squamous cell carcinoma — a type of skin cancer

squeeze machine — a machine that shapes sand in a mold by ramming the pattern into the sand

stabilizer — a compound added to plastic to protect it from the destructive effects of sunlight

stable isotopes — isotopes of an element that do not disintegrate radioactively

stain tube — a device for estimating gas or vapor concentration in the air by the length of a colored reaction product produced in a tube through which the air sample has passed

static pressure — air pressure in a duct connected to a fan

STEL — Short Term Exposure Limit: the TLV for a limited period of time, typically for 15 minutes

stratum corneum — the keratin layer of the epidermis

strip mine (surface mine, open pit mine) — a mine in which minerals are removed from a hole in the surface of the ground

styrene sickness — nausea, vomiting and fatigue resulting from styrene exposure

subacute testing or dosing — daily dosing of a test animal for a moderate period of time, usually 90 days

surface mine (open pit mine, strip mine) — a mine in which minerals are removed from a hole in the surface of the ground

surrogate — one species substituting for another in testing

Sv (sievert) — 100 rem

sweat glands — glands in the skin at the base of the pore that produce sweat

systemic toxin — a poison causing immediate damage, usually cell death and usually in a specific target organ

T_A — dry bulb (ordinary thermometer) temperature

T_G — globe temperature, which measures the effect of convective heat transfer

T_{NWB} — natural wet bulb temperature: evaluates cooling by evaporation using a wet bulb thermometer exposed to natural air movement

tackyfier — a compound used to make an elastomer more sticky

target population — the group of individuals exposed to a particular agent

temporary threshold shift — a temporary drop in hearing acuity due to exposure to loud noise

tennis elbow (epicondylitis) — inflammation of the elbow due to excessive rotation of the radius bone in the forearm

tenosynovitis — inflammation of tendons due to abrasion resulting from excessive use

teratogen — a compound that causes birth defects when the developing fetus is exposed to it

thermoluminescent detector (TLD) — a personal dosage detector that stores energy on exposure to radiation, then emits light proportionately when discharged

thermoplastic — a linear polymer which can soften and melt on heating

thermoset plastic — a three dimensional polymer which does not soften on heating

threshold toxicity — the dose that is at the interface of the lowest producing an effect and the highest not producing an effect

third octave bands — a dividing of the range of audible sound into intervals, each one third of an octave, to discuss hearing problems

TLD (thermoluminescent detector) — a personal dosage detector that stores energy on exposure to radiation, then emits light proportionately when discharged

TLV — Threshold Limit Value — a recommendation by the ACGIH for the highest safe level of exposure to a chemical

TLV-C — Ceiling Threshold Limit Value — an exposure level that should not be exceeded even for a very short time

tracers — in research, the fate of particular atoms may be determined or followed by use of radioactive isotopes of those atoms

trachea — the upper passage of the respiratory between the nasal cavity and the splitting out of the bronchial tubes

transport velocity — the air velocity needed to keep particulate contamination suspended in a duct

trigger finger — tenosynovitis from repetitive finger action on the trigger switch of a power tool

TSCA — Toxic Substances Control Act: gives the EPA authority to act to prevent injury from chemicals

tumor — a mass of cells displaying an alteration of expected pattern and unusually rapid growth

TWA — Time Weighted Average — refers to the practice of sampling exposure levels at several times and averaging the exposure over the day used in monitoring compliance with TLV or PEL values

ulna (anatomy) — the rigid bone of the forearm forming one side of the elbow

upper flammability limit (combustion rich limit mixture) — the highest concentration of a gas or vapor able to sustain continuous combustion

upper respiratory tract irritant — a gas which is so irritating that we only allow it to be inhaled into the upper part of the lung

UV spectrophotometer — an analytical instrument that measures the absorbance of specific frequencies of ultraviolet light by a compound under study

vacuum forming — forming a plastic sheet by evacuating the space between the sheet and the mold

Van der Waals forces — the electron clouds around atoms vibrate temporarily giving the atoms a positively and negatively charged end. Two atoms very close together are weakly attracted to one another by these charges.

vapor — a normally liquid chemical in the gas phase due to its evaporation

vapor degreasing — dipping an object in solvent vapors to remove grease

velocity pressure — pressure due to air movement in a duct

visible spectrophotometer — an analytical instrument that measures the absorbance of specific frequencies of visible light by a compound under study

volatile — the property of a liquid with a low boiling point such that it evaporates readily

vulcanization — a process by which elastomers are crosslinked to produce harder, more abrasion resistant products

warning properties of a gas or vapor — the smell or irritation experienced on exposure to a gas or vapor

water manometer — a curved tube containing water used to measure the difference between static pressure in a duct and atmospheric pressure

wavelength (sound) — when sound is considered as a wave, wavelength is the distance from one peak height to the next

WBGT (wet bulb globe temperature index) — the most used method to evaluate the heat stress potential of the workplace

wet scrubbers — water sprays used to remove gases and vapors from the air

white finger disease (Raynard's disease) — damage to nerves and blood vessels in hands due to the vibration of hand tools

Williams-Steiger Occupational Safety and Health Act — the act of 1970 that created OSHA

X-rays — high energy electromagnetic radiation generated by bombarding a metal target with a high voltage electric arc

Appendix A.
Simple Statistical
Calculations

Standard Deviation

The most common means of expressing the confidence limits on a set of experimental values is called the standard deviation. This is expressed by following the average value for the data plus or minus a number. If you had a large sample of data with a random error in the determination, you would expect 68% of the values to fall in a range from a value of the average minus the standard deviation to a value of the average plus the standard deviation. To obtain the standard deviation, follow these steps:

1. Average all of the experimental values.
2. Find the difference between the average value and each experimental value.
3. Square each of these differences.
4. Total the squares of the differences.
5. Divide the total of the squares of the differences by 1 less than the total number of data items.
6. Take the square root of the number obtained in step 5. This is the standard deviation.

95% Confidence Limit

95% confidence limits are often expressed in toxicology. We are expressing now the likelihood that the averages obtained will fall inside a certain range. At the 95% limit, we are stating that 95% of a large number of averages taken of sets of data on a particular measurement will fall inside the stated range. Notice that we are not talking about the individual measurements, but of the averages of sets of data, a much more dependable set of numbers, falling within a stated range. 95% confidence limits are obtained as follows:

1. Take the standard deviation of the data as outlined above.
2. Divide the standard deviation by the square root of the number of items of data.
3. Multiply this value by the appropriate number (called the Student t value) from Table A.1. This is the desired confidence limit.

Table A.1   Student t Values.

Number of Items of Data	t Value
2	12.70
3	4.30
4	3.18
5	2.78
6	2.57
7	2.45
8	2.36
9	2.31
10	2.26
11	2.23
16	2.13
21	2.09
31	2.04
∞	1.96

Appendix B.
Material Safety Data Sheet

BENZENE
BENZENE
BENZENE

MATERIAL SAFETY DATA SHEET

FISHER SCIENTIFIC
CHEMICAL DIVISION
1 REAGENT LANE
FAIR LAWN NJ 07410
(201) 796-7100

EMERGENCY NUMBER: (201) 796-7100
CHEMTREC ASSISTANCE: (800) 424-9300

THIS INFORMATION IS BELIEVED TO BE ACCURATE AND REPRE-
SENTS THE BEST INFORMATION CURRENTLY AVAILABLE TO US.
HOWEVER, WE MAKE NO WARRANTY OF MERCHANTABILITY OR
ANY OTHER WARRANTY, EXPRESS OR IMPLIED, WITH RESPECT TO
SUCH INFORMATION, AND WE ASSUME NO LIABILITY RESULTING
FROM ITS USE. USERS SHOULD HAKE THEIR OWN INVESTIGATIONS
TO DETERMINE THE SUITABILITY OF THE INFORMATION FOR THEIR
PARTICULAR PURPOSES.

SUBSTANCE IDENTIFICATION

SUBSTANCE: **BENZENE**   CAS-NUMBER 71-43-2

TRADE NAMES/SYNONYMS:
   BENZOL; CYCLOHEXATRIENE; BENZOLE; PHENE; PYROBENZOL;
   PYROBENZOLE; CARBON OIL; COAL TAR NAPHTHA; PHENYL
   HYDRIDE; BENZOLENE; BICARBURET OF HYDROGEN; COAL
   NAPHTHA; MOTOR BENZOL; ANNULENE; (6)ANNULENE; NITRA-

TION BENZENE; MINERAL NAPHTHA; B426; 13065; B243; B245S; B245; 8411; B414; RCRA U019; UN 1114; STCC 4908110; C6H6;

CHEMICAL FAMILY:
Hydrocarbon, aromatic

MOLECULAR FORMULA: C6-H6

MOLECULAR WEIGHT: 78.11

CERCLA RATINGS (SCALE 0–3):   HEALTH=3 FIRE=3 REACTIVITY=0
                             PERSISTENCE=1
NFPA RATINGS (SCALE 0–4):   HEALTH-2 FIRE=3 REACTIVITY=O

COMPONENTS AND CONTAMINANTS

COMPONENT: BENZENE   PERCENT: >99
             CAS# 71-43-2

OTHER CONTAMINANTS: 0.15% NON-AROMATICS; 1 PPM THIOPHENE

EXPOSURE LIMITS:
BENZENE:
  1 ppm OSHA TWA; 5 ppm OSHA 15 minute STEL; 0.5 ppm OSHA action
  level 10 ppm (30 mg/m3) ACGIH TWA;
  ACGIH A2-Suspected Human Carcinogen
  (Notice of Intended Changes 1990–91)
  0.1 ppm (0.32 mg/m3) NIOSH recommended 8 hour TWA;
  1 ppm (3.2 mg/m3) NIOSH recommended 15 minute ceiling

  Measurement method: Charcoal tube; carbon disulfide; gas chromatography
  with flame ionization detection (NIOSH Vol. III #1500 Hydrocarbons)
  10 pounds CERCLA Section 103 Reportable Quantity
  Subject to SARA Section 313 Annual Toxic Chemical Release Reporting
  Subject to California Proposition 65 cancer and/or reproductive toxicity
    warning and release requirements- (February 27, 1987)

PHYSICAL DATA

DESCRIPTION: Colorless to light yellow liquid with an aromatic odor

BOILING POINT: 176 F (80 C)   MELTING POINT: 42 F (6 C)

SPECIFIC GRAVITY: 0.8765 @ 20 C   VISCOSITY: 0.6468 cP @ 20 C

VOLATILITY: 100%   VAPOR PRESSURE: 75 mmHg @ 20 C

EVAPORATION RATE: (butyl acetate=1) 5.1

SOLUBILITY IN WATER: 0.18% @ 25 C

ODOR THRESHOLD: 4.68 ppm   VAPOR DENSITY: 2.8

SOLVENT SOLUBILITY: Soluble in acetone, alcohol, carbon disulfide, acetic acid, carbon tetrachloride, chloroform, ether, oils

----
FIRE AND EXPLOSION DATA

FIRE AND EXPLOSION HAZARD:
Dangerous fire hazard when exposed to heat or flame.

Moderate explosion hazard when exposed to heat or flame.

Vapor-air mixtures are explosive above flash point.

Vapors are heavier than air and may travel a considerable distance to a source of ignition and flash back.

Due to low electroconductivity of the substance, flow or agitation may generate electrostatic charges resulting in sparks with possible ignition.

FLASH POINT: 12 F (-11 C) (CC)   UPPER EXPLOSIVE LIMIT 7.8%

LOWER EXPLOSIVE LIMIT: 1.2% AUTOIGNITION TEMP.: 928 F (498 C)

FLAMMABILITY CLASS(OSHA): IB

FIREFIGHTING MEDIA:
Dry chemical, carbon dioxide, water spray or regular foam
(1990 Emergency Response Guidebook, DOT P 5800.5)

For larger fires, use water spray, fog or regular foam
(1990 Emergency Response Guidebook, DOT P 5800.5)

FIREFIGHTING:
Move container from fire area if you can do it without risk. Apply cooling water to sides of containers that are exposed to flames until well after fire is out. Stay away from ends of tanks. For massive fire in cargo area, use unmanned

hose holder or monitor nozzles; if this is impossible, withdraw from area and let fire burn. Withdraw immediately in case of rising sound from venting safety device or any discoloration of tank due to fire. Isolate for 1/2 mile in all directions if tank, rail car or tank truck is involved in fire (1990 Emergency Response Guidebook, DOT P 5800.5, Guide Page 27).

Extinguish only if flow can be stopped. Use water in flooding quantities as a fog; solid streams may spread fire. Cool containers with flooding amounts of water; apply from as far a distance as possible. Avoid breathing hazardous materials; keep upwind. Evacuate to a radius of 1500 feet for uncontrollable fires. Consider evacuation of downwind area if material is leaking.

Water may be ineffective (NFPA 325M, Fire Hazard Properties of Flammable Liquids, Gases, and Volatile Solids, 1991)

Fire fighting phases: Dry chemical, alcohol foam or carbon dioxide. Water may be ineffective. Use water to keep fire-exposed containers cool. If a leak or spill has not ignited, use water spray to disperse the vapors and to provide protection for the men attempting to stop the leak. Water spray may be used to flush spills away from exposures (NFPA 49, Hazardous Chemicals Data, 1975).

TRANSPORTATION DATA

U.S. DEPARTMENT OF TRANSPORTATION SHIPPING NAME-ID NUMBER, 49 CFR 172.101:
Benzene-UN 1114

U.S. DEPARTMENT OF TRANSPORTATION HAZARD CLASS OR DIVISION, 49 CFR 172.101:
3 - Flammable liquid

U.S. DEPARTMENT OF TRANSPORTATION PACKING GROUP, 49 CFR 172.101:
PG II

U.S. DEPARTMENT OF TRANSPORTATION LABELING REQUIREMENTS, 49 CFR 172.101 AND SUBPART E:
Flammable liquid

U.S. DEPARTMENT OF TRANSPORTATION PACKAGING AUTHORIZATIONS:
EXCEPTIONS: 49 CFR 173.150
NON-BULK PACKAGING: 49 CFR 173.202

BULK PACKAGING: 49 CFR 173.242

U.S. DEPARTMENT OF TRANSPORTATION QUANTITY LIMITATIONS
49 CFR 172.101:
PASSENGER AIRCRAFT OR RAILCAR: 5 l
CARGO AIRCRAFT ONLY: 60 l

TOXICITY

BENZENE:
IRRITATION DATA: 20 mg/24 hours skin-rabbit moderate; 15 mg/24 hours
open skin-rabbit mild; 88 mg eye-rabbit moderate; 2 mg/24 hours eye-rabbit
severe.
TOXICITY DATA: 2000 ppm/5 minutes inhalation-human LCLo; 2 ppm/5
minutes inhalation-human LCLo; 65 mg/m3/5 years inhalation-human LCLo;
100 ppm inhalation-human TCLo; 150 ppm/1 year intermittent inhalation-
man TCLo; 20,000 ppm/5 minutes inhalation-mammal LCLo; 10,000 ppm/7
hours inhalation-rat LC50; 30 ppm/6 hours/13 weeks intermittent inhalation-
rat TCLo; 9980 ppm inhalation-mouse LC50; 300 ppm/6 hours/13 weeks
intermittent inhalation-mouse TCLo; 103 ppm/6 hours/5 days intermittent
inhalation-mouse TCLo; 10 ppm/6 hours/10 weeks intermittent inhalation-
mouse TCLo; 302 ppm/6 hours/26 weeks intermittent inhalation-mouse
TCLo; 4680 ppm/8 hours/4 days intermittent inhalation-mouse TCLo;
146,000 mg/m3 inhalation-dog LCLo; 170,000 mg/m3 inhalation-cat LCLo;
>8263 mg/kg skin-rabbit LD50; >8263 mg/kg skin-guinea pig LD50; 50
mg/kg oral-man LDLo; 930 mg/kg oral-rat LD50; 6600 mg/kg/27 weeks
intermittent oral-rat TDLo; 4700 mg/kg oral-mouse LD50; 2000 mg/kg oral-
dog LDLo; 88 mg/kg intravenous-rabbit LDLo; 2890 µg/kg intraperitoneal-rat
LD50; 340 mg/kg intraperitoneal-mouse LD50; 527 mg/kg intraperitoneal-
guinea pig LDLo; 1500 mg/kg intraperitoneal-mammal LDLo; 1400 mg/kg
subcutaneous-frog LDLo; 194 mg/kg unreported-man LDLo; mutagenic data
(RTECS); reproductive effects data (RTECS); tumorigenic data (RTECS).
CARCINOGEN STATUS: OSHA Carcinogen; Known Human Carcinogen
(NTP); Human Sufficient Evidence, Animal Sufficient Evidence (IARC
Group-1). Numerous case reports and series have suggested a relationship
between exposure to benzene and the occurrence of various types of leukemia. Several case-control studies have also shown increased odds ratios for
exposure to benzene, but mixed exposure patterns and poorly defined exposures render their interpretation difficult. Three independent cohort studies
have demonstrated an increased incidence of acute nonlymphocytic leukemia
in workers exposed to benzene.
LOCAL EFFECTS: Irritant-inhalation, skin and eye.

ACUTE TOXICITY LEVEL: Moderately toxic by inhalation and ingestion; slightly toxic by dermal absorption.

TARGET EFFECTS: Central nervous system depressant; bone marrow depressant. Poisoning may also affect the immune system and the heart.

AT INCREASED RISK FROM EXPOSURE: Persons with certain immunological tendencies, poor nutrition, anemia and drug or chemically induced agranulocytemia.

ADDITIONAL DATA: Use of alcoholic beverages may enhance the toxic effects. Use of stimulants such as epinephrine may cause cardiac arrhythmias. May cross the placenta. Interactions with medications have been reported.

HEALTH EFFECTS AND FIRST AID

INHALATION:

BENZENE:

IRRITANT/NARCOTIC/BONE MARROW DEPRESSANT/CARCINOGEN.

ACUTE EXPOSURE - Concentrations of 300 ppm may cause respiratory tract irritation; more severe exposures may result in pulmonary edema. Systemic effects are mainly on the central nervous system and depend on exposure time and concentrations. No effects were noted at 25 ppm for 8 hours; signs of intoxication began at 50–150 ppm within 5 hours; at 500–15000 ppm, within 1 hour; were severe at 7500 ppm, within 30–60 minutes; and 20,000 ppm was fatal within 5–10 minutes. Effects may include nausea, vomiting, headache, dizziness, drowsiness, weakness, sometimes preceded by a brief period of exhilaration or euphoria, irritability, malaise, confusion, ataxia, staggering, weak, rapid pulse, chest pain and tightness with breathlessness, pallor, cyanosis of the lips and fingertips, and tinnitus. In severe exposures there may be blurred vision, shallow, rapid breathing, delirium cardiac arrhythmias, unconsciousness, deep anesthesia, paralysis, and coma characterized by motor restlessness, tremors and hyperreflexia, sometimes preceded by convulsions. Recovery depends on the severity of exposure. Polyneuritis may occur and there may be persistent nausea, anorexia, muscular weakness, headache, drowsiness, insomnia, and agitation. Nervous irritability, breathlessness, and unsteady gait may persist for 2–3 weeks; a peculiar skin color and cardiac distress for 4 weeks. Liver and kidney effects may occur, but are usually mild, temporary impairments. Chromosomal damage has been found after exposure to toxic levels. Although generally hematotoxicity is not a significant concern in acute exposure, delayed hematological effects, including anemia and thrombocytopenia, have been reported, as have petechial hemorrhages, spontaneous internal bleeding and secondary infections. In fatal exposures, death may be due to asphyxia, central nervous system depression, cardiac or respiratory failure and circulatory collapse, or occasionally, sudden ventricular fibrillation. It may occur within a few minutes to several hours, or cardiac arrhythmia may occur at anytime with 24 hours. Also, death from central nervous system, respiratory or hemorrhagic complications may occur up to 5 days after exposure. Pathologic findings have included respiratory inflammation with edema and hemorrhage of the lungs, renal congestion, cerebral edema, and extensive petechial hemorrhages in the brain, pleurae, pericardium, urinary tract, mucous membranes, and skin.

CHRONIC EXPOSURE - Long-term exposure may cause symptoms referable to the central nervous, hematopoietic and immune systems. Early effects are vague and varied and may include headache, light-headedness, dizziness, nausea, anorexia, abdominal discomfort, and fatigue. Sore, dry throat, weakness, lethargy, malaise, drowsiness, nervousness, and irritability have also been reported. Later there may be dyspnea, pallor, slightly increased, temperature, decreased blood pressure, rapid pulse, palpitations, and visual disturbances. Dizziness when cold water is placed in the ear and associated with ataxia, tremors and emotional lability. Workers exposed to benzene in combination with other solvents have exhibited polyneuritis. Several case reports, one of them an acute exposure, suggest the possibility that systemic exposure may be associated with retrobulbar or optic neuritis. Occasionally hemorrhages in retina and conjunctiva occur and rarely neuroretinal edema and papilledema have accompanied the retinal hemorrhages. Hematological effects vary widely and may appear after a few weeks or many years of exposure or even many years after exposure has ceased. The degree of exposure below which no blood effects will occur cannot be established with certainty. In the early stages, there may be blood clotting defects due to morphological, functional and quantitative platelet alteration with resultant bleeding from the nose and gums, easy bruising and petechiae; leukopenia with predominant lymphocytopenia or neutropenia; and anemia which may be normochromic or macrocytic and hypochromic. Extramedullary hematopoiesis, splenomegaly, circulating immature marrow cells, and an initial increase in leukocytes, erythrocytes and platelets have also been reported. The bone marrow may be hyper-, hypo- or normoplastic and does not always correlate with the peripheral blood picture. Also, the symptoms do not always parallel the laboratory findings. If treated at this stage, the effects appear reversible, although recovery may be protracted and there may be relapses. Decreased erythrocyte survival, hemolysis, capillary fragility, internal hemorrhages, iron metabolism disturbances, and hyperbilirubinemia have also been reported. Exposure to high levels for longer periods may also result in aplasia and fatty degeneration of the bone marrow with pancytopenia. The most serious cases of aplastic anemia may be fatal due to hemorrhage and infection; death may occur within 3 months of diagnosis. Enormous variability in individual response, including non-dose dependent aplasia, and the finding of eosinophilia suggests that, in some cases, the blood dyscrasia may partially be an allergic reaction. Numerous case reports and series have suggested a relationship between exposure to benzene and the occurrence of various types of leukemia. Several case-control studies have also shown increased odds ratios for exposure to benzene, but mixed exposure patterns and poorly defined exposures render their interpretation difficult. Three independent cohort studies have demonstrated an increased incidence of acute nonlymphocytic leukemia in workers exposed to benzene. Several studies have also suggested a link between occupational exposures and multiple myeloma and lymphoma, both Hodgkin's and increased's. Although aplastic anemia is probably the more likely consequence of long-term exposure, it is not uncommon for an individual surviving this, to go through a preleukemic phase into frank leukemia. Conversely, leukemia without precedent aplastic anemia can occur. In one study the range of time from the start of the exposure to the diagnosis of leukemia was 3–24 years. It has been suggested that the chromosomal aberrations which can arise in peripheral blood and bone marrow cells and persist for a long time after exposure ceases, may be associated with the increased incidence of leukemia. The immunosuppressive effects has also ben suggested as being associated with the leukemogenesis. Adverse effects on the immunological system have been shown to make rabbits more susceptible to tuberculosis and pneumonia and may explain why the terminal event in some cases of benzene intoxication may be overwhelming infection. Exposed mice exhibited a tendency toward induction of lymphoid neoplasms. Rats exhibited an increased incidence of neoplasms, mainly carcinomas, at various sites. Menstrual disturbances have been reported more frequently in exposed women. Testicular damage has been reported in rats, rabbits and guinea pigs. Some animal studies have demonstrated embryo/fetotoxicity, sometimes at levels as low as 10 ppm and the potential for teratogenic effects such as decreased body weight and skeletal variants, have also been shown. Other studies have not produced any abnormalities or embryolethality.

FIRST AID - Remove from exposure area to fresh air immediately. If breathing has stopped, give artificial respiration. Maintain airway and blood pressure and administer oxygen if available. Keep affected person warm and at rest. Treat symptomatically and supportively. Administration of oxygen should be performed by qualified personnel. Get medical attention immediately.

SKIN CONTACT:
BENZENE:
IRRITANT.
ACUTE EXPOSURE - Direct contact may cause irritation. Effects may include erythema, a burning sensation, and with prolonged contact, blister-

ing and edema. Under normal conditions, significant signs of systemic toxicity are unlikely from skin contact alone due to the slow rate of absorption; it may however, contribute to the toxicity from inhalation. Application to guinea pigs resulted in increased dermal permeability.

CHRONIC EXPOSURE - Repeated or prolonged contact defats the skin and may result in dermatitis with erythema, scaling, dryness, vesiculation, and fissuring, possibly accompanied by paresthesias of the fingers which may persist several weeks after the dermatitis subsides. Peripheral neuritis has also been reported. Secondary infections may occur. Tests on guinea pigs indicate sensitization is possible. Although animal studies have failed to establish a relationship between skin contact and a carcinogenic effect, most of the studies were inadequate; some papillomas and hematopoietic effects have been reported.

FIRST AID - Remove contaminated clothing and shoes immediately. Wash affected area with soap or mild detergent and large amounts of water until no evidence of chemical remains (approximately 15–20 minutes). Get medical attention immediately.

EYE CONTACT:
BENZENE:
IRRITANT.

ACUTE EXPOSURE - May cause irritation. Vapor concentrations of 3000 ppm are very irritating, even on brief exposure. Droplets cause a moderate burning sensation, but only a slight, transient corneal epithelial injury with rapid recovery.

CHRONIC EXPOSURE - Repeated or prolonged exposure may cause conjunctivitis. 50% of rats exposed to 50 ppm for more than 600 hours developed cataracts.

FIRST AID - Wash eyes immediately with large amounts of water or normal saline, occasionally lifting upper and lower lids, until no evidence of chemical remains (approximately 15–20 minutes). Get medical attention immediately.

INGESTION:
BENZENE:
NARCOTIC/CARCINOGEN.

ACUTE EXPOSURE - May cause local irritation and burning sensation in the mouth, throat and stomach, and hemorrhagic inflammatory lesions of the mucous membranes in contact with the liquid. Signs of symptoms of systemic intoxication may include nausea, vomiting, headache, dizziness, weakness, staggering, chest pain and tightness, shallow, rapid pulse and respiration, breathlessness, pallor followed by flushing, and a fear of impending death. There may be visual disturbances, tremors, convulsions, ventricular irregularities, and paralysis. Excitement, euphoria or delirium may precede weariness, fatigue, sleepiness and followed by stupor and unconsciousness, coma and death from respiratory failure. Those who survive the central nervous system effects may develop bronchitis, pneumonia, pulmonary edema, and intrapulmonary hemorrhage. Aspiration may cause immediate pulmonary edema and hemorrhage. The usual lethal dose in humans is 10–15 milliliters, but smaller amounts have been reported to cause death. A single exposure may produce long-term effects with pancytopenia persisting up to a year.

CHRONIC EXPOSURE - Daily administration to humans of 2–5 grams in olive oil caused headache, vertigo, bladder irritability, impotence, gastric disturbances, and evidence of renal congestion. In female rats treated with 132 single daily doses over 187 days, no effects were observed at 1 mg/kg; slight leukopenia at 10 mg/kg; and both leukopenia and anemia at 50 and 100 mg/kg. Oral administration to rats and mice at various dose levels induced neoplasms at multiple sites in males and females. In a one year gavage study, rats given 50 or 250 mg/kg, 4–5 days/week for 52 weeks did not exhibit acute or subacute toxic effects, but a dose correlated increase of leukemias and mammary carcinomas was observed; some other tumor types were also reported. Reproductive effects have been reported in animals.

FIRST AID - Extreme care must be used to prevent aspiration. Gastric lavage with a cuffed endotracheal tube in place to prevent further aspiration should be done within 15 minutes. In the absence of depression or convulsions or impaired gag reflex, emesis can also be induced using syrup of ipecac without increasing the hazard of aspiration (Dreisbach, Handbook of Poisoning, 12th Ed.). Treat symptomatically and supportively. Gastric lavage should be performed by qualified medical personnel. Get medical attention immediately.

ANTIDOTE:
No specific antidote. Treat symptomatically and supportively.

REACTIVITY

REACTIVITY:
Stable under normal temperatures and pressures.

INCOMPATIBILITIES:

BENZENE:
  ACIDS (STRONG): Incompatible.
  ALLYL CHLORIDE WITH DICHLOROETHYL ALUMINUM OR ETHYL-
    ALUMINUM SESQUICHLORIDE: Possible explosion.
  ARSENIC PENTAFLUORIDE + POTASSIUM METHOXIDE: Explosive
    interaction.
  BASES (STRONG): Incompatible.
  BROMINE + IRON: Incompatible.
  BROMINE PENTAFLUORIDE: Fire and explosion hazard.
  BROMINE TRIFLUORIDE: Possible explosion or ignition.
  CHLORINE: Explosion in the presence of light.
  CHLORINE TRIFLUORIDE: Violent reaction with possible explosion.
  CHROMIC ANHYDRIDE (POWERED): Ignition.
  DIBORANE: Spontaneously explosive reaction in air.
  DIOXYGEN DIFLUORIDE: Ignition, even at reduced temperatures.
  DIOXYGENYL TETRAFLUOROEBORATE: Ignition reaction.
  INTERHALOGEN COMPOUNDS: Ignition or explosion.
  IODINE HEPTAFLUORIDE: Ignition on contact.
  IODINE PENTAFLUORIDE: Violent interaction above 50 C.
  NITRIC ACID: Violent or explosive unless properly agitated and cooled.
  NITRYL PERCHLORATE: Explosive interactions.
  OXIDIZERS (STRONG): Fire and explosion hazard.
  OXYGEN (LIQUID): Explosive mixture.
  OZONE: Formation of explosive gelatinous ozonide.
  PERCHLORATES (METAL): Formation of explosive complex.
  PERCHLORYL FLUORIDE + ALUMINUM CHLORIDE: Formation of
    shock sensitive compound.
  PERMANGANATES + SULFURIC ACID: Possible explosion.
  PERMANGANIC ACID: Explosive hazard.
  PEROXODISULFURIC ACID: Explosion hazard.
  PEROXOMONOSULFURIC ACID: Explosive interaction.
  POTASSIUM PEROXIDE: Ignition.
  SILVER PERCHLORATE: Formation of explosive complex.
  SODIUM PEROXIDE + WATER: Ignition.
  URANIUM HEXAFLUORIDE: Violent reaction.

DECOMPOSITION:
Thermal decomposition products may include toxic oxides of carbon.

POLYMERIZATION:
Hazardous polymerization has not been reported to occur under normal temperatures and pressures.

STORAGE AND DISPOSAL

Observe all federal, state and local regulations when staring or disposing of this substance.

Storage

Store in accordance with 29 CFR 1910.106.

Bonding and grounding: Substances with low electroconductivity, which may be ignited by electrostatic sparks, should be stored in containers which meet the bonding and grounding guidelines specified in NFPA 77-1983, Recommended Practice on Static Electricity.

Protect against physical damage. Outside or detached storage is preferable. Inside storage should be in a standard flammable liquids storage room or cabinet. Separate from oxidizing materials (NFPA 49, hazardous chemicals data, 1975).

Store away from incompatible substances.

Disposal

Disposal must be in accordance with standards applicable to generators of hazardous waste, 40 CFR 262. EPA Hazardous Waste Number U019.

Benzene - Regulatory level: 0.5 mg/L (TCLP-40 CFR 261 Appendix II) materials which contain the above substance at or above the TCLP regulatory level meet the EPA toxicity characteristic, and must be disposed of in accordance with 40 CFR part 262. EPA Hazardous Waste Number 0018.

****

CONDITIONS TO AVOID

Avoid contact with heat, sparks, flames, or other sources of ignition. Vapors may be explosive. Avoid overheating of containers; containers may violently rupture in heat of fire. Avoid contamination of water sources.

****

SPILL AND LEAK PROCEDURES

SOIL SPILL:
Dig holding area such as lagoon, pond or pit for containment.

Dike flow of spilled material using soil or sandbags or foamed barriers such as polyurethane or concrete.

Use cement powder, fly ash, sawdust or commercial sorbent to absorb bulk liquid.

Reduce vapor and fire hazard with fluorocarbon water foam.

AIR SPILL:
Knock down vapors with water spray. Keep upwind.

WATER SPILL:
Limit spill motion and dispersion with natural barriers or ail spill control booms.

Apply detergents soaps alcohols or another surface active agent to thicken spilled material.

Apply universal gelling agent to immobilize trapped spill and increase efficiency of removal.

If dissolved, apply activated carbon at ten times the spilled amount in the region of 10 ppm or greater concentration.

Use suction hoses to remove trapped spill material.

Use dredges or lifts to extract immobilized masses of pollution and precipitates.

OCCUPATIONAL SPILL:
Shut off ignition sources. Stop leak if you can do it without risk. Use water spray to reduce vapors. For small spills, take up with sand or other absorbent material and place into containers for later disposal. For larger spills, dike far ahead of spill for later disposal. No smoking, flames or flares in hazard area. Keep unnecessary people away; isolate hazard area and restrict entry.

Reportable Quantity (RQ): 1000 pounds
The Superfund Amendments and Reauthorization Act (SARA) Section 304 requires that a release equal to or greater than the reportable quantity for this substance be immediately reported to the local emergency planning committee and the state emergency response commission (40 CFR 355.40). If the release of this substance is reportable under CERCLA Section 103, the National Response Center must be notified immediately at (800) 424-8802 or (202) 426-2675 in the metropolitan Washington, D.C. area (40 CFR 302.6).

PROTECTIVE EQUIPMENT

VENTILATION:
Provide local exhaust or process enclosure ventilation to meet the published exposure limits. Ventilation equipment should be explosion-proof if explosive concentrations of dust, vapor or fume are present.

Benzene:
Ventilation should meet the requirements in 29 CFR 1910.1028(f).

RESPIRATOR:
The following respirators are the minimum legal requirements as set forth by the Occupational Safety and Health Administration found in 29 CFR 1910, Subpart Z.

BENZENE:

Concentration:	Required respirator:
Less than or equal to 50 ppm-	Half-mask air-purifying respiratory with organic vapor cartridge.
Less than or equal to 50 ppm-	Full facepiece respiratory with organic vapor cartridges. Full facepiece gas mask with chin style canister.
Less than or equal to 100 ppm-	Full facepiece powered air-purifying respiratory with organic vapor canister.
Less than or equal to 1000 ppm-	Supplied air respirator with full facepiece in positive-pressure mode.
Greater than 1000 ppm or unknown concentration-	Self-contained breathing apparatus with full facepiece in positive-pressure mode. Full facepiece positive-pressure supplied-air respiratory with auxiliary self-contained air supply.
Escape-	Any organic vapor gas mask. Any self-contained breathing apparatus with full facepiece.
Firefighting-	Full facepiece self-contained breathing apparatus in positive-pressure mode.

The following respirators and maximum use concentrations are recommendations by the U.S. Department of Health and Human Services, NIOSH pocket guide to chemical hazards or NIOSH criteria documents.
The specific respirator selected must be based on contamination levels found in the work place and be jointly approved by the National Institute of Occupational Safety and Health and the Mine Safety and Health Administration.

At any detectable concentration:

 Self-contained breathing apparatus with full facepiece operated in pressure-demand or other positive pressure mode.
 Supplied-air respirator with full facepiece operated in pressure-demand or other positive pressure mode in combination with an auxiliary self-con-

tained breathing apparatus operated in pressure-demand or other positive pressure mode.

Escape-Air-purifying full facepiece respirator (gas mask) with a chin-style or front- or back-mounted organic vapor canister.
Escape-type self-contained breathing apparatus.

FOR FIREFIGHTING AND OTHER IMMEDIATELY DANGEROUS TO LIFE OR HEALTH CONDITIONS:

Any self-contained breathing apparatus that has a full facepiece and is operated in a pressure-demand or other positive-pressure mode.

Any supplied-air respirator that has a full facepiece and is operated in a pressure-demand or other positive-pressure mode in combination with an auxiliary self-contained breathing apparatus operated in pressure-demand or other positive-pressure mode.

CLOTHING:
Employee must wear appropriate protective (impervious) clothing and equipment to prevent repeated or prolonged skin contact with this substance.

BENZENE:
Protective clothing should meet the requirements for personal protective equipment in 29 CFR 1910.1028(h).

GLOVES:
Employee must wear appropriate protective gloves to prevent contact with this substance.

BENZENE: Protective gloves should meet the requirements for personal protective equipment in 29 CFR 1910.1028(h).

EYE PROTECTION:
Employee must wear splash-proof or dust-resistant safety goggles to prevent eye contact with this substance.

Emergency eye wash: Where there is any possibility that an employee's eyes may be exposed to this substance the employer should provide an eye wash fountain within the immediate work area for emergency use.

BENZENE:
Protective eye equipment should meet the requirements for protective clothing and equipment in 29 CFR 1910.1028(h).

AUTHORIZED - FISHER SCIENTIFIC, INC.
CREATION DATE: 10/11/84   REVISION DATE: 01/15/94

-ADDITIONAL INFORMATION-

THIS INFORMATION IS BELIEVED TO BE ACCURATE AND REPRE-
SENTS THE BEST INFORMATION CURRENTLY AVAILABLE TO US.
HOWEVER, WE MAKE NO WARRANTY OF MERCHANTABILITY OR
ANY OTHER WARRANTY, EXPRESS OR IMPLIED, WITH RESPECT TO
SUCH INFORMATION AND WE ASSUME NO LIABILITY RESULTING
FROM ITS USE. USERS SHOULD MAKE THEIR OWN INVESTIGATIONS
TO DETERMINE THE SUITABILITY OF THE INFORMATION FOR THEIR
PARTICULAR PURPOSES.

Note: This MSDS has been provided by Fisher Scientific, whose MSDSs are
created, maintained, and licensed by Occupational Health Services, Inc.;
New York, NY.

Appendix C. Selection of Detection Tubes

Gas or Vapor to be measured	Chemical Formula	TLV-TWA (ACGIH)[a]	Tube No.	Detector Tube to be used	Measurable Concentration Range	Pump Strokes taken[b]	Shelf Life (year)	SEI Tube
Acetaldehyde	CH_3CHO	100ppm	*92	Acetaldehyde	4–750ppm	1–4	1.5	
			*91L	Formaldehyde-(LR)	1–12ppm	2,4	1	
Acetic Acid	CH_3COOH	10ppm	81	Acetic Acid	1–80ppm	½–2	3	
Acetic Anhydride	$(CH_3CO)_2O$	C, 5ppm	81	Acetic Acid	0.5–40ppm	½–2	3	
Acetone	$(CH_3)_2CO$	750ppm	151	Acetone	0.01–2.0%	1–2	3	
Acetone Cyanohydrin	$(CH_3)_2C(OH)CN$	—	12L	Hydrogen Cyanide-(LR)	5–70ppm	1	2	
Acetonitrile (Methyl Cyanide)	CH_3CN	40ppm	11L (PYR)	Nitrogen Oxides-(LR)	2–145ppm	2	2	
Acetylene	C_2H_2	—	171	Acetylene	0.1–4.0%	½–2	3	
			103	Hydrocarbons, Low Class	0.075–36%	½–2	2	
Acrolein	$CH_2{:}CHCHO$	0.1ppm	*93	Acrolein	3–800ppm	2–4	2	
Acrylonitrile	CH_2CHCN	2ppm	+191	Acrylonitrile	2–360ppm	1–4	2	
			+191L	Acrylonitrile-(LR)	0.125–15ppm	2–7	2	
Allyl chloride	CH_2CHCH_2Cl	1ppm	+131La	Vinyl Chloride	25–270ppm	1	2	
Amines	Varied	—	180	Amines	0.25–60	1–2	3	
Ammonia	NH_3	25ppm	3H	Ammonia-(HR)	0.2–32%	½–5	3	
			**3M	Ammonia-(MR)	10–1000ppm	½–5	3	
			3La	Ammonia-(LR)	1–200ppm	½–2	3	√
			3L	Ammonia-(LR)	1–60ppm	½–2	3	
n-Amyl Acetate	$CH_3COOC_2H_4CH(CH_3)_2$	100ppm	142	Butyl Acetate	0.01–0.9%	2	3	
Aniline	$C_6H_5NH_2$	2ppm	*181	Aniline	1.25–60ppm	2–5	1	
Arsine	AsH_3	0.05ppm	19L	Arsine-(LR)	0.05–5ppm	10–20	3	

Gas or Vapor to be measured	Chemical Formula	TLV-TWA (ACGIH)[a]	Tube No.	Detector Tube to be used	Measurable Concentration Range	Pump Strokes taken[b]	Shelf Life (year)	SEI Tube
Benzene	C_6H_6	10ppm	+121S	Benzene	10–400ppm	1–4	3	
			**121	Benzene	2.5–120ppm	1–4	3	√
			+121L	Benzene-(LR)	0.125–60ppm	1–10	3	
Benzyl Bromide	$C_6H_5CH_2Br$	—	+136	Methyl Bromide	25–850ppm	1	2	
Benzyl Chloride	$C_6H_5CH_2Cl$	1ppm	*132L	Trichloroethylene-(LR)	0.5–25ppm	1	2	
Boron Trichloride	BCl_3	—	12L	Hydrogen Cyanide-(LR)	1–14ppm	2	2	
Bromine	Br_2	0.1ppm	9L	Nitrogen Dioxide-(LR)	2–23ppm	2	2	
Bromoform (Tribromomethane)	$CHBr_3$	0.5ppm	+136	Methyl Bromide	0.5–50ppm	1–2	2	
Butadiene	$CH_2CHCHCH_2$	10ppm	174	Butadiene	50–800ppm	1	3	
			174L	Butadiene-(LR)	2.5–100ppm	4–8	2	
Butane	C_4H_{10}	800ppm	104	n-Butane	25–1400ppm	1	3	
			103	Hydrocarbons, Low Class	0.035–1.68%	½–2	2	
n-Butyl Acetate	$CH_3COOC_4H_9$	150ppm	142	Butyl Acetate	0.01–0.8%	2	3	
			142L	Butyl Acetate-(LR)	10–300ppm	2	2	
n-Butyl Alcohol (n-Butanol)	C_4H_9OH	C, 50ppm	112	Ethyl Alcohol (Ethanol)	100–1500ppm	4	3	
			114	n-Butyl Alcohol	5–150ppm	3,6	2	
sec-Butyl Alcohol	$C_2H_4(OH)C_2H_5$	100ppm	112	Ethyl Alcohol (Ethanol)	100–3500ppm	4	3	
			115	sec-Butyl Alcohol	5–150ppm	4,8	2	
tert-Butyl Alcohol	$(CH_3)_3COH$	100ppm	102L	n-Hexane-(LR)	1000–8900ppm	2	3	
			104	n-Butane	1000–10,000ppm	2	3	
n-Butyl Amine	$C_4H_9NH_2$	C, 5ppm	180	Amines	1–60ppm	½–2	3	

Gas or Vapor to be measured	Chemical Formula	TLV-TWA (ACGIH)[a]	Tube No.	Detector Tube to be used	Measurable Concentration Range	Pump Strokes taken[b]	Shelf Life (year)	SEI Tube
sec-Butyl Amine	$CH_3CHNH_2C_2H_5$	C, 5ppm	180	Amines	1–60ppm	½–2	3	
tert-Butyl Mercaptan	$(CH_3)_3\ CSH$	0.5ppm	75	tert-Butyl Mercaptan	2.5–30mg/m³	2	2	
			75L	tert-Butyl Mercaptan	0.5–30mg/m³	½–2	2	
Carbon Dioxide	CO_2	5000ppm	2HH	Carbon Dioxide-(HR)	2.5–40%	½–1	3	
			2H	Carbon Dioxide-(HR)	0.5–20%	½–2	3	
			**2L	Carbon Dioxide-(LR)	0.13–6.0%	½–2	3	✓
			2LL	Carbon Dioxide-(XLR)	300–5000ppm	1	3	
Carbon Disulfide	CS_2	10ppm	+13M	Carbon Disulfide-(MR)	20–4000ppm	½–2	3	
			+13	Carbon Disulfide	0.5–125ppm	½–4	3	
Carbon Monoxide	CO	50ppm	1HH	Carbon Monoxide-(XHR)	1–40%	½–1	2	
			1H	Carbon Monoxide-(HR)	0.1–10%	½–2	2	
			1M	Carbon Monoxide-(MR)	0.05–4.0%	½–2	2	
			1L	Carbon Monoxide-(LR)	5–2000ppm	½–5	3	
			1La	Carbon Monoxide-(LR)	8–1000ppm	½–3	2	✓
			1LL	Carbon Monoxide-(XLR)	5–50ppm	2	2	
Carbon Tetrachloride	CCL_4	5ppm	*+134	Carbon Tetrachloride	1–60ppm	1–5	1	
Carbonyl Sulfide	COS	—	*+21	Carbonyl Sulfide	5–200ppm	½–2	2	
Chlorine	Cl_2	1ppm	8HH	Chlorine-(XHR)	0.25–10%	½–2	2	
			8H	Chlorine	25–1000ppm	½–2	2	
			8La	Chlorine-(LR)	0.05–16ppm	½–10	2	✓
Chlorine Dioxide	ClO_2	0.1ppm	8La	Chlorine-(LR)	0.33–16ppm	½–3	2	

Gas or Vapor to be measured	Chemical Formula	TLV-TWA (ACGIH)ᵃ	Tube No.	Detector Tube to be used	Measurable Concentration Range	Pump Strokes takenᵇ	Shelf Life (year)	SEI Tube
Chlorobenzene (Monochlorobenzene)	C_6H_5Cl	75ppm	121	Benzene	5–350ppm	1–2	3	
Chlorobromomethane	CH_2ClBr	200ppm	+136	Methyl Bromide	8–80ppm	1	2	
			+135	Methyl Chloroform	20–350ppm	1	1	
Chlorodifluoromethane (Freon 22) (TN)	$CHClF_2$	1000ppm	50	Freons (PYR)	175–1400	3	2	
Chloroform	$CHCl_3$	10ppm	+137	Chloroform	4–400ppm	3–7	2	
Chloropicrin (Trichloronitromethane)	CCl_3NO_2	0.1ppm	*+134	Carbon Tetrachloride	1–60ppm	1–5	1	
m-Cresol	$C_6H_4OHCH_3$	5ppm	*61	Cresol	1–28ppm	2	2	
o-Cresol	$C_6H_4OHCH_3$	5ppm	*61	Cresol	0.4–62.5ppm	1–4	2	
p-Cresol	$C_6H_4OHCH_3$	5ppm	*61	Cresol	1–30ppm	2	2	
Cumene	$C_6H_5CH(CH_3)_2$	50ppm	122	Toluene	5–2400ppm	½–2	3	
Cyclohexane	C_6H_{12}	300ppm	102H	Hexane-(HR)	0.015–1.2%	½–2	3	
Cyclohexanol	$C_6H_{11}O_6$	50ppm	118	Cyclohexanol	5–100ppm	5, 10	2	
Cyclohexanone	$C_6H_{10}O$	25ppm	*154	Cyclohexanone	2–60ppm	2–4	2	
Cyclohexene	C_6H_{10}	300ppm	151	Acetone	100–1000ppm	1	3	
Cyclohexylamine	$C_6H_{13}N$	10ppm	180	Amines	1–60ppm	½–2	3	
Cyclohexyl Chloride	$C_6H_{11}Cl$	—	102L	n-Hexane-(LR)	40–1200ppm	2	3	
n-Decane	$CH_3(CH_2)_8CH_3$	—	105	Hydrocarbons, High Class	200–6000ppm	1,2	2	
Diacetone Alcohol	$CH_3COCH_2C(CH_3)_2OH$	50ppm	154	Cyclohexanone	5–150ppm	1,2	2	
1,1-Dibromoethane (Ethylidine Dibromide)	$CHBr_2CH_2$	—	+136	Methyl Bromide	7–70ppm	1	2	

Gas or Vapor to be measured	Chemical Formula	TLV-TWA (ACGIH)[a]	Tube No.	Detector Tube to be used	Measurable Concentration Range	Pump Strokes taken[b]	Shelf Life (year)	SEI Tube
1,2-Dibromoethane (Ethylene Dibromide)	CH_2BrCH_2Br	—	+136	Methyl Bromide	6–80ppm	1	2	
o-Dichlorobenzene	$C_6H_4Cl_2$	C, 50ppm	121	Benzene	50–70ppm	1	3	
			+121L	Benzene-(LR)	3–70ppm	3	2	
			127	Dichlorobenzene	5–300ppm	2	3	
Dichlorodifluoromethane (Freon 12) (TN)	CCl_2F_2	1000ppm	50	Freons (PYR)	50–400ppm	3	2	
1,1-Dichloroethane	Cl_2CHCH_3	200ppm	+135	Methyl Chloroform	30–1300ppm	1–4	1	
1,2-Dichloroethane (Ethylene Dichloride)	$CH_2ClCHCl$	10ppm	+135	Methyl Chloroform	200–2600ppm	3	1	
1,2-Dichloroethylene (Acetylene Dichloride)	$CHClCHCl$	200ppm	*132H	Trichloroethylene-(HR)	10–450ppm	1	2	
			139	Dichloroethylene	10–450ppm	1	2	
1,2-Dichloropropane (Propylene Dichloride)	$CH_3CHClCH_2Cl$	75ppm	+131La	Vinyl Chloride-(LR)	40–600ppm	2	2	
			+135	Methyl Chloroform	30–1500ppm	4	1	
Dichloroetetrafluoroethane (Freon 114) (TN)	$CClF_2CClF_2$	1000ppm	50	Freons (PYR)	125–1000ppm	3	2	
Diethanolamine	$(HOCH_2CH_2)_2NH$	3ppm	180	Amines	1–60ppm	½–2	3	
Diethyl Amine	$(C_2H_5)_2NH$	10ppm	180	Amines	1–60ppm	½–2	3	
Diethyl Benzene	$C_6H_4(CH_3)_2$	—	122	Toluene	25–2400ppm	2	3	
Diethylenetriamine	$C_4H_{13}N_3$	1ppm	180	Amines	1–60ppm	½–2	3	
Diisobutyl Ketone	$(CH_3)_2CHCH_2COCH_2CH(CH_3)_2$	25ppm	*91L	Formaldehyde-(LR)	1–36ppm	4	1	
Diisopropylamine	$C_6H_{15}N$	5ppm	180	Amines	1–60ppm	½–2	3	
Dimethyl Acetamide	$CH_3CON(CH_3)_2$	10ppm	184	Dimethyl Acetamide	1.5–240ppm	1–4	2	

Gas or Vapor to be measured	Chemical Formula	TLV-TWA (ACGIH)[a]	Tube No.	Detector Tube to be used	Measurable Concentration Range	Pump Strokes taken[b]	Shelf Life (year)	SEI Tube
Dimethyl Amine	$(CH_3)_2NH$	10ppm	180	Amines	1–60ppm	½–2	3	
Dimethyl Aniline (N-Dimethyl Aniline)	$C_6H_5(CH_3)_2$	5ppm	*181	Aniline	2.5–45ppm	3	1	
Dimethyl Ether (Methyl Ether)	$(CH_3)_2O$	—	161	Ethyl Ether (Ether)	0.03–0.9%	1	3	
Dimethylformamide	$HCON(CH_3)_2$	10ppm	183	Dimethyl Formamide	1.0–90ppm	½–2	2	
Dioxane	(C_2H_4O)	25ppm	163	Ethylene Oxide	0.6–5.6%	1	2	
1,3-Dioxolan	$C_6H_{12}O_3$	—	113L	Isopropyl Alcohol-(LR)	20–320ppm	1	2	
Divinyl Benzene	$C_6H_4(CHCH_2)_2$	10ppm	124L	Styrene-(LR)	1–20ppm	3	3	
Epichlorohydrin	C_3H_5ClO	2ppm	163L	Ethylene Oxide-(LR)	0.5–130ppm	2	1	
Ethanolamine	C_2H_7NO	3ppm	180	Amines	1–60ppm	½–2	3	
Ethyl Acetate	$CH_3COOC_2H_5$	400ppm	141	Ethyl Acetate	0.04–1.5%	1	3	
			141L	Ethyl Acetate-(LR)	25–800ppm	2	2	
Ethyl Acrylate	$CH_8{:}CHCOOCH_2CH_3$	5ppm	151	Acetone	0.05–0.65%	2	3	
Ethyl Alcohol (Ethanol)	C_2H_5OH	1000ppm	112	Ethyl Alcohol	0.02–7.5%	½–2	3	
			112L	Ethyl Alcohol-(LR)	25–2000ppm	1,2	2	
Ethyl Benzene	$C_6H_5C_2H_5$	100ppm	122	Toluene	7–700ppm	½–2	3	
Ethylbenzyl Chloride	$ClCH_2C_6H_4C_2H_5$	—	+131La		6–42.5ppm	2	2	
Ethyl Bromide	C_2H_5Br	200ppm	+136	Methyl Bromide	10–90ppm	1	2	
Ethyl Chloride	C_2H_5Cl	1000ppm	138	Methylene Chloride	250–10,000ppm	3–5	1	
Ethyl Chloroformate	$ClCOOC_2H_5$	—	+131La	Vinyl Chloride-(LR)	15–140ppm	2	2	
Ethylene (Ethene)	C_2H_4	—	172	Ethylene	50–800ppm	1	3	
			172L	Ethylene-(LR)	0.2–50ppm	4	2	

Gas or Vapor to be measured	Chemical Formula	TLV-TWA (ACGIH)[a]	Tube No.	Detector Tube to be used	Measurable Concentration Range	Pump Strokes taken[b]	Shelf Life (year)	SEI Tube
Ethylenediamine	C₂H₄(NH₂)₂	10ppm	180	Amines	2–120ppm	½–2	3	
Ethylene Glycol	C₂H₆O₂	C, 50ppm	165	*+Ethylene Glycol	50–300mg/m³	3	2	
Ethylene Glycol Monobutyl Ether (Butyl Cellosolve)	C₄H₉OCH₂CH₂OH	25ppm	113L	Isopropyl Alcohol-(LR)	60–640ppm	1	2	
Ethylene Glycol Monoethyl Ether (Cellosolve)	C₂H₅OCH₂CH₂OH	5ppm	113L	Isopropyl Alcohol-(LR)	100–1200ppm	1	2	
Ethylene Glycol Monomethyl Ether (Methyl Cellosolve)	CH₃OCH₂CH₂OH	5ppm	113L	Isopropyl Alcohol-(LR)	30–900ppm	1	2	
Ethylene Glycol Monomethyl Ether Acetate (Methyl Cellosolve Acetate)	C₂H₅OCH₂CH₂O(O)CCH₃	5ppm	113L	Isopropyl Alcohol-(LR)	50–700ppm	1	2	
Ethylene Imine	C₂H₅N	0.5ppm	180	Amines	1–60ppm	½–2	3	
Ethylene Oxide	(CH₂)₂O	1ppm	163	Ethylene Oxide	0.05–3.0%	1	3	
			*+163L	Ethylene Oxide-(LR)	0.4–100ppm	2–4	1	
Ethyl Ether (Ether)	(C₂H₅)₂	400ppm	161	Ethyl Ether (Ether)	0.04–1.0%	1	3	
Ethyl Mercaptan	C₂H₅SH	0.5ppm	72	Ethyl Mercaptan	0.5–120ppm	1–10	3	
Formaldehyde	HCHO	1ppm	+91	Formaldehyde	2–20ppm	2	3	
			*91L	Formaldehyde-(LR)	0.1–5ppm	5	1	
Formic Acid	HCOOH	5ppm	81	Acetic Acid	1–80ppm	½–2	3	
Freons (TN)	Varied	—	50	Freon-(PYR)	See data on page 3			
Furfural	C₅H₄O₂	2ppm	*154	Cyclohexanone	2–60ppm	2–4	2	
Gasoline	Mixture	300ppm	101	Gasoline	0.015–1.2%	½–2	3	
			101L	Gasoline-(LR)	30–1000ppm	2	3	

Gas or Vapor to be measured	Chemical Formula	TLV-TWA (ACGIH)ᵃ	Tube No.	Detector Tube to be used	Measurable Concentration Range	Pump Strokes takenᵇ	Shelf Life (year)	SEI Tube
Heptamethylene Diamine	$NH_2(CH_2)_7NH_2$	—	180	Amines	0.5–30ppm	½–2	3	
n-Heptane	C_7H_{16}	40ppm	101	Gasoline	0.015–1.2%	½–2	3	
			105	Hydrocarbons, High Class	90–2700ppm	1,2	2	
Hexamethylene Diamine	$NH_2(CH_2)_6NH_2$	—	180	Amines	0.5–30ppm	½–2	3	
n-Hexane	C_6H_{14}	50ppm	102H	Hexane-(HR)	0.015–1.2%	½–2	3	
			102L	n-Hexane-(LR)	50–1200ppm	1	3	
			105	Hydrocarbons, High Class	80–2400ppm	1,2	2	
			106	Hydrocarbons, Petroleum Distillates	300–4000ppm	1	3	
Hexylamine	$CH_3(CH_2)_5NH_2$	—	180	Amines	1–60ppm	½–2	3	
Hydrazine	N_2H_4	0.1ppm	185	Hydrazine	0.05–2ppm	5–10	2	
			180	Amines	2–120ppm	½–2	3	
Hydrocarbons, High Class (n-Octane, Nonane, Decane)	Varied	—	105	Hydrocarbons, High Class	100–3000ppm	1	3	
Hydrocarbons, Low Class (Propane, Butane)	Varied	—	+103	Hydrocarbons, Low Class	0.05–2.4%	½–2	1	
Hydrocarbons, Petroleum Distillates	Varied	—	106	Hydrocarbons, Petroleum Distillates	0.5–28mg/l mg/m³	½–2	3	
Hydrogen	H_2	—	30	Hydrogen (Color Intensity Tube)	0.5–2.0%	1	3	
Hydrogen Bromide	HBr	C, 3ppm	14L	Hydrogen Chloride-(LR)	2.5–27.5ppm	4	3	

Gas or Vapor to be measured	Chemical Formula	TLV-TWA (ACGIH)[a]	Tube No.	Detector Tube to be used	Measurable Concentration Range	Pump Strokes taken[b]	Shelf Life (year)	SEI Tube
Hydrogen Chloride (Hydrochloric Acid)	HCl	C, 5ppm	14M	Hydrogen Chloride-(MR)	10–1000ppm	½–2	3	
			14L	Hydrogen Chloride-(LR)	0.1–40ppm	½–10	3	
Hydrogen Cyanide (Hydrocyanic Acid)	HCN	C, 10ppm	12H	Hydrogen Cyanide-(HR)	0.05–2.0%	1	3	
			12M	Hydrogen Cyanide-(MR)	17–2400ppm	½–2	2	
			**12L	Hydrogen Cyanide-(LR)	0.5–120ppm	½–5	2	√
Hydrogen Fluoride	HF	C, 3ppm	17	Hydrogen Fluoride	0.25–20ppm	4–7	3	
Hydrogen Sulfide	H₂S	10ppm	4HT	Hydrogen Sulfide	1–40%	½–2	3	
			4HH	Hydrogen Sulfide-(XHR)	0.1–4.0%	½–1	3	
			4H	Hydrogen Sulfide-(HR)	10–3200ppm	1–10	3	
			4M	Hydrogen Sulfide-(MR)	12.5–500ppm	½–2	3	
			4L	Hydrogen Sulfide-(LR)	1–240ppm	½–10	3	
			**4LL	Hydrogen Sulfide-(XLR)	0.25–60ppm	1–10	3	√
Hydrogen Sulfide/Sulfur Dioxide (Total Sulfur Tube)	H_2S/SO_2	—	45H	Hydrogen Sulfide/Sulfur Dioxide	0.02–8.0%	½–1	2	
Iodine	I₂	C, 0.1ppm	9L	Nitrogen Dioxide-(LR)	0.5–12ppm	2	2	
Isobutyl Acetate	$CH_3COOC_2H_3(CH_3)_2$	150ppm	142	Butyl Acetate	0.005–0.66%	2	3	
Isobutyl Alcohol	$(CH_3)_2CHCH_2OH$	50ppm	112	Ethyl Alcohol (Ethanol)	100–3000ppm	2–4	3	
			116	Isobutyl Alcohol	5–150ppm	3–6	2	
Iso-Octane	$(CH_3)_3C_3H_3(CH_3)_2$	—	101	Gasoline	0.015–1.2%	½–	3	
Isopentyl Alcohol (Primary Amyl Alcohol)	$(CH_3)_2CHCH_2CH_2OH$	—	117	Isopentyl Alcohol	10–300ppm	1–4	2	
Isophorone	$C_9H_{14}O$	C, 5ppm	154	Cyclohexanone	2–60ppm	2,4	2	
Isopropyl Acetate	$CH_3COOCH(CH_3)_2$	250ppm	151	Acetone	0.05–0.75%	2	3	

Gas or Vapor to be measured	Chemical Formula	TLV-TWA (ACGIH)[a]	Tube No.	Detector Tube to be used	Measurable Concentration Range	Pump Strokes taken[b]	Shelf Life (year)	SEI Tube
Isobutyl Alcohol	$(CH_3)_2CHCH_2OH$	50ppm	112	Ethyl Alcohol (Ethanol)	100–3000ppm	2–4	3	
			116	Isobutyl Alcohol	5–150ppm	3–6	2	
Iso-Octane	$(CH_3)_3C_3H_3(CH_3)_2$	—	101	Gasoline	0.015–1.2%	½–	3	
Isopentyl Alcohol (Primary Amyl Alcohol)	$(CH_3)_2CHCH_2CH_2OH$	—	117	Isopentyl Alcohol	10–300ppm	1–4	2	
Isophorone	$C_9H_{14}O$	C, 5ppm	154	Cyclohexanone	2–60ppm	2,4	2	
Isopropyl Acetate	$CH_3COOCH(CH_3)_2$	250ppm	151	Acetone	0.05–0.75%	2	3	
Isopropyl Alcohol (Isopropanol)	$(CH_3)_2CHOH$	400ppm	113	Isopropyl Alcohol (Isopropanol)	0.02–5.0%	½–2	3	
			113L	Isopropyl Alcohol-(LR)	25–800ppm	1–2	2	
Isopropylamine	C_3H_9N	5ppm	180	Amines	1–60ppm	½–2	3	
Kerosene	Hydrocarbon Mixture	—	121	Benzene	250–4000ppm	2	3	
			106	Petroleum Distillates Hydrocarbon	0.5–28mg/L	½–2	3	
L.P. Gas	Hydrocarbon Mixture	1000ppm	100A	L.P. Gas	0.02–0.8%	1	3	
Mercaptans, Total	R-SH	—	70	Total Mercaptans (See also Ethyl, Methyl & Butyl Mercaptan)	0.5–120ppm	1–10	3	
Mercury Vapor	Hg	$0.05mg/m^3$	40	Mercury Vapor	$0.05–13.2mg/m^3$	1	2	
Methacrylonitrile	$CH_2{:}C(CH_3)CN$	1ppm	+192	Methacrylonitrile	0.2–32ppm	1–4	2	
Methyl Acrylate	$CH_2{:}CHCOOCH_3$	10ppm	151	Acetone	0.07–0.75%	2	3	
Methyl Alcohol (Methanol)	CH_3OH	200ppm	111	Methyl Alcohol (Methanol)	0.004–4.5%	½–2	3	
			111L	Methyl Alcohol-(LR)	20–1000ppm	1,2	2	

Gas or Vapor to be measured	Chemical Formula	TLV-TWA (ACGIH)a	Tube No.	Detector Tube to be used	Measurable Concentration Range	Pump Strokes takenb	Shelf Life (year)	SEI Tube
Methyl Bromide	CH_3Br	5ppm	+136	Methyl Bromide	2-200ppm	½-4	2	
			+136H	Methyl Bromide-(HR)	10-600ppm	½-2	3	
2-Methyl, 3-Butenitrile	$CH_2CHCH(CH_3)CN$	—	+191L	Acetonitrile	1.2-11.5ppm	4	2	
Methyl Chloroform (1,1,1-Trichloroethane)	CH_3CCl_3	350ppm	+135	Methyl Chloroform	100-500ppm	1	1	
			+135L	Methyl Chloroform-(LR)	20-200ppm	1	1	
Methyl Chloroformate	$ClCOOCH_3$	—	+131La	Vinyl Chloride-(LR)	70-1050ppm	5	2	
Methyl Cyclohexanone	$CH_3C_5H_9CHO$	50ppm	155	Methyl Cyclohexanone	2-100ppm	2,3	2	
Methylene Chloride (Dichloromethane) (Freon 30) TN	CH_2Cl_2	100ppm	+138	Methylene Chloride	25-1500ppm	3-7	1	
			50	Freons (PYR)	50-400ppm	3	2	
Methyl Ethyl Ketone	$CH_3COC_2H_5$	200ppm	152	Methyl Ethyl Ketone	0.02-0.6%	2	3	
Methyl Isobutyl Ketone	$CH_3COCH_2CH(CH_3)_2$	50ppm	153	Methyl Isobutyl Ketone	0.01-0.6%	2	3	
Methyl Mercaptan	CH_3SH	0.5ppm	71	Methyl Mercaptan	0.25-70ppm	1-10	3	
Methyl Methacrylate	$CH_2C(CH_3)COOCH_3$	100ppm	141	Ethyl Acetate	0.005-0.275%	4	3	
Methyl Propionate	$CH_3CH_2COOCH_3$	—	81	Acetic Acid	0.1-2.3%	5	3	
Mineral Spirits	Hydrocarbon Mixture	—	121	Benzene	55-1000ppm	1	3	
Monoethyl Amine (Ethyl Amine)	$C_2H_5NH_2$	10ppm	180	Amines	1-60ppm	½-2	3	
Monomethyl Amine (Methyl Amine)	CH_3NH_2	10ppm	180	Amines	1-60ppm	½-2	3	
Monomethyl Aniline	$C_6H_5NHCH_3$	0.5ppm	*181	Aniline	2.5-45ppm	3	1	
Morpholine	C_4H_9NO	20ppm	180	Amines	1-60ppm	½-2	3	
Nickel Carbonyl	$Ni(CO)_4$	0.05ppm	20L	Nickel Carbonyl	10-800ppm	1-4	3	
Nitric Acid	HNO_3	2ppm	15L	Nitric Acid-(LR)	0.1-40ppm	½-10	3	

Gas or Vapor to be measured	Chemical Formula	TLV-TWA (ACGIH)[a]	Tube No.	Detector Tube to be used	Measurable Concentration Range	Pump Strokes taken[b]	Shelf Life (year)	SEI Tube
Nitric Oxide	NO	25ppm	** +10	Nitric Oxide	NO:2–200ppm / NO2:5–200ppm	1–5	1	
			11	Total Nitrogen Oxides (NO + NO2)	25–600ppm	1	1	
Nitrogen Dioxide	NO_2	3ppm	**9L	Nitrogen Dioxide-(LR)	0.5–100ppm	1–2	2	
			+10	Nitric Oxide	NO2:5–200ppm / NO2:2–200ppm	1–5	1	
Nitrogen Oxides	NO + NO2	NO:25ppm NO2:3ppm	+10	Nitric Oxide	NO:2–200ppm / NO2:5–200ppm	1–5	1	
			11L	Nitrogen Oxides-(LR)	0.08–5ppm	2,4	2	
			11	Nitrogen Oxides	25–600ppm	1	1	
			11H	Nitrogen Oxides-(HR) (NO + NO2)	50–2000ppm	1	1	
n-Nonane	$CH_3(CH_2)_7CH_3$	200ppm	105	Hydrocarbons, High Class	130–3900ppm	1,2	2	
Octane	C_8H_{18}	300ppm	101	Gasoline	0.015–1.2%	½–2	3	
			105	Hydrocarbons, High Class	100–3000ppm	1,2	2	
Oxygen	O_2	—	+31	Oxygen	6–24%	½	2	
Ozone	O_3	0.1ppm	18L	Ozone-(LR)	0.025–3ppm	1–10	3	
			18M	Ozone-(MR)	4–400ppm	½–5	2	
Pentamethyl Diamine	$NH_2(CH_2)_5NH_2$	—	180	Amines	0.5–30ppm	½–2	3	
Pentane	C_5H_{12}	600ppm	104	n-Butane	110–1600ppm	1	3	
			105	Hydrocarbons, High Class	90–2700ppm	1,2	2	

Gas or Vapor to be measured	Chemical Formula	TLV-TWA (ACGIH)[a]	Tube No.	Detector Tube to be used	Measurable Concentration Range	Pump Strokes taken[b]	Shelf Life (year)	SEI Tube
Pentenenitrile	$CH_3CH_2CH{:}CHCN$	—	193	Pentenenitrile	0.5–15ppm	2.4	1.5	
Perchloroethylene	$CCl_2{:}CCl_2$	50ppm	*133M	Perchloroethylene-(MR)	2–250ppm	½–2	2	
			132Ha	Trichloroethylene-(HR)	20–1300ppm	½–2	2	
Petroleum Ether	Hydrocarbon Mixture (C_5 to C_7)	—	161	Ethyl Ether	80–2000mg/m³	1	3	
			106	Hydrocarbons, Petroleum Distillate	0.5–2.8mg/L	½–2	3	
Phenol	C_6H_5OH	5ppm	*60	Phenol	0.4–62.5ppm	1–4	2	
Phosgene	$COCl_2$	0.1ppm	*16	Phosgene	0.05–20ppm	1–10	1	
Phosphine	PH_3	0.3ppm	7J	Phosphine-(HR)	0.00025–0.1% (2.5–1000ppm)	½–10	3	
			7	Phosphine-(HR)	2.5–100ppm	1–4	3	
			7L	Phosphine-(LR)	0.15–5ppm	5–10	3	
Polytec	Varied	—	107	Polytec	Qualitative	3	3	
Propane	C_3H_8	—	100B	Propane†	0.1–2%	20mL/ 2min.	3	
			103	Hydrocarbons, Low Class	0.05–2.4%	½–2	2	
Propionitrile (Ethyl Cyanide)	CH_3CH_2CN	5mg/M³	*191	Acrylonitrile	100–1000ppm	2,4	2	
n-Propyl Acetate	$CH_3COOC_3H_7$	200ppm	151	Acetone	0.06–0.9%	2	3	
n-Propyl Alcohol	C_3H_7OH	200ppm	112	Ethyl Alcohol (Ethanol)	0.02–0.8%	2	3	
			113	Isopropyl Alcohol	0.55–2.65%	1	3	
			113L	Isopropyl Alcohol-(LR)	80–800ppm	1	2	
Propyl Amine	$C_3H_7NH_2$	—	180	Amines	1–60ppm	½–2	3	
Propylene	C_3H_6	—	100A	L.P. Gas (Propylene)	0.02–0.8%	1	3	

Gas or Vapor to be measured	Chemical Formula	TLV-TWA (ACGIH)[a]	Tube No.	Detector Tube to be used	Measurable Concentration Range	Pump Strokes taken[b]	Shelf Life (year)	SEI Tube
Propyleneimine	C_3H_7N	2ppm	180	Amines	1–60ppm	½–2	3	
Propylene Oxide	C_3H_6O	20ppm	163	Ethylene Oxide	0.3–3.6%	1	3	
			163L	Ethylene Oxide-(LR)	0.4–100ppm	2,4	1	
Pyridine	C_5H_5N	5ppm	182	Pyridine	0.2–35ppm	½–2	3	
Smoke Tubes	—	—	501	For use in Air Flow Indicator Kit & Qualitative Respirator Fit Testing				
Stoddard Solvent	Hydrocarbon Mixture	100ppm (575mg/M³)	128	Stoddard Solvent	100–8000mg/M³	1	3	
Styrene (Monostyrene)	$C_6H_5CHCH_2$	50ppm	124	Styrene	10–1000ppm	½–2	3	
			124L	Styrene-(LR)	2–100ppm	1–4	3	
Sulfur Dioxide	SO_2	2ppm	5H	Sulfur Dioxide-(HR)	0.05–8.0%	½–10	3	
			5M	Sulfur Dioxide-(MR)	20–3600ppm	½–4	3	
			**5L	Sulfur Dioxide-(LR)	1.25–200ppm	½–4	3	
			5La	Sulfur Dioxide-(LR)	0.5–60ppm	1–8	3	
			5Lb	Sulfur Dioxide-(LR)	0.05–10ppm	1–8	3	
1,1,2,2-Tetrachloroethane	$Cl_2CHCHCl_2$	1ppm	+131La	Vinyl Chloride-(LR)	3–75ppm	4	3	√
Tetrahydrofuran	C_4H_8O	200ppm	161	Ethyl Ether (Ether)	0.01–0.8%	2,5	2	
			159	Tetrahydrofuran	20–800ppm	1,2	3	
Tetramethylenediamine	$NH_2(CH_2)_4NH_2$	—	180	Amines	0.5–30ppm	½–2	2	
Thionyl Chloride	$SOCl_2$	C, 1ppm	5La	Sulfur Dioxide-(LR)	1–25ppm	2	3	
Toluene	$C_6H_5CH_3$	100ppm	**161	Ethyl Ether (Ether)	0.02–0.85%	1	3	
			122	Toluene	5–600ppm	½–2	3	

Gas or Vapor to be measured	Chemical Formula	TLV-TWA (ACGIH)ᵃ	Tube No.	Detector Tube to be used	Measurable Concentration Range	Pump Strokes takenᵇ	Shelf Life (year)	SEI Tube
O-Toluene	$CH_3C_6H_4NH_2$	2ppm	*181	Aniline	2.5-35ppm	3	1	
1,1,2-Trichloroethane	$CH_2ClCHCl_2$	10ppm	+135	Methyl Chloroform	220-1900ppm	1-2	1	
Trichloroethylene	$CHCl{:}CCl_2$	50ppm	*132M	Trichloroethylene-(MR)	2-250ppm	½-2	2	
			*132L	Trichloroethylene-(LR)	2-50ppm	½-1	2	
			132Ha	Trichloroethylene-(HR)	20-1300ppm	½-2	2	
Trichlorofluoromethane (Freon 11) (TN)	CCl_3F	C, 1000ppm	50	Freons (PYR)	38-300ppm	3	2	
Trichlorofluoroethane (Freon 113) (TN)	$C_2Cl_3F_3$	1000ppm	50	Freons (PYR)	38-300ppm	3	2	
Triethylamine	$(C_2H_5)_3N$	10ppm	3M	Ammonia-(MR)	3.5-140ppm	2-4	3	
Trimethylamine	$(CH_3)_3N$	10ppm	3M	Ammonia-(MR)	2.5-100ppm	2-4	3	
			180	Amines	1-60ppm	½-2	3	
Vinyl Acetate	$CH_2CHCOOCH_3$	10ppm	141	Ethyl Acetate	0.01-1.0%	1	3	
			+143	Vinyl Acetate	5-250ppm	1-4	3	
Vinyl Chloride	CH_2CHCl	5ppm	**131	Vinyl Chloride	0.025-2.0%	½-2	3	
			+131La	Vinyl Chloride-(LR)	0.25-54ppm	½-4	2	
			+131L	Vinyl Chloride-(LR)	0.1-8.8ppm	2-7	2	
Vinylidene Chloride	$CH_2{:}CCl_2$	5ppm	+131La	Vinyl Chloride-(LR)	0.3-30ppm	½-2	2	
Water Vapor	H_2O	—	6	Water Vapor	0.5-32mg/L	½-2	3	
			6L	Water Vapor-(LR)	0.1-2.0mg/L	½-1	3	
			6LP	Pipeline "Dewpoint" Water Vapor-(LR)	6-80LB/MMCF	1-2	3	
Xylene	$C_6H_4(CH_3)_2$	100ppm	123	Xylene	10-500ppm	½-2	3	

Legend

[a] Threshold Limit Value for 1985–86, adopted by American Conference of Governmental Industrial Hygienists. 1 ppm = 0.0001%

[b] One stroke of the sampling pump draws a 100-mL air sample. One-half pump stroke draws a 50-mL air sample.

√ Indicates a Safety Equipment Institute (SEI)-certified tube.

* Tubes to be stored at 0°C (32°F) or below.

** NIOSH-certified tube. (The NIOSH Certification Program has been discontinued.)

+ Twin tubes to be combined with primary and analyzer tubes, 5 tests.

† Must be used with propane kit.

LR = low range.

MR = mid range

HR = high range.

XLR = extra low range

XHR = extra high range

PYR = requires pyrolyzer

TN = trade name

Appendix D. Conversion Factors.

To convert from	To	Multiply by
calories	Btu	3.968×10^{-3}
centimeters	inches	0.3937
cubic centimeters	cubic inches	0.06102
cubic feet	liters	28.32
cubic inches	cubic meters	1.639×10^{-5}
cubic inches	liters	0.01639
cubic meters	cubic feet	35.31
feet	centimeters	30.482
gallons	cubic centimeters	3785
gallons	liters	3.785
grams	ounces	0.03527
grams	pounds	0.002205
inches	centimeters	2.540
kilograms	pounds	2.205
kilograms	tons (short)	1.102×10^{-3}
kilograms	metric tons	0.001000
liters	cubic centimeters	1000
liters	cubic feet	0.03532
liters	cubic inches	61.025
liters	gallons	0.2642
meters	inches	39.37
meters	feet	3.281
milligrams	ounces	3.527×10^{-5}
milligrams	pounds	2.205×10^{-6}
milliliters	cubic centimeters	1.000
milliliters	cubic inches	0.06103
milliliters	ounces (fluid)	0.03381
millimeters	inches	0.03937
ounces	grams	28.35
ounces (fluid)	cubic centimeters	29.57
ounces (fluid)	liters	0.02957
pounds	grams	453.6
pounds	kilograms	0.4536
quarts	liters	0.9463
tons (short)	kilograms	907.2
tons (short)	tons (metric)	0.9072

Source: Lange's Handbook of Chemistry, Handbook Publishers, Inc., Sandusky, OH.

Appendix E. Safety Phone Numbers

OSHA–Health Standards	1-202-523-7075
NIOSH	1-800-356-4674
National Safety Council	1-312-527-4800
National Response Center (to report releases)	1-800-424-8802
Chemical Emergency Preparedness Hotline (CERCLA)	1-800-535-0202
Department of Transportation Hotline	1-202-366-4488
Chemical Transportation Emergency Center	1-800-424-9300
EPA (RCRA) Hazardous Waste Hotline (Emergency Response)	1-800-424-9346
Chemical Manufacturers Association (Chemical Referral Center)	1-800-262-8200
Sustance Identification	1-800-848-6538

Index